Clinical Manual of Drug Interaction Principles for Medical Practice

Gary H. Wynn, M.D.

Jessica R. Oesterheld, M.D.

Kelly L. Cozza, M.D.

Scott C. Armstrong, M.D.

D0931968

American Psychiatric Publishing, Inc.

Washington, DC
London, England

Note: The authors have worked to ensure that all information in this book is accurate at the time of publication and consistent with general psychiatric and medical standards, and that information concerning drug dosages, schedules, and routes of administration is accurate at the time of publication and consistent with standards set by the U.S. Food and Drug Administration and the general medical community. As medical research and practice continue to advance, however, therapeutic standards may change. Moreover, specific situations may require a specific therapeutic response not included in this book. For these reasons and because human and mechanical errors sometimes occur, we recommend that readers follow the advice of physicians directly involved in their care or the care of a member of their family.

Books published by American Psychiatric Publishing, Inc., represent the views and opinions of the individual authors and do not necessarily represent the policies and opinions of APPI or the American Psychiatric Association.

The opinions or assertions contained herein are the private views of the authors and are not to be construed as official or as reflecting the views of the Department of the Army or the Department of Defense.

Copyright © 2009 American Psychiatric Publishing, Inc.
ALL RIGHTS RESERVED

Manufactured in the United States of America on acid-free paper
12 11 10 09 08   5 4 3 2 1
First Edition
Typeset in Adobe's Formata and AGaramond.

American Psychiatric Publishing, Inc.
1000 Wilson Boulevard
Arlington, VA 22209-3901
www.appi.org

To purchase 25–99 copies of this or any other APPI title at a 20% discount, please contact APPI Customer Service at appi@psych.org or 800-368-5777. If you wish to buy 100 or more copies of the same title, please e-mail us at bulksales@psych.org for a price quote.

Library of Congress Cataloging-in-Publication Data
Clinical manual of drug interaction principles for medical practice / Gary H. Wynn ... [et al.].
-- 1st ed.
    p. ; cm.
Includes bibliographical references and index.
ISBN 978-1-58562-296-2 (alk. paper)
1. Drug interactions. 2. Pharmacokinetics. I. Wynn, Gary H., 1958-
II. American Psychiatric Publishing.
[DNLM: 1. Drug Interactions. 2. Pharmacogenetics. 3. Pharmacokinetics. QV 38 C6398 2009]
RM302.C555 2009
615′.7045--dc22

                                                                                            2008035657

British Library Cataloguing in Publication Data
A CIP record is available from the British Library.

Contents

PART I
Introduction and Basic Pharmacology
of Metabolic Drug Interactions
Gary H. Wynn, M.D., Editor

1 **Introduction to Drug Interactions**
Gary H. Wynn, M.D.
Scott C. Armstrong, M.D., D.F.A.P.A., F.A.P.M.

Gary H. Wynn, M.D.
Scott C. Armstrong, M.D., D.F.A.P.A., F.A.P.M.

Michael H. Court, B.V.Sc., D.A.C.V.A., Ph.D.

Jessica R. Oesterheld, M.D.

PART II
P450 Enzymes
Jessica R. Oesterheld, M.D., Editor

PART III
Drug Interactions by Medical Specialty
Gary H. Wynn, M.D., Editor

Contributors

Scott C. Armstrong, M.D., D.F.A.P.A., F.A.P.M.
Medical Director, Tuality Center for Geriatric Psychiatry; Associate Clinical Professor of Psychiatry, Oregon Health and Science University, Forest Grove, Oregon

David M. Benedek, M.D., D.F.A.P.A.
Associate Professor and Assistant Chair, Department of Psychiatry; Associate Director and Senior Scientist, Center for the Study of Traumatic Stress, Uniformed Services University School of Medicine, Bethesda, Maryland

Michael H. Court, B.V.Sc., D.A.C.V.A., Ph.D.
Assistant Professor, Comparative and Molecular Pharmacogenomics Laboratory, Department of Pharmacology and Experimental Therapeutics, Tufts University School of Medicine, Boston, Massachusetts

Kelly L. Cozza, M.D., F.A.P.M., F.A.P.A.
Associate Professor, Uniformed Services University of the Health Sciences, Bethesda, Maryland; Consulting Psychiatrist, Department of Psychiatry, Walter Reed Army Medical Center, Washington, D.C.

Justin M. Curley, M.D.
Internal Medicine and Psychiatry Resident, Walter Reed Army Medical Center, Washington, D.C.; Assistant Professor, Uniformed Services University School of Medicine, Bethesda, Maryland

Javier Muniz, M.D.
Medical Director, Outpatient Mental Health Services, Malcolm Grow Medical Center, Andrews Air Force Base, Maryland; Assistant Professor, Uniformed Services University School of Medicine, Bethesda, Maryland

Jessica R. Oesterheld, M.D.
Acting Director, Spurwink, Department of Psychiatry, Maine Medical Center; Clinical Associate Professor, University of Vermont College of Medicine, Burlington, Vermont

Neil Sandson, M.D.
Staff Psychiatrist, Veterans Affairs of Maryland Health Care System; Clinical Associate Professor, Department of Psychiatry, University of Maryland School of Medicine, Baltimore, Maryland

Miia R. Turpeinen, M.D., Ph.D.
Senior Lecturer, Department of Pharmacology and Toxicology, University of Oulu, Oulu, Finland

Scott G. Williams, M.D.
Internal Medicine and Psychiatry Resident, Walter Reed Army Medical Center, Washington, D.C.; Assistant Professor, Uniformed Services University School of Medicine, Bethesda, Maryland

Joseph E. Wise, M.D.
Psychiatry Resident, Walter Reed Army Medical Center, Washington, D.C.

Gary H. Wynn, M.D.
Staff Psychiatrist, Walter Reed Army Medical Center, Washington, D.C.; Assistant Professor, Uniformed Services University School of Medicine, Bethesda, Maryland

Disclosure of Interests
The contributors to this book have indicated a financial interest in or other affiliation with a commercial supporter, a manufacturer of a commercial product, a provider of a commercial service, a nongovernmental organization, and/or a government agency, as follows:

Jessica R. Oesterheld, M.D., has a financial relationship with Genelex, a genetic testing firm. The other contributors affirm that they have no financial interest or affiliation relevant to this work.

Preface

It has been five years since we last presented this topic as a Concise Guide. Since the previous edition we have seen a multitude of changes, ranging from new medications entering the market to a vast increase in the understanding of pharmacokinetics and pharmacogenetics. In light of this growing body of knowledge and ever-expanding scope of research, this text has been upgraded to a Clinical Manual. There are a number of changes in this edition, including the addition of a section on addictions treatments and the reconfiguring of the P-glycoprotein (now ABCB1) chapter into a broader discussion of transporters. Additionally, the section on HIV has been expanded, specifics about pharmacogenetics and transporters have been integrated into each chapter, and all of the chapters and tables have been updated. Finally, the Concise Guide's pocket reference has been folded into the text as an easy-to-use appendix. We hope you find this Clinical Manual both convenient and clinician-friendly.

For this edition, we have also seen some changes to our authorship. Dr. Wynn moved from his role as contributing author to senior author and editor, and Drs. Oesterheld, Cozza, and Armstrong continued their work in that role from previous editions. Dr. Oesterheld has continued to expand our understanding of transporters, aided by Dr. Scherrmann, who graciously assisted us in this effort. Dr. Court has brought his expertise to phase II metabolism, and Dr. Turpeinen has added an international dimension to this edition. Drs. Muniz and Sandson helped update the topic of psychiatric medications and contributed a new section on addictions treatments. Dr. Benedek returns to continue his commentary on the legal ramifications of drug-drug interactions. Our young authors, Drs. Williams, Wise, and Curley, with their energy and astute questioning, continually remind us why we work on this topic.

As in the first two editions, references in each chapter support both the text and the tables. Most references are cited in the text; those not cited in the text are included because they support general concepts or data found in the tables.

The work contained in this manual was not funded or supported in any way by pharmaceutical companies or distributors.

We hope that the expanded coverage of pharmacokinetic drug interactions to include pharmacogenetics and transporters will allow the reader to quickly and confidently predict and identify possible drug interactions so as to be able to provide the highest quality care to patients.

—G.H.W.

Acknowledgments

For Grammy, who always understood. —*G.H.W.*

Thanks to my husband, Mark, who has always supported my intellectual work, and to my grandchildren: Nathaniel, Henry, William, Marla, Anna, and Annabel, who anchor me in the future. —*J.R.O.*

Thanks yet again to my husband, Steve, and to my children, Vincent and Cecilia, for their love and understanding during the preparation of this book. A special thanks to all of my mentors and patients. —*K.L.C.*

Thanks to my parents, Patricia and Richard Armstrong, to my spouse, JoAnn, and to my children, Joseph, Kaitlyn, and Caleb. And thanks to all my patients, who motivate me to do my best for them. —*S.C.A.*

PART I

Introduction and Basic Pharmacology of Metabolic Drug Interactions

Gary H. Wynn, M.D.
Editor

Introduction to Drug Interactions and to This Clinical Manual

Gary H. Wynn, M.D.

Scott C. Armstrong, M.D., D.F.A.P.A., F.A.P.M.

An understanding of drug interactions has become essential to the practice of medicine. The increasing pharmacopoeia, coupled with prolonged human life spans, makes polypharmacy commonplace. The better-studied metabolism phase, phase I, includes the P450 system (in this manual, we use the term *P450* rather than *cytochrome P450* or *CYP450*). The need for a thorough understanding of the P450 system became apparent in the 1980s, when the combination of terfenadine and antibiotics or of selective serotonin reuptake inhibitors and tricyclic antidepressants resulted in arrhythmias and sudden death. In recent years, research on glucuronidation and other phase II reactions and on efflux transporters such as ABCB1 (P-glycoprotein) has increased the understanding of drug-drug interactions. Despite burgeoning data and clear manufacturer warnings, patients are still at risk for these interactions.

Why Is It Important to Understand Pharmacokinetic Drug Interactions?

The following case demonstrates the importance of understanding drug-drug interactions:

> A 64-year-old woman with chronic schizophrenia required acute admission to a geriatric psychiatry unit from a group home setting because of an iatrogenic drug interaction. The patient had responded well for several years to clozapine 125 mg twice a day. Two months before acute admission, clozapine therapy was stopped because she appeared to be having a toxic reaction; a clozapine level (1,043 ng/mL; usual range, 350–475 ng/mL) confirmed the clinical presentation. Her primary psychiatrist then initiated ziprasidone therapy, thinking the patient could not tolerate or metabolize clozapine because of her age. The patient's psychosis worsened, necessitating the acute hospitalization on the geriatric psychiatry unit. The inpatient psychiatrist (S.C.A.) asked for the list of medications that the patient had been taking at the time of the toxicity. It was discovered that the patient had been taking ciprofloxacin for a mild urinary tract infection. The inpatient psychiatrist determined that the ciprofloxacin increased the serum level of clozapine through inhibition of the P450 enzymes 1A2 and 3A4. This information permitted titration to the patient's original, therapeutic clozapine dose. She was discharged back to her group home. Subsequent outpatient clozapine levels ranged from 500 to 600 ng/mL.

If the patient's initial care providers had known about the ciprofloxacin-clozapine interaction, a different antibiotic could have been chosen or the clozapine dose could have been reduced or temporarily withheld, allowing the patient to maintain her usual level of functioning. Knowledge of this drug interaction would have led to better patient care and reduced health care costs.

Drug interactions have become an important preventable iatrogenic complication. As many as 5% of hospitalizations per year are due to drug-drug interactions (Becker et al. 2007) with as many as half of those interactions involving a drug that requires outpatient monitoring (Budnitz et al. 2006). Those individuals on polypharmacy regimens and the elderly are at highest risk for a drug-drug interaction requiring hospitalization. The amount spent on these hospitalizations can exceed $1 billion a year (Johnson and Bootman 1995). In one large study involving Medicaid patients (n=315,084), the risk of hospitalization greatly increased when patients were prescribed azole antifungals (odds ratio, 3.43) or rifamycins (odds ratio, 8.07), two drug classes that have significant P450 drug-drug interaction profiles (Hamilton et al. 1998). In addition, a study by the Institute of Medicine in the United States indicates that adverse medication reactions—including drug-drug interactions—account for up to 7,000 deaths annually in the United States (Kohn et al. 2000). Clearly, drug interactions are a problem of which clinicians need to remain aware.

These data also seem to underrepresent the situation. Many interactions are not reported by patients to physicians or by physicians to monitoring agencies. Some bothersome side effects may in fact be effects of drug interactions (e.g., a patient's serum caffeine levels may be affected by newly administered medications, resulting in "jitters"). Adverse drug reactions such as tardive dyskinesia may be more pronounced in patients who are poor metabolizers at some P450 enzymes; some of these poor metabolizers are genetically predisposed, and others are made poor metabolizers via drug interactions. We suspect that most drug interactions are less than lethal and go unrecognized, leading to ineffective or aborted drug therapy and/or increased care costs.

Why Does the Metabolic System Exist?

Metabolic enzymes may have originated from a common ancestral gene or protein in plants and animals billions of years ago. Initially, these enzymes may have helped maintain cell membrane integrity through their contribution to steroid metabolism. As animals evolved and ate plants to survive, the plants that developed toxins survived. Animals then needed to be able to detoxify these chemicals, and thus an elaborate detoxification system developed (Jefferson and Greist 1996).

P450 enzymes are oxidase systems (see Figure 1–1). These enzymes oxidize endogenous and exogenous compounds and usually render them less active (see Chapter 2, "Definitions and Phase I Metabolism"), preparing them for further transformation by phase II reactions such as glucuronidation or sulfation, or for elimination from the body. The P450 system's primary role is to metabolize endogenous compounds such as steroids and neuropeptides. A secondary role, especially for enzymes in the gastrointestinal and hepatic system, is to detoxify ingested chemicals. These exogenous, or xenobiotic, chemicals are compounds in foods, medicines, smoke, and any other ingested organic molecule. Medications are one group of "toxins" in the modern human environment (others include pesticides and organic solvents), and the role of the P450 system in metabolizing drugs is a relatively recent phenomenon in human history. P450 enzymes are indiscriminate in their activity; that is, these enzymes will metabolize an ancient food chemical as readily as they will metabolize a modern drug. P450 enzymes are also multifunctional, with each enzyme capable of metabolizing many different compounds.

More than 40 individual P450 enzymes have been identified in humans. However, many of these enzymes play minor roles in drug metabolism. Six enzymes are responsible for more than 90% of human drug oxidation: 1A2, 3A4, 2C9, 2C19, 2D6, and 2E1 (Guengerich 1997). The enzymes 2A6, 2B6, and 2C8 play clinically relevant but smaller roles in human metabolism. Each of these enzymes is discussed in this text. The 3A subfamily, including 3A3, 3A5, and 3A7, are discussed in conjunction with 3A4.

The six major P450 enzymes (and perhaps all human P450 enzymes) exhibit genetic variability via polymorphisms. Phase II enzymes can also exhibit genetic variability or polymorphisms, though understanding of phase II genetic variability remains far behind that of phase I metabolism. It is speculated that the variety of and differences in human genotypes are the result of isolated populations' evolving different abilities to survive local stressors—in this case, to metabolize indigenous organic compounds. As ethnic isolation has become replaced by diversity within the modern world, genotypic variability has become less predictable. The evolutionary advantage of genetic variability has disappeared for modern humans. Variability in drug response may now be an obstacle rather than an advantage.

Gender may also play a role in the variability of metabolic enzyme systems. It is difficult to establish gender differences, however, because research evaluating these differences can be obscured by large interindividual variation in enzyme quantity and efficiency. Furthermore, to uncover metabolic enzyme-based gender distinctions in *in vivo* studies, one must factor out gender-based differences. It is difficult and costly to control factors related to weight and volume of distribution, ethnicity and polymorphism, smoking and alcohol consumption, obesity, age, cotherapy (including hormone replacement therapy and oral contraceptives), and gender differences in pharmacodynamics.

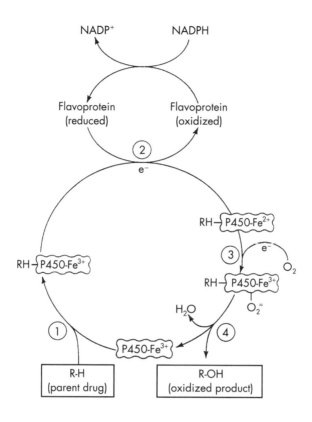

FIGURE 1–1. P450 cycle in drug oxidations.

e^-=electron; $NADP^+$=oxidized form of nicotinamide adenine dinucleotide phosphate; NADPH=reduced form of NADP; R-H=parent drug. R-OH= oxidized metabolite.

Source. Reprinted from Holford NHG: "Pharmacokinetics and Pharmacodynamics: Rational Dosing and the Time Course of Drug Action," in *Basic and Clinical Pharmacology,* 8th Edition. Edited by Katzung BG. New York, Lange Medical Books/McGraw-Hill, 2001, p. 53. Used with permission.

How to Use This Manual

This Clinical Manual is designed to be a practical guide—a reference at the bedside, on rounds, or in the office—and as such, this text is divided into several sections for ease of use.

Part I, "Introduction and Basic Pharmacology of Metabolic Drug Interactions," is a succinct review of pharmacology, written with clinicians in mind. We recommend reading this part first and then returning to particular chapters in the section as needed when consulting later chapters or reviewing other current literature. Phase I and phase II metabolism are fully introduced in Part I, as are transporters.

Part II, "P450 Enzymes," contains pertinent tables, short reviews, and carefully chosen clinical and research illustrations for individual P450 enzymes. This section will perhaps be frequently consulted once the reader is familiar with the nomenclature and the pertinent literature to date. We and many of our mentors carry lists such as these to refer to when we cannot remember where in the system a particular drug is metabolized or which drug inhibits or induces metabolism at which enzyme site. These chapters conclude with case vignettes, to provide an opportunity for study and to better illustrate P450-mediated drug interactions in clinical practice.

Part III, "Drug Interactions by Medical Specialty," covers the drugs commonly used in medicine that have clinically significant pharmacokinetic phase I–, phase II–, or transporter-related drug interactions and are arranged by specialty (and listed in tables), with pertinent clinical and research data to support and clarify the issues. Some of the data in this section's tables are the same as in Parts I and II, so that each part can stand alone.

Part IV, "Practical Matters," includes suggestions about prescribing multiple drugs and monitoring for drug interactions, a discussion of legal issues, and a chapter on how to review the literature. In that chapter, we share the approaches we have developed in culling the enormous and sometimes confusing amount of information available in this area. We hope that this section will assist clinicians in becoming drug interaction experts in their own field as they identify useful references and develop practice-specific drug lists of their own.

Throughout this Clinical Manual we provide tables of P450, UGT, transporter, and medical subspecialty information for easy and ready access to clinically valuable information. These tables are collated in the Appendix to this manual, which is intended to be a rapid reference for clinicians.

Finally, we remind the reader that drugs interact in many ways. An adverse drug reaction may occur without the involvement of the P450 system, glucuronidation, or transporters. Nevertheless, hepatic and gut wall P450-mediated interactions are the majority. We do not review all types of interactions here (absorption, protein-binding, renal excretion and elimination, and pharmacodynamic interactions are excluded), and we refer the reader elsewhere for information on those specific interactions.

References

Becker ML, Kallewaard M, Caspers PW, et al: Hospitalisations and emergency department visits due to drug-drug interactions: a literature review. Pharmacoepidemiol Drug Saf 16:641–651, 2007

Budnitz DS, Pollock DA, Weidenbach KN, et al: National surveillance of emergency department visits for outpatient adverse drug events. JAMA 296:1858–1866, 2006

Guengerich FP: Role of cytochrome P450 enzymes in drug-drug interactions. Adv Pharmacol 43:7–35, 1997

Hamilton RA, Briceland LL, Andritz MH: Frequency of hospitalization after exposure to known drug-drug interactions in a Medicaid population. Pharmacotherapy 18:1112–1120, 1998

Jefferson JW, Greist JH: Brussels sprouts and psychopharmacology: understanding the cytochrome P450 enzyme system, in The Psychiatric Clinics of North America Annual of Drug Therapy, Vol 3. Edited by Greist JH, Jefferson JW. Philadelphia, PA, WB Saunders, 1996, pp 205–222

Johnson JA, Bootman JL: Drug-related morbidity and mortality: a cost-of-illness model. Arch Intern Med 155:1949–1956, 1995

Kohn L, Corrigan J, Donaldson M (eds): To Err Is Human: Building a Safer Health System. Washington, DC, National Academy Press, 2000

2

Definitions and Phase I Metabolism

Gary H. Wynn, M.D.
Scott C. Armstrong, M.D., D.F.A.P.A., F.A.P.M.

Drug interactions reflect a shift in drug activity or effect in the body as a result of another chemical's presence or activity. Drug interactions are usually considered either pharmacodynamic or pharmacokinetic. In this chapter, we outline pharmacodynamic and pharmacokinetic principles. Pharmacokinetics, specifically phase I metabolism, is then fully described.

Drug Interactions Definitions

Pharmacodynamic Interactions

Pharmacodynamic interactions are interactions due to one drug's influence on another drug's effect at the latter's intended receptor site or end organ. These interactions or alterations in drug function are not due to a change in absorption, distribution, metabolism, or elimination. When two drugs act at a receptor site in the brain, causing a combined, typically unwanted effect, or negating a wanted effect, a pharmacodynamic interaction is said to have occurred. The potentially dangerous monoamine oxidase inhibitor–tricyclic antidepressant serotonin syndrome is an example of a pharmacodynamic interaction.

Pharmacokinetic Interactions

Pharmacokinetic interactions are interactions due to one drug's effect on the movement of another drug through the body. These interactions are alterations in the way the body would normally process a drug towards eventual elimination (including the way it is metabolized). Pharmacokinetic interactions may result in delayed onset of effect, decreased or increased effect, toxicity, or altered excretion, and directly affect the concentration of drug that reaches the target site. Pharmacokinetic interactions encompass alterations in absorption, distribution, metabolism, and excretion.

Absorption interaction: An alteration due to one drug's effect on another drug's route of entry into the body. Most absorption interactions occur in the gut; some examples of these interactions are those due to altered gastric pH, food coadministration, mechanical blockade or chelation, and loss of gut flora.

Distribution interaction: An alteration in how a drug travels throughout the body. These interactions are typically the result of alterations in plasma protein binding. Drug effect is directly related to the amount of free drug available to the target site. More or less drug is available if another drug affects the protein-bound fraction of a drug. Warfarin is an example of a drug that is sensitive to protein-binding displacement by many other drugs.

Metabolism interaction: An alteration in the biotransformation of a compound into active drug or excretable inactive compounds. These interactions usually result in a change in drug concentration due to an increase or decrease in enzymatic activity and thus metabolic rate. In this Clinical Manual, metabolism is covered in great depth.

Excretion interaction: An alteration in the ability to eliminate an unaltered drug or metabolite from the body. An example of this type of interaction is the effect of sodium concentration or diuretics on lithium retention by the kidneys.

Metabolism

The act of metabolism may have developed to assist humans in removing endogenous substances no longer needed (e.g., catecholamines, steroids, bilirubin). Over time, metabolism came to include biotransformation of exogenous substances, such as food, chemicals, environmental toxins, and drugs. Drugs are usually lipophilic, thus allowing them to enter cell membranes and exert an effect on target organs and tissues. Lipophilic compounds are generally difficult to eliminate from the body. Metabolism—or biotransformation of these compounds into more polar, inactive metabolites that are more water soluble—is necessary for eventual elimination. These metabolites can then more readily exit the body via urine, bile, or stool. The workhorses of biotransformation are the metabolic enzymes.

Biotransformation occurs throughout the body, with the greatest concentration of activity in the liver and the gut wall. At the cellular level, biotransformation occurs in the endoplasmic reticulum. With many drugs, there is a *first-pass effect* as the drug crosses the gut wall and again as the drug passes through the liver before reaching the systemic circulation and its target sites. Most drugs lose at least some portion of their functional activity during "first pass" metabolism as the body begins the process of preparing the drug for elimination. This first-pass effect can limit the oral availability of drugs and is a factor in determining whether a drug should be administered parenterally or orally. Several processes contribute to the first-pass effect, including phase I metabolism in the gut wall (with CYP 3A4 responsible for 70% of intestinal P450 activity) (DeVane 1998), metabolism in the liver, and phase II metabolic reactions in both gut and liver, and by active transporters. The brain, kidneys, lungs, and skin also have significant metabolic activity, but any cell with endoplasmic reticulum has the capacity for some metabolic reactions.

Drug biotransformation is generally accomplished in two phases: phase I and phase II. Some drugs undergo these processes in order, some undergo them simultaneously, and some undergo only phase I or phase II. In the remainder of this chapter, phase I is reviewed. The reader is referred to Chapter 3 ("Metabolism in Depth: Phase II") for a full discussion of phase II.

Phase I Metabolism

Phase I reactions typically add small, polar groups to a parent drug by adding or exposing a functional group on a compound or drug through oxidative reactions such as *N*-dealkylation, *O*-dealkylation, hydroxylation, *N*-oxidation, *S*-oxidation, or deamination (Table 2–1). The resulting compounds may then lose all pharmacological activity. These compounds are then ready for reactions to form highly water-soluble conjugates with subsequent elimination by means of phase II metabolism. As mentioned in the previous paragraph, many drugs are not metabolized in this linear fashion, and some may undergo conjugation reactions of phase II before phase I, or these reactions may occur simultaneously. Some drugs do not lose pharmacological activity at all and instead have similar or increased activity after metabolism. Some drugs (e.g., terfenadine) are pro-drugs and must be metabolized to their active metabolite. The cytochrome P450 monooxygenase system is a phase I system.

Cytochrome P450 Monooxidase System ("P450")

More than 200 P450 enzymes exist in nature, and a collection of at least 40 enzymes is found in humans (Figure 2–1). Six enzymes are responsible for about 90% of all the metabolic activity of P450 enzymes—1A2, 3A4, 2C9, 2C19, 2D6, and 2E1—and all play an important role in xenobiotic oxidative metabolism. Minor but clinically relevant enzymes include 2A6, 2B6, and 2C8. All of these enzymes are on the smooth endoplasmic reticulum of hepatocytes and the luminal epithelium of the small intestine. These proteins are closely associated with nicotinamide adenine dinucleotide phosphate (reduced form) (NADPH) cytochrome P450 reductase, which donates or is the source of the electrons needed for oxidation (Wilkinson 2001). (For details on the enzymatic reactions, see Table 2–1 and the references at the end of this chapter.)

The P450 enzymes contain red-pigmented heme, and when bound to carbon monoxide they absorb light at a wavelength of 450 nm. In the term *cytochrome P450*, *cyto* stands for *microsomal vesicles*, *chrome* for *colored*, *P* for *pigmented*, and *450* for *450 nanometers*. Each enzyme is encoded by one particular gene—one gene, one enzyme. The enzymes are grouped into families and subfamilies according to the similarity of their amino acid sequences. Enzymes in the same family are homologous for 40%–55% of amino acid sequences, and enzymes within the same subfamily are homologous for more than 55%. (See Table 2–2 for more on P450 nomenclature.)

Until the early 1990s, it was standard to label a family by using a Roman numeral; an enzyme might be labeled *CYPIID6*. Today, Arabic numerals are

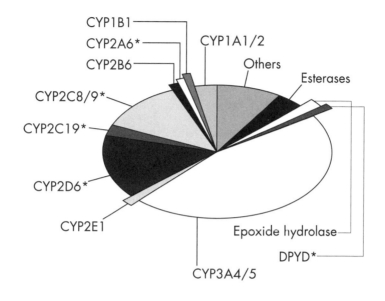

FIGURE 2–1. Proportion of drugs metabolized by the major phase I enzymes.

The relative size of each pie section indicates the estimated percentage of phase I metabolism that each enzyme contributes to the metabolism of drugs based on literature reports. Enzymes that have functional allelic variants are indicated by an asterisk (*). In many cases, more than one enzyme is involved in a particular drug's metabolism. CYP=cytochrome P450; DPYD= dihydropyrimidine dehydrogenase.

Source. Adapted from Wilkinson GR: "Pharmacokinetics: The Dynamics of Drug Absorption, Distribution, and Elimination," in *Goodman & Gilman's The Pharmacological Basis of Therapeutics,* 10th Edition. Edited by Hardman JG, Limbird LE, Gilman AG. New York, McGraw-Hill, 2001, p. 15. Used with permission of the McGraw-Hill Companies.

TABLE 2–1. Major reactions involved in phase I drug metabolism

Reaction	Phase I	Examples
Oxidative reactions		
N-Dealkylation	$RNHCH_3 \longrightarrow RNH_2 + CH_2O$	Imipramine, diazepam, codeine, erythromycin, morphine, tamoxifen, theophylline, caffeine
O-Dealkylation	$ROCH_3 \longrightarrow ROH + CH_2O$	Codeine, indomethacin, dextromethorphan
Aliphatic hydroxylation	$RCH_2CH_3 \longrightarrow RCHCH_3$ with OH	Tolbutamide, ibuprofen, pentobarbital, meprobamate, cyclosporine, midazolam
Aromatic hydroxylation	(aromatic ring $R \longrightarrow$ epoxide $\longrightarrow$ phenol R–OH)	Phenytoin, phenobarbital, propranolol, phenylbutazone, ethinyl estradiol, amphetamine, warfarin
N-Oxidation	$RNH_2 \longrightarrow RNHOH$	Chlorpheniramine, dapsone, meperidine
	$R_1R_2NH \longrightarrow R_1R_2N\text{—OH}$	Quinidine, acetaminophen
S-Oxidation	$R_1R_2S \longrightarrow R_1R_2S{=}O$	Cimetidine, chlorpromazine, thioridazine, omeprazole
Deamination	$RCHCH_3 \rightarrow R\text{—C}\text{—}CH_3 \rightarrow R\text{—C}\text{—}CH_3 + NH_2$ (with OH, NH$_2$, O)	Diazepam, amphetamine

TABLE 2–1. Major reactions involved in phase I drug metabolism *(continued)*

Reaction	Phase I	Examples
Hydrolysis reactions		
	$R_1COR_2 \xrightarrow{} R_1COOH + R_2OH$ (C=O with O above)	Procaine, aspirin, clofibrate, meperidine, enalapril, cocaine
	$R_1CNR_2 \xrightarrow{} R_1COOH + R_2NH_2$ (C=O with O above)	Lidocaine, procainamide, indomethacin

Source. Adapted from Wilkinson GR: "Pharmacokinetics: The Dynamics of Drug Absorption, Distribution, and Elimination," in *Goodman and Gilman's The Pharmacological Basis of Therapeutics*, 10th Edition. Edited by Hardman JG, Limbird LE, Gilman AG. New York, McGraw-Hill, 2001, p. 14. Used with permission of the McGraw-Hill Companies.

TABLE 2–2.   P450 nomenclature

Category	Addition(s) to *P450*	Examples
Family	Arabic numeral	P4501, P4502
Subfamily	Arabic numeral+ uppercase letter	P4501A, P4502D
Single gene or protein	Arabic numeral+ uppercase letter+ Arabic numeral	P4501A2, P4502D6

used; the same enzyme is now labeled *CYP2D6* or simply *2D6*. The P450 system is also referred to as the *cytochrome P450 system*, *CYP450 system*, and *P450 system*. For brevity, we use *P450 system* throughout this book. Individual enzymes are designated several ways in the literature; a particular enzyme might be labeled *cytochrome P450 2D6*, *CYP2D6*, or *2D6*. In this book, we use the shortest forms, labeling individual P450 enzymes *2D6* or *3A4*, for example. Finally, the P450 enzymes are often called *isoenzymes* or *isozymes* in the literature, but these terms are misnomers because they refer to enzymes that catalyze one reaction. To avoid confusion, the term *enzyme* is used throughout this text.

Other Phase I Systems

Other phase I metabolism systems include the alcohol dehydrogenase, flavin monooxygenase, esterase, and amidase systems. The monoamine oxidase system is also considered a part of phase I (but not of the P450 system). This group includes monoamine oxidases, which have been studied in relation to depression and Parkinson's disease.

Metabolic Variability: Inhibition, Induction, and Pharmacogenetics

Drug response varies greatly across groups and individuals. This variability is due to many pharmacological factors. In phase I and phase II metabolism, this variability may be a reflection of enzyme inhibition, enzyme induction, and/or genetic differences.

Inhibition

Drugs metabolized at a common human enzyme will at times be competing for or sharing metabolic sites within the endoplasmic reticulum in the liver and gut. A drug's affinity for an enzyme is called its *inhibitory potential*, or *Ki*. These values are routinely determined *in vitro* via human liver microsome studies (see Chapter 23, "How to Retrieve and Review the Literature," for a discussion of *in vitro* studies). Drugs with little affinity for an enzyme have a high Ki and probably will not bind. Drugs with a low Ki, or great affinity for enzyme binding, are very likely to bind and may compete with other drugs for the same site. Drugs with a Ki less than 2.0 μM are typically considered *potent* inhibitors. When drugs are coadministered, the drug with the greater affinity (lower Ki) will competitively inhibit the binding of the drug with lower affinity (high Ki) (Owen and Nemeroff 1998) (see Figure 2–2). Some drugs bind to an enzyme and inhibit its activity without needing the enzyme for their own metabolism. Some drugs may be both substrates of an enzyme (i.e., may require that enzyme for metabolism) and inhibitors of the same enzyme. Drugs that inhibit an enzyme may slow down the enzyme's activity or block the activity needed for metabolism of other drugs there, which will result in increased levels of any drug dependent on that enzyme for biotransformation. This inhibition leads to prolonged pharmacological effect and may result in drug toxicity. Inhibition is immediate in its effect, and when treatment with the offending drug is discontinued, the enzyme quickly returns to normal function. Effects resulting from drug inhibition have rapid onset and disappear quickly.

The P450 system is greatly affected by competitive inhibition. Ketoconazole has a low Ki for 3A4, or a high affinity for enzyme binding at 3A4. When coadministered with terfenadine, this drug prevents the metabolism of the parent compound of terfenadine, leading to an increased serum level of toxic terfenadine, which causes arrhythmias (Monahan et al. 1990). Without coadministration of a potent inhibitor, terfenadine is rapidly metabolized at 3A4 to its nontoxic but pharmacodynamically active compound fexofenadine. Terfenadine was removed from the market due to this interaction. Each human P450 enzyme is fully discussed in Part II of this manual, with tables of known inhibitors provided.

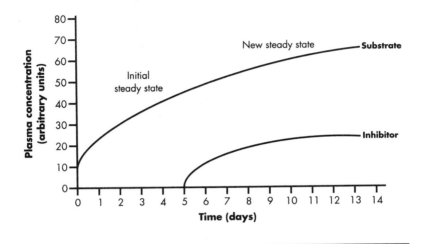

FIGURE 2–2.   Alteration in plasma kinetics in the presence of a potent *inhibitor* for that P450 enzyme and substrate.

Induction

Some xenobiotics (drugs and environmental substances such as cigarette smoke) increase the synthesis of P450 proteins, actually increasing the number of sites available for biotransformation. When more sites are available, the metabolic activity of an enzyme increases. This induction process may lead to decreases in the amount of parent drug administered and increases in the amount of metabolites produced. An inducer may cause the level of a coadministered drug to decrease to below the level needed for therapeutic effect, resulting in loss of clinical efficacy (see Figure 2–3). For example, coadministration of the potent inducer rifampin with methadone has led to opiate withdrawal (see Chapter 18, "Pain Management II: Narcotic Analgesics"). Drugs that require P450 metabolism for activation may have increased side effects or even become toxic if the active metabolites are produced more quickly than expected. Such a phenomenon is hypothesized to occur with the induction of valproic acid: the production of toxic metabolites is increased, leading to hepatotoxicity (see Chapter 15, "Neurology"). 3A4, 2D6, 1A2, 2C9, 2C19, and 2E1 may all be induced.

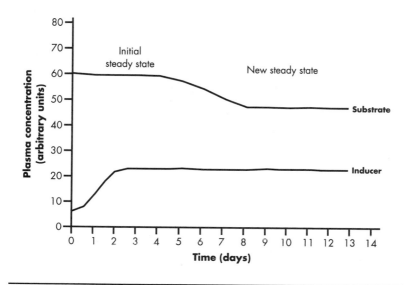

FIGURE 2–3. Alteration in plasma kinetics in the presence of a potent *inducer* for that P450 enzyme and substrate.

Pharmacogenetics: Polymorphisms

Clinicians have noted for decades that drug metabolism varies greatly across racial groups. The concept of racially polymorphic enzymes makes evolutionary sense. Enzymes exist in part to metabolize exogenous substances in human environments, and as a result different populations developed different enzymatic capabilities based on exposure to their particular environmental stressors.

Each person has two copies (or *alleles*) of each gene. The most common pair of alleles is usually referred to as the "wild type." Some individuals will have variations to the wild-type alleles. These variations are called *genetic polymorphisms*. Genetic polymorphisms can arise from a variety of alterations of the alleles, including minor base pair changes, missense mutations, entire allelic deletions, and extra alleles. Individuals with nonworking or missing alleles are considered *poor metabolizers* (PMs): their metabolism is slower, meaning they are less able to biotransform a compound at a specific enzyme site compared with the rest of a population. *Ultraextensive metabolizers* (UEMs) biotransform a compound at a much higher rate than the rest of the population and frequently require more drug than expected to achieve therapeutic effect. They typically have extra allelic copies of the wild-type genome. *Extensive metabolizers* (EMs) biotransform at a rate typical of the population. EMs, however, can be "converted" into PMs by adding an inhibitor resulting in slower metabolism, effectively making an individual a phenotypic PM despite his or her genetic status. Most P450 and some phase II enzymes are polymorphic. Further information on specific enzymatic polymorphisms is provided throughout this text.

Summary

Drug interactions occur for many reasons. Pharmacokinetic interactions are due to an alteration in drug absorption, distribution, metabolism, or elimination. Most drug interactions are due to alterations in enzymatic processes of phase I and phase II metabolism. Much of this variability is due to alterations in the P450 system through the environmental and genetic influences of inhibition, induction, and genetic variability. Despite this, an understanding of phase I metabolism alone does not explain why there is so much variability in patient response to drugs.

References

DeVane CL: Principles of pharmacokinetics and pharmacodynamics, in The American Psychiatric Press Textbook of Psychopharmacology, 2nd Edition. Edited by Schatzberg AF, Nemeroff CB. Washington, DC, American Psychiatric Press, 1998, pp 155–169

Monahan BP, Ferguson CL, Killeavy ES, et al: Torsades de pointes occurring in association with terfenadine use. JAMA 264:2788–2790, 1990

Owen JR, Nemeroff CB: New antidepressants and the cytochrome P450 system: focus on venlafaxine, nefazodone and mirtazapine. Depress Anxiety 7 (suppl 1):24–32, 1998

Wilkinson GR: Pharmacokinetics: the dynamics of drug absorption, distribution, and elimination, in Goodman and Gilman's The Pharmacological Basis of Therapeutics, 10th Edition. Edited by Hardman JG, Limbird LE, Gilman AG. New York, McGraw-Hill, 2001, pp 3–29

3

Metabolism in Depth: Phase II

Michael H. Court, B.V.Sc., D.A.C.V.A., Ph.D.

In phase II or conjugation reactions, water-soluble molecules are added to a drug, usually making an inactive, easily excreted metabolite. Covalent linkages are made by specific enzymes between the drug and glucuronic acid, sulfate, acetate, amino acids, and/or glutathione (Wilkinson 2001) (Table 3–1).

Until recently, phase II enzymes were not well understood because most research focused on the P450 or phase I enzymes. Phase I enzymes came to researchers' attention first because these enzymes are more apt to produce active or toxic metabolites. The vast majority of clinically important drug interactions involve P450. Nevertheless, the attention paid to phase II enzymes has increased, and it has become clear that this system also has the potential for significant metabolic drug interactions. What follows is an overview of the three most thoroughly studied phase II enzymatic processes: glucuronidation, sulfation, and methylation (see Table 3–1 and Figure 3–1).

TABLE 3–1.   Major reactions involved in phase II drug metabolism

Reaction	Examples
Conjugation reactions	**Phase II**
Glucuronidation	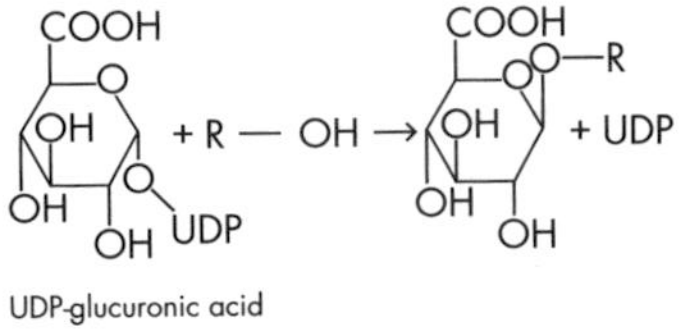 Acetaminophen, morphine, oxazepam, lorazepam

UDP-glucuronic acid

| Sulfation | 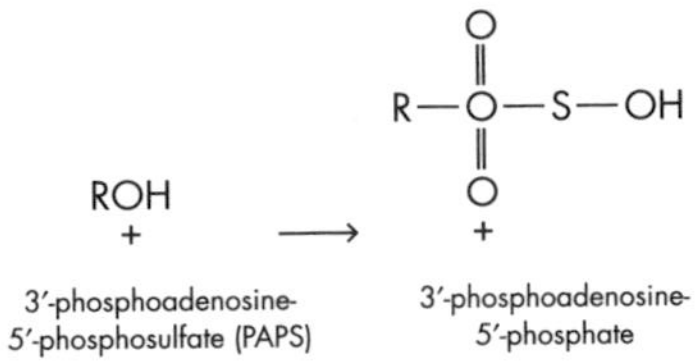 Acetaminophen, steroids, methyldopa |

ROH

3'-phosphoadenosine-
5'-phosphosulfate (PAPS)

3'-phosphoadenosine-
5'-phosphate

| Acetylation | 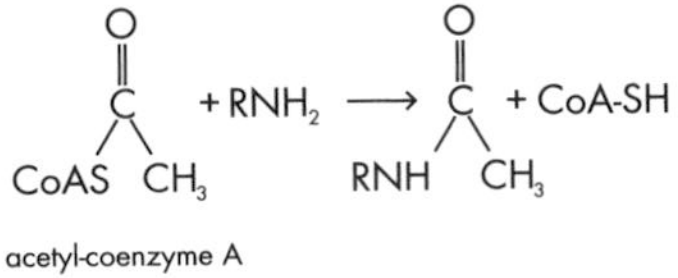 Sulfonamides, isoniazid, dapsone, clonazepam |

acetyl-coenzyme A

Source.   Adapted from Wilkinson GR: "Pharmacokinetics: The Dynamics of Drug Absorption, Distribution, and Elimination," in *Goodman and Gilman's The Pharmacological Basis of Therapeutics*, 10th Edition. Edited by Hardman JG, Limbird LE, Gilman AG. New York, McGraw-Hill, 2001, p. 14. With permission of the McGraw-Hill Companies.

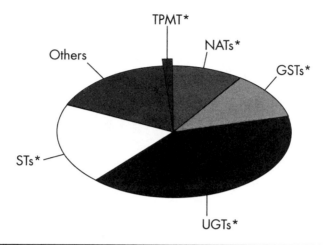

FIGURE 3–1. Proportion of drugs metabolized by the major phase II enzymes.

The relative size of each pie section indicates the estimated percentage of phase II metabolism that each enzyme contributes to the metabolism of drugs based on literature reports. Enzymes that have allelic variants with altered function are indicated by an asterisk (*). In many cases, more than one enzyme is involved in a particular drug's metabolism. GST=glutathione *S*-transferase; NAT=*N*-acetyltransferase; ST=sulfotransferase; TPMT=thiopurine methyltransferase; UGT=uridine 5′ diphosphate glucuronosyltransferase.

Source. Adapted from Wilkinson GR: "Pharmacokinetics: The Dynamics of Drug Absorption, Distribution, and Elimination," in *Goodman & Gilman's The Pharmacological Basis of Therapeutics,* 10th Edition. Edited by Hardman JG, Limbird LE, Gilman AG. New York, McGraw-Hill, 2001, p. 15. With permission of the McGraw-Hill Companies.

Glucuronidation

The most abundant phase II enzymes belong to the family of uridine 5′-diphosphate glucuronosyltransferases (UGTs). This enzyme family constitutes a major detoxification system, and the liver is the main site of glucuronidation (McGurk et al. 1998). UGTs are also found throughout the gastrointestinal tract, where they are an integral part of prehepatic first-pass metabolism (Tukey and Strassburg 2000). In addition, UGTs are found in the kidneys, brain, placenta, and other important sites in the body. UGTs are membrane proteins found in the endoplasmic reticulum and the nuclear membrane (Radominska-Pandya et al. 2002). In terms of both quantity of substrates and variety of substrates conjugated, UGTs are the most important phase II enzymes. In addition to drugs and exogenous compounds, many endogenous products, such as bilirubin, bile acids, thyroxine, and steroids, are substrates of the UGTs.

After glucuronidation, the glucuronidated drugs are typically excreted back into the intestine via the hepatobiliary system. These glucuronides are then broken down by intestinal bacterial β-glucuronidases, releasing the unconjugated drug, which is then reabsorbed into the systemic circulation. This enterohepatic recirculation system effectively "recycles" used drug, increasing plasma half-life and substantially increasing duration of action of drugs metabolized this way.

The vast majority of glucuronides are inactive regarding pharmacological or potential adverse effects, but there are several notable exceptions. One is morphine-6-glucuronide, a metabolite of morphine that is about 20 times more potent as an analgesic compound than morphine. Other major exceptions are the drug glucuronides that have an unstable ester linkage between the drug and glucuronide group (acyl glucuronides). Such glucuronides can bind to plasma proteins and produce hypersensitivity reactions. In 1993, zomepirac was withdrawn from sale in the United States because of a high incidence of hypersensitivity reactions, probably caused by its unstable acyl glucuronide (Spielberg 1994).

Many drugs are metabolized first by phase I metabolism and then by glucuronidation, but some drugs are directly conjugated by UGTs. Although most benzodiazepines are metabolized by the P450 system and then glucuronidated (see Chapter 19, "Psychiatry"), lorazepam, oxazepam, and temazepam are directly glucuronidated. Because UGTs are less affected than P450 enzymes by chronic hepatic disease, these three drugs are the preferred benzodiazepines for use in patients with chronic hepatic disease. More than 1,000 exogenous compounds—including chemicals, carcinogens, flavonoids, and drugs—are substrates of UGTs (Tukey and Strassburg 2000). Drugs primarily handled by UGTs include lamotrigine, valproate, nonsteroidal anti-inflammatory drugs, zidovudine, and most opioids (e.g., morphine and codeine). Like the P450 enzymes, UGTs have both unique and overlapping drug substrates and are vulnerable to drug inhibition and induction. Competitive inhibition between substrates of the same UGT may occur, and UGT inhibitors and inducers may be metabolized by other UGTs or through other drug metabolism systems.

In general, UGTs are considered less prone to drug-drug inhibition interactions than the P450 enzymes, in part because they tend to have lower affinity for drug substrates (and inhibitors) and have substantial overlap in substrate specificities (Williams et al. 2004). Researchers at the National Institutes of Health recently discovered a novel indirect molecular mechanism for drug-drug interaction that appears (so far) to be unique to the UGTs. Specifically, they discovered that to be fully active, the UGT enzymes must be phosphorylated at certain amino acid sites on the enzyme (Basu et al. 2003). Consequently, inhibition of the kinase enzyme (protein kinase C) that phosphorylates these amino acids on the UGTs has been shown to inhibit enzyme activity completely in various cell model systems. Most recently, it was shown that curcumin (a turmeric spice derivative, herbal medication, and kinase inhibitor) enhances the bioavailability and immune suppressant effects of mycophenolic acid in a mouse model, largely through inhibition of mycophenolic acid glucuronidation in the gut and liver (Basu et al. 2007). It remains unclear, however, as to whether these effects might occur in people given drugs or herbals that are kinase inhibitors.

All human UGT genes have been identified and sequenced, and, as in the case of P450 enzymes, a nomenclature system based on similarity of the predicted amino acid sequence of the enzyme has been developed: the root symbol (UGT) is followed by the family (Arabic numeral), subfamily (uppercase letter), and individual gene (Arabic numeral)—for example, *UGT1A1*. The Web site address for the committee responsible for UGT nomenclature is http://som.flinders.edu.au/FUSA/ClinPharm/UGT.

Specific probe substrates (drugs metabolized by one UGT isoenzyme) are starting to be identified for each UGT (Court 2005; Miners et al. 2006) (Tables 3–2 and 3–3). So far, however, there is little consensus among researchers on specific UGT inducers and inhibitors. Furthermore, most current information about UGT-based drug interactions was obtained through *in vitro* experiments, which may not be directly relevant to living individuals. For example, there is *in vitro* evidence that diclofenac significantly inhibits codeine glucuronidation (Ammon et al. 2000), but *in vivo* testing has failed to support this interaction (Ammon et al. 2002). Some drugs currently known to be substrates of the seven best-characterized hepatic UGTs are listed in Tables 3–2 and 3–3.

UGT subfamily 1A and UGT family 2 enzymes are the most important in drug metabolism, and the liver is the most abundant site for these enzymes. Subfamily UGT1A enzymes found in the liver include UGT1A1, UGT1A3, UGT1A4, UGT1A6, and UGT1A9. They are all products of a single complex gene located on chromosome 2. They have in common exons 2–5 encoding the C-terminal half of the enzyme protein, but exon 1 encoding the N-terminal half of the protein is unique to each enzyme. Bilirubin is glucuronidated exclusively by UGT1A1, and variants of the gene encoding UGT1A1 are responsible for the rare congenital unconjugated hyperbilirubinemias, Crigler-Najjar syndrome type I (completely inactivated enzyme) and Crigler-Najjar syndrome type II (partially active enzyme). Individuals with Gilbert syndrome (affecting 8%–12% of Caucasian populations) have a fluctuating mild unconjugated hyperbilirubinemia caused by a dinucleotide insertion polymorphism (*UGT1A1*28*) in the UGT1A1 gene promoter that decreases gene transcription.

Like slow metabolizers of the P450 enzyme 2D6, individuals with the congenital unconjugated hyperbilirubinemias have increased blood concentrations of other substrates of UGT1A1. Because UGT1A1 is inducible by phenobarbital, this drug has been used to reduce levels of bilirubin in patients with partially inactive UGT1A1 (Jansen 1999; Sugatani et al. 2001). Individuals with Gilbert syndrome who are given the anticancer drug irinotecan have been shown to have an increased risk of toxicity because UGT1A1 is the major metabolic pathway for its active metabolite, SN-38 (Ando et al. 2002; Iyer et al. 1999, 2002). Genotyping individuals for UGT1A1 polymorphisms before treatment with irinotecan has been advocated to decrease side effects (diarrhea and immunosuppression) and enhance efficacy, although such an approach has yet to be validated by prospective clinical trials (Rauchschwalbe et al. 2002). A preliminary study suggests that oral pretreatment of patients with the dietary bioflavonoid chrysin decreases the incidence of irinotecan diarrhea, presumably through induction of UGT1A1 in the gastrointestinal mucosa (Tobin et al. 2006). Unconjugated hyperbilirubinemia without hepatocellular injury is a common side effect (6%–25% incidence) associated with treatment of HIV infection with the protease inhibitors indinavir and atazanavir (Rotger et al. 2005). This effect results from inhibition of UGT1A1-mediated bilirubin glucuronidation by these drugs, which is more profound in patients with Gilbert's syndrome (Rotger et al. 2005; Zucker et al. 2001).

Because UGT1A4 catalyzes many of the amine-containing drugs, many psychotropic agents (e.g., amitriptyline, chlorpromazine, imipramine, lamotrigine, olanzapine, clozapine, and promethazine) are substrates. Probenecid, a broadly potent UGT inhibitor, affects the clearance of olanzapine but not of risperidone, the latter being handled primarily by P450 enzymes (Markowitz et al. 2002).

Unlike UGT1A, the UGT2B enzymes, including UGT2B4, UGT2B7, UGT2B10, UGT2B11, UGT2B15, and UGT2B17, are produced from separate genes, all located on chromosome 4. The best characterized of these with regard to drug metabolism are UGT2B7 and UGT2B15. Other findings suggest UGT2B10 may be the principal enzyme responsible for glucuronidation of nicotine in the liver (Kaivosaari et al. 2007).

UGT2B7 has the widest substrate spectrum of all the UGTs; drugs catalyzed include nonsteroidal anti-inflammatory drugs (e.g., naproxen, ibuprofen, ketoprofen), zidovudine, epirubicin, chloramphenicol (Barbier et al. 2000; Radominska-Pandya et al. 2001), and most of the opioid drugs, including morphine and codeine (see Table 3–2). UGT2B7 has a common allelic variant form in about 50% of Caucasians that differs only by a single amino acid (H268Y, tyrosine substituted for a histidine at position 268). However, results of *in vitro* and clinical studies suggest the effects of this variant on enzyme function are relatively small at best. Decreased clearance of naproxen in subjects coadministered valproic acid results in part from inhibition of naproxen glucuronidation by UGT2B7 (Addison et al. 2000). Similarly, fluconazole has been shown to inhibit the clearance of coadministered zidovudine by UGT2B7, although the extent of the interaction was relatively small (about a 50% increase in area under the plasma concentration-time curve [AUC]) (Brockmeyer et al. 1997).

TABLE 3–2. Some UGT1A substrates

	1A1	1A3	1A4	1A6	1A9
Chromosome	2	2	2	2	2
Polymorphism*	UGT1A1*28	T31C, G81A, T140C	P24T, L48V	T181A, R184S	C3Y, M33T
In vitro effect?	Yes	Yes	Yes	Yes	Yes
In vivo effect?	Yes	Unknown	Unknown	Equivocal	Unknown
Some endogenous substrates	Bilirubin Estriol	Estrones	Androsterone Progestins	Serotonin	2-Hydroxyestradiol Thyroxine
Some drug substrates	Acetaminophen Atorvastatin Buprenorphine Cerivastatin Ciprofibrate Clofibrate Ethinyl estradiol Etoposide Gemfibrozil *Nalorphine* *Naltrexone* Simvastatin SN-38	*Amitriptyline* Atorvastatin Buprenorphine Cerivastatin *Chlorpromazine* *Clozapine* Cyproheptadine Diclofenac Diflunisal *Diphenhydramine* *Doxepin* Fenoprofen Gemfibrozil	*Amitriptyline* Chlorpromazine Clozapine Cyproheptadine Diphenhydramine Doxepin 4-hydroxytamoxifen Lamotrigine Loxapine Meperidine Olanzapine Promethazine Retigabine	Acetaminophen Entacapone Ketoprofen Naftazone Salicylate	Acetaminophen Clofibric acid Dapsone Diclofenac Diflunisal *Ethinyl estradiol* Estrone Flavonoids *Furosemide* *Ibuprofen* *Ketoprofen* Laberalol *Mefenamic acid*

TABLE 3–2.   Some UGT1A substrates *(continued)*

1A1	1A3	1A4	1A6	1A9
Telmisartan	*4-Hydroxytamoxifen*	Trifluoperazine		*Naproxen*
Troglitazone	Ibuprofen			R-Oxazepam
	Imipramine			Propofol
	R-Lorazepam			Propranolol
	Losartan			Retinoic acid
	Loxapine			*SN-38*
	Morphine			Tolcapone
	Nalorphine			*Valproate*
	Naloxone			
	Naltrexone			
	Naproxen			
	Naringenin			
	Norbuprenorphine			
	Promethazine			
	Simvastatin			
	SN-38			
	Tripelennamine			
	Valproate			

Note. *Italics* indicate that the UGT is a minor pathway for the substrate. <u>Underline</u> indicates that this compound may be a specific probe substrate for the UGT. CN-I=Crigler-Najjar syndrome type I; CN-II=Crigler-Najjar syndrome type II; GS=Gilbert syndrome; UGT=uridine 5'-diphosphate glucuronosyltransferase.   *See updated listing maintained at http://galien.pha.ulaval.ca/alleles/alleles.html.

TABLE 3–3.   Some UGT2B substrates

	2B7	2B10	2B15
Chromosome	4	4	4
Polymorphism*	H268Y	Yes	D85Y
In vitro effect?	Equivocal	Yes	Yes
In vivo effect?	Equivocal	Unknown	Yes
Some endogenous substrates	Androsterone Bile acids	Arachidonic and linoleic acid metabolites	Catechol estrogens Testosterone
Some substrate drugs	Chloramphenicol Clofibric acid Codeine Cyclosporine Diclofenac *Entacapone* Epirubicin *Fenoprofen* Hydromorphone *Ibuprofen* Ketoprofen Lorazepam Losartan	Nicotine	Dienestrol *Entacapone* *S–Lorazepam* *S–Oxazepam* *Tolcapone*

TABLE 3–3. Some UGT2B substrates *(continued)*

2B7	2B10	2B15
Morphine		
Nalorphine		
Naloxone		
Naltrexone		
Naproxen		
Norcodeine		
R-Oxazepam		
Oxycodone		
Tacrolimus		
Temazepam		
Tolcapone		
Valproate		
Zidovudine		
Zomepirac		

Note. *Italics* indicate that the UGT is a minor pathway for the substrate. Underline indicates that this compound may be a specific probe substrate for the UGT. CN-I=Crigler-Najjar syndrome type I; CN-II=Crigler-Najjar syndrome type II; UGT=uridine 5′-diphosphate glucuronosyltransferase.

*See updated listing maintained at http://galien.pha.ulaval.ca/alleles/alleles.html.

UGT2B15 is the main enzyme responsible for glucuronidation of the 3-hydroxybenzodiazepines oxazepam and lorazepam (and probably temazepam) (Court 2005). This enzyme is polymorphic; 50% of Caucasians have a tyrosine substituted for an aspartate amino acid at position 85 (Court et al. 2004). Several studies show this polymorphism results in reduced enzyme activity for various substrates *in vitro* and also reduced clearance and effects of lorazepam administered to volunteer subjects (Chung et al. 2005; Court et al. 2004). There is also evidence this enzyme is more highly expressed in the livers of men than women (Court et al. 2004). Catalytic inhibition of UGT2B15 by valproic acid accounts for the decreased clearance of lorazepam by glucuronidation reported for subjects coadministered these drugs (Samara et al. 1997). This interaction also appears to be clinically significant in that coma attributed to increased lorazepam levels was reported in a patient receiving both of these drugs (S.A. Lee at al. 2002).

Some UGT levels are diminished in neonates (Burchell et al. 1989; de Wildt et al. 1999), and some decrease in the elderly (Sonne 1993). Some UGTs are more efficient in men than in women, which may account for differences in concentrations of acetaminophen and other drugs that are substrates of the UGTs (Meibohm et al. 2002; Morissette et al. 2001). Like P450 enzymes, glucuronidation is affected by many environmental factors—smoking (Benowitz and Jacob 2000; Yue et al. 1994), foods (e.g., watercress [Hecht et al. 1999]), and disease states (e.g., hypothyroidism [Sonne 1993])—although specifics regarding individual UGTs are scarce.

Sulfation

Sulfation, or sulfonation (an older term), is another important phase II drug conjugation mechanism. The process involves the transfer of a sulfuryl group from a donor substrate, 3′-phosphoadenosine 5′-phosphosulfate (PAPS), to an acceptor compound via sulfotransferase (SULT). Sulfation is dependent on the availability of PAPS, as PAPS can be easily depleted in human liver in a matter of minutes (Klaassen and Boles 1997), although more can be made quickly. This "turnover" of PAPS, along with sulfation and desulfation (by sulfatase enzyme), creates an active recycling process within tissues.

A nomenclature system has been proposed that is similar to the systems used for the UGT and P450 enzymes, including a cytosolic superfamily of

10 SULTs that are tissue-specific and substrate-specific (Blanchard et al. 2004). There are two families of important drug-conjugating human SULTs: SULT1 and SULT2. Subfamilies SULT1A, SULT1B, SULT1C, SULT1E, SULT2A, and SULT2B handle small endogenous compounds (e.g., steroids, thyroid hormones, catechol estrones, neurotransmitters) and exogenous compounds (e.g., ethinyl estradiol, acetaminophen, budesonide [SULT2A1 {Meloche et al. 2002}]). A major step in troglitazone's metabolism is sulfation (SULT1A1 [Honma et al. 2002]). Although sulfation of many drugs produces water-soluble, less-toxic metabolites, other conjugates are active (e.g., minoxidil), and some are reactive and can "attach" to DNA or RNA. Additionally, sulfation often produces promutagens (Glatt 2000).

Less information is available about specific SULT inhibitors, inducers, and substrates than UGT inhibitors, inducers, and substrates, but environmental influences are known to occur. One of the more interesting is the apparent inhibition of SULTs by components such as vanillin, wine, tartrazine (a synthetic coloring agent), tea, and coffee (Burchell and Coughtrie 1997; Coughtrie et al. 1998). Several SULTs are known to be polymorphic as well (e.g., SULT1A1), but the clinical significance is unknown (Raftogianis et al. 2000).

Methylation

Enzyme subfamilies that conjugate through methylation are almost too numerous to count. The effects of S-adenosylmethionine (SAMe, or "sammy"), is used to treat illnesses including depression, may be related to methylation enzymes, because SAMe is a methyl donor compound (Bottiglieri and Hyland 1994). Three of the better-studied and more interesting methyltransferases are catechol O-methyltransferase (COMT), histamine N-methyltransferase (HNMT), and thiopurine methyltransferase (TPMT).

Catechol O-Methyltransferase

The best appreciated of the methyltransferases, COMT has been known for more than 70 years (Kopin 1994). COMT is one of two means by which catecholamines are metabolized, the other being by monoamine oxidases (a non-P450 phase I system). COMTs also metabolize steroid catechols. COMTs are located in virtually all regions, the two most important of which may be the brain and the liver. There are two main forms of COMT: membrane-bound COMT and soluble COMT (Ellingson et al. 1999), and genetic polymorphisms have been noted (Kuchel 1994). Three COMT genes have been identified that differ in activity level (M.S. Lee et al. 2002). Many studies have examined the genetic diversity of COMT as a factor in illnesses, including breast cancer, hypertension, obsessive-compulsive disorder, bipolar disorder, and schizophrenia.

Selective and reversible inhibition of COMT was developed as a strategy for enhancing dopamine availability in patients with Parkinson's disease (Goldstein and Lieberman 1992). Tolcapone and entacapone are COMT-inhibiting drugs, but how they inhibit COMT remains unclear.

Histamine *N*-Methyltransferase

HNMT metabolizes histamine by adding a methyl group to histamine; then monoamine oxidase further catabolizes the methylhistamine. HNMT is located in many regions, particularly the liver and the kidney (De Santi et al. 1998). Its gene is located on chromosome 2, and HNMT has genetic polymorphic variability, but the clinical significance of the polymorphisms is currently unclear (Preuss et al. 1998). A common group of inhibitors, the 4-aminoquinolones (e.g., chloroquine and hydroxychloroquine), are believed to be inhibitors of HNMT, and this inhibition may account for their usefulness in treating rheumatoid arthritis, systemic lupus erythematosus (Cumming et al. 1990), and other inflammatory disorders. Another group of inhibitors of HNMTs are steroidal neuromuscular relaxants, such as pancuronium and vecuronium. Inhibition of HNMT may explain instances of flushing and hypotension with the use of these drugs (Futo et al. 1990).

Thiopurine Methyltransferase

TPMT is best known for its rare polymorphic character; a complete absence of this enzyme occurs in about 10% of Caucasian populations (McLeod and Siva 2002). In fact, TPMT's routine phenotyping in patients with leukemia who require drugs such as 6-mercaptopurine is perhaps the first use of phenotype testing to enter general practice (Weinshilboum et al. 1999). Other drugs known to be conjugated by TPMT are thioguanine and azathioprine (Nishida et al. 2002). In women, TPMT has a slightly lower activity (Meibohm et al. 2002).

References

Addison RS, Parker-Scott SL, Hooper WD, et al: Effect of naproxen co-administration on valproate disposition. Biopharm Drug Dispos 21: 235–242, 2000

Ammon S, von Richter O, Hofmann U, et al: In vitro interaction of codeine and diclofenac. Drug Metab Dispos 28:1149–1152, 2000

Ammon S, Marx C, Behrens C, et al: Diclofenac does not interact with codeine metabolism in vivo: a study in healthy volunteers. BMC Clin Pharmacol 2:2, 2002

Ando Y, Ueoka H, Sugiyama T, et al: Polymorphisms of UDP-glucuronosyltransferase and pharmacokinetics of irinotecan. Ther Drug Monit 24: 111–116, 2002

Barbier O, Turgeon D, Girard C, et al: 3'-Azido-3'-deoxythimidine (AZT) is glucuronidated by human UDP-glucuronosyltransferase 2B7 (UGT2B7). Drug Metab Dispos 28:497–502, 2000

Basu NK, Kole L, Owens IS: Evidence for phosphorylation requirement for human bilirubin UDP-glucuronosyltransferase (UGT1A1) activity. Biochem Biophys Res Commun 303:98–104, 2003

Basu NK, Kole L, Basu M, et al: Targeted inhibition of glucuronidation markedly improves drug efficacy in mice: a model. Biochem Biophys Res Commun 360:7–13, 2007

Benowitz NL, Jacob P 3rd: Effects of cigarette smoking and carbon monoxide on nicotine and cotinine metabolism. Clin Pharmacol Ther 67:653–659, 2000

Blanchard RL, Freimuth RR, Buck J, et al: A proposed nomenclature system for the cytosolic sulfotransferase (SULT) superfamily. Pharmacogenetics 14:199–211, 2004

Bottiglieri T, Hyland K: S-Adenosylmethionine levels in psychiatric and neurological disorders: a review. Acta Neurol Scand Suppl 154:19–26, 1994

Brockmeyer NH, Tillmann I, Mertins L, et al: Pharmacokinetic interaction of fluconazole and zidovudine in HIV-positive patients. Eur J Med Res 2:377–383, 1997

Burchell B, Coughtrie MW: Genetic and environmental factors associated with variation of human xenobiotic glucuronidation and sulfation. Environ Health Perspect 105 (suppl 4):739–747, 1997

Burchell B, Coughtrie M, Jackson M, et al: Development of human liver UDP-glucuronosyltransferases. Dev Pharmacol Ther 13:70–77, 1989

Chung JY, Cho JY, Yu KS, et al: Effect of the UGT2B15 genotype on the pharmacokinetics, pharmacodynamics, and drug interactions of intravenous lorazepam in healthy volunteers. Clin Pharmacol Ther 77:486–494, 2005

Coughtrie MW, Sharp S, Maxwell K, et al: Biology and function of the reversible sulfation pathway catalysed by human sulfotransferases and sulfatases. Chem Biol Interact 109:3–27, 1998

Court MH: Isoform-selective probe substrates for in vitro studies of human UDP-glucuronosyltransferases. Methods Enzymol 400:104–116, 2005

Court MH, Duan SX, Guillemette C, et al: Stereoselective conjugation of oxazepam by human UDP-glucuronosyltransferases (UGTs): S-oxazepam is glucuronidated by UGT2B15, while R-oxazepam is glucuronidated by UGT2B7 and UGT1A9. Drug Metab Dispos 30:1257–1265, 2002

Court MH, Hao Q, Krishnaswamy S, et al: UDP-glucuronosyltransferase (UGT) 2B15 pharmacogenetics: UGT2B15 D85Y genotype and gender are major determinants of oxazepam glucuronidation by human liver. J Pharmacol Exp Ther 310:656–665, 2004

Cumming P, Reiner PB, Vincent SR: Inhibition of rat brain histamine-N-methyltransferase by 9-amino-1,2,3,4-tetrahydroacridine (THA). Biochem Pharmacol 40:1345–1350, 1990

De Santi C, Donatelli P, Giulianotti PC, et al: Interindividual variability of histamine N-methyltransferase in the human liver and kidney. Xenobiotica 28:571–577, 1998

de Wildt SN, Kearns GL, Leeder JS, et al: Glucuronidation in humans: pharmacogenetic and developmental aspects. Clin Pharmacokinet 36:439–452, 1999

Ellingson T, Duppempudi S, Greenberg BD, et al: Determination of differential activities of soluble and membrane-bound catechol-O-methyltransferase in tissues and erythrocytes. J Chromatogr B Biomed Sci Appl 729:347–353, 1999

Futo J, Kupferberg JP, Moss J: Inhibition of histamine N-methyltransferase (HNMT) in vitro by neuromuscular relaxants. Biochem Pharmacol 39:415–420, 1990

Glatt H: Sulfotransferases in the bioactivation of xenobiotics. Chem Biol Interact 129:141–170, 2000

Goldstein M, Lieberman A: The role of the regulatory enzymes of catecholamine synthesis in Parkinson's disease. Neurology 42:8–14, 41–48, 1992

Hecht SS, Carmella SG, Murphy SE: Effects of watercress consumption on urinary metabolites of nicotine in smokers. Cancer Epidemiol Biomarkers Prev 8:907–913, 1999

Honma W, Shimada M, Sasano H, et al: Phenol sulfotransferase, ST1A3, as the main enzyme catalyzing sulfation of troglitazone in human liver. Drug Metab Dispos 30:944–952, 2002

Iyer L, Hall D, Das S, et al: Phenotype-genotype correlation of in vitro SN38 (active metabolite of irinotecan) and bilirubin glucuronidation in human liver tissue with UGT1A1 promoter polymorphism. Clin Pharmacol Ther 65:576–582, 1999

Iyer L, Das S, Janisch L, et al: UGT1A1*28 polymorphism as a determinant of irinotecan disposition and toxicity. Pharmacogenomics J 2:43–47, 2002

Jansen PL: Diagnosis and management of Crigler-Najjar syndrome. Eur J Pediatr 158 (suppl 2):S89–S94, 1999

Kaivosaari S, Toivonen P, Hesse LM, et al: Nicotine glucuronidation and the human UDP-glucuronosyltransferase UGT2B10. Mol Pharmacol 72:761–768, 2007

Klaassen CD, Boles JW: Sulfation and sulfotransferases 5: the importance of 3'-phosphoadenosine 5'-phosphosulfate (PAPS) in the regulation of sulfation. FASEB J 11:404–418, 1997

Kopin IJ: Monoamine oxidase and catecholamine metabolism. J Neural Transm Suppl 41:57–67, 1994

Kuchel O: Clinical implications of genetic and acquired defects in catecholamine synthesis and metabolism. Clin Invest Med 17:354–373, 1994

Lee MS, Kim HS, Cho EK, et al: COMT genotype and effectiveness of entacapone in patients with fluctuating Parkinson's disease. Neurology 58:564–567, 2002

Lee SA, Lee JK, Heo K: Coma probably induced by lorazepam-valproate interaction. Seizure 11:124–125, 2002

Markowitz JS, DeVane CL, Liston HL, et al: The effects of probenecid on the disposition of risperidone and olanzapine in healthy volunteers. Clin Pharmacol Ther 71:30–38, 2002

McGurk KA, Brierley CH, Burchell B: Drug glucuronidation by human renal UDP-glucuronosyltransferases. Biochem Pharmacol 55:1005–1012, 1998

McLeod HL, Siva C: The thiopurine S-methyltransferase gene locus—implications for clinical pharmacogenomics. Pharmacogenomics 3:89–98, 2002

Meibohm B, Beierle I, Derendorf H: How important are gender differences in pharmacokinetics? Clin Pharmacokinet 41:329–342, 2002

Meloche CA, Sharma V, Swedmark S, et al: Sulfation of budesonide by human cytosolic sulfotransferase, dehydroepiandrosterone-sulfotransferase (DHEA-ST). Drug Metab Dispos 30:582–585, 2002

Miners JO, Knights KM, Houston JB, et al: In vitro-in vivo correlation for drugs and other compounds eliminated by glucuronidation in humans: pitfalls and promises. Biochem Pharmacol 71:1531–1539, 2006

Morissette P, Albert C, Busque S, et al: In vivo higher glucuronidation of mycophenolic acid in male than in female recipients of a cadaveric kidney allograft and under immunosuppressive therapy with mycophenolate mofetil. Ther Drug Monit 23:520–525, 2001

Nishida A, Kubota T, Yamada Y, et al: Thiopurine S-methyltransferase activity in Japanese subjects: metabolic activity of 6-mercaptopurine 6-methylation in different TPMT genotypes. Clin Chim Acta 323:147–150, 2002

Preuss CV, Wood TC, Szumlanski CL, et al: Human histamine N-methyltransferase pharmacogenetics: common genetic polymorphisms that alter activity. Mol Pharmacol 53:708–717, 1998

Radominska-Pandya A, Little JM, Czernik PJ: Human UDP-glucuronosyltransferase 2B7. Curr Drug Metab 2:283–298, 2001

Radominska-Pandya A, Pokrovskaya ID, Xu J, et al: Nuclear UDP-glucuronosyltransferases: identification of UGT2B7 and UGT1A6 in human liver nuclear membranes. Arch Biochem Biophys 399:37–48, 2002

Raftogianis R, Creveling C, Weinshilboum R, et al: Estrogen metabolism by conjugation. J Natl Cancer Inst Monogr 27:113–124, 2000

Rauchschwalbe SK, Zuhlsdorf MT, Schuhly U, et al: Predicting the risk of sporadic elevated bilirubin levels and diagnosing Gilbert's syndrome by genotyping UGT1A1*28 promoter polymorphism. Int J Clin Pharmacol Ther 40:233–240, 2002

Rotger M, Taffe P, Bleiber G, et al: Gilbert syndrome and the development of antiretroviral therapy-associated hyperbilirubinemia. J Infect Dis 192:1381–1386, 2005

Samara EE, Granneman RG, Witt GF, et al: Effect of valproate on the pharmacokinetics and pharmacodynamics of lorazepam. J Clin Pharmacol 37:442–450, 1997

Sonne J: Factors and conditions affecting the glucuronidation of oxazepam. Pharmacol Toxicol 73 (suppl 1):1–23, 1993

Spielberg TE: Examining product risk in context: the case of zomepirac (letter). JAMA 272:1252, 1994

Sugatani J, Kojima H, Ueda A, et al: The phenobarbital response enhancer module in the human bilirubin UDP-glucuronosyltransferase UGT1A1 gene and regulation by the nuclear receptor CAR. Hepatology 33:1232–1238, 2001

Tobin PJ, Beale P, Noney L, et al: A pilot study on the safety of combining chrysin, a non-absorbable inducer of UGT1A1, and irinotecan (CPT-11) to treat metastatic colorectal cancer. Cancer Chemother Pharmacol 57:309–316, 2006

Tukey RH, Strassburg CP: Human UDP-glucuronosyltransferases: metabolism, expression, and disease. Annu Rev Pharmacol Toxicol 40:581–616, 2000

Weinshilboum RM, Otterness DM, Szumlanski CL: Methylation pharmacogenetics: catechol O-methyltransferase, thiopurine methyltransferase, and histamine N-methyltransferase. Annu Rev Pharmacol Toxicol 39:19–52, 1999

Wilkinson GR: Pharmacokinetics: the dynamics of drug absorption, distribution, and elimination, in Goodman & Gilman's The Pharmacological Basis of Therapeutics, 10th Edition. Edited by Hardman JG, Limbird LE, Gilman AG. New York, McGraw-Hill, 2001, pp 3–29

Williams JA, Hyland R, Jones BC, et al: Drug-drug interactions for UDP-glucurono-syltransferase substrates: a pharmacokinetic explanation for typically observed low exposure (AUCI/AUC) ratios. Drug Metab Dispos 32:1201–1208, 2004

Yue QY, Tomson T, Sawe J: Carbamazepine and cigarette smoking induce differentially the metabolism of codeine in man. Pharmacogenetics 4: 193–198, 1994

Zucker SD, Qin X, Rouster SD, et al: Mechanism of indinavir-induced hyperbiliru-binemia. Proc Natl Acad Sci USA 98:12671–12676, 2001

4

Transporters

Jessica R. Oesterheld, M.D.

The previous chapters of this section discussed the two major detoxification systems of the body: phase I metabolism such as that performed by the P450 cytochromes, and phase II metabolism such as that involving the uridine 5′-diphosphate glucuronosyltransferases (UGTs). Understanding the basics of both is essential to mastering a majority of drug interactions. Beyond the basics of phase I and phase II metabolism, in recent years drug-drug interactions have been discovered that are based on the inhibiting or inducing of specialized membrane-bound proteins that act as transporters. As the potentially important nature of these transporters has become apparent, some authors have started labeling these transporters "phase 0" or "phase III" (Decleves et al. 2006).

Until 20 years ago, it was believed that only lipid-soluble drugs diffused across the apical border of the intestine, whereas some small hydrophilic ionized drugs could pass through paracellular junctions, crossing the cytosol and passing through the basolateral border to the portal circulation (see Figure 4–1). Further study has shown that non–lipid-soluble drugs are indeed transported (both influxed and effluxed) across both the apical and the basolateral borders by membrane-embedded transporters. Transporters are present not only in the intestine, where they provide a barrier to absorption, but also in hepatocytes, renal tubule cells, blood-brain barrier, and many other sites. In these locations, transporters affect the transfer of drugs in and out of organs, into blood and target cells, and through the placenta and blood-brain barrier. Transporters also play an important role in drug elimination through biliary and urinary excretion.

These transporters are members of two "superfamilies": the ATP-binding cassette (ABC) transporters and the solute-linked carrier (SLC) transporters. The ABC transporters are a single family with a shared structural element and ontologic history. The SLC transporter families are a diverse group that share no common structural elements and transport a wide array of exogenous and endogenous products. In general, ABC transporters are effluxing or exporting transporters and SLC transporters are influxing or importing transporters.

The nomenclature for both superfamilies can be found at http:// www.genenames.org/cgi-bin/hgnc_search.pl. As is the case for P450 enzymes and UGTs, the nomenclature follows the family, subfamily, and isoform convention in which an integer represents the family (e.g., 1, 2); a letter, the subfamily (e.g., A, B, C); and an integer, the isoform (e.g., 1, 2). Because they are all in a single family, members of the ABC family all have the same ABC designation, and the first integer is skipped. The ABC designation is followed by the subfamily letter and the isoform integer (e.g., ABCB1). Because members of the SLC superfamily are not in a single family, their designation, SLC, is followed by an integer (family), letter (subfamily), and integer (isoform). There is a single exception in the SLC nomenclature. The SLC subfamily 21 is designated as SLCO instead of SLC (e.g., SLCO1A2). In addition to their official names, previous symbols and previous names or "aliases" can also be found at the Web address above. In this chapter, both the official name and the common alias will be provided except for ABCB1.

This chapter is intended to provide clinicians with basic information necessary to understand transporter-based drug-drug interactions (DDIs), and thus it begins with an overview of the ABC and the SLC transporter families. An individual transporter from each superfamily will be reviewed in some detail to provide a model of understanding transporter-based DDIs. For the ABCs, ABCB1 will be highlighted, and for the SLCs, SLCO1B1 (OATP1B1), a hepatic influxing transporter, will be reviewed.

Because transporters have different organ placements and unique membrane siting, either apical or basolateral, a diagram of where they are located is essential (Figure 4–1). Study of the anatomical sites of transporters is continuing, and thus the provided diagrams are simplified and should be viewed only as an approximation. Diagrammatic representations such as these may require substantial revision as more becomes known about the sites of transporters.

ABC Transporters

By the 1970s, it was known that certain cancers (e.g., leukemias, breast cancer) could acquire resistance to previously effective oncology drugs. This process was named multidrug resistance (MDR), and the question of why many drugs become ineffective at the same time has been gradually answered. The *MDR1* gene on chromosome 7 of cancer cells was found to overexpress a protein that can actively transport some cancer drugs (e.g., vinca alkaloids, paclitaxel, and others) across membranes in association with an energy source, adenosine triphosphate (ATP). The membrane-based glycoprotein was named P-glycoprotein (P-gp; *P* designates permeability). MDR develops when the *MDR1* gene directs P-gp to multiply and then to extrude all its substrates from or to "bounce" all substrates out of the cell. MDR develops because many P-gp substrates (i.e. oncological drugs) have been affected at the same time.

P-gp, as mentioned earlier, is one of a superfamily of transporters with a distinctive sequence of ATP binding domains (nucleotide-binding folds), and the family was thus named the ATP-binding cassette (ABC) transporter family. Although a few are influx transporters, most are efflux transporters. A Web site of ABC transporters that describes what is known of the subfamilies and their functions has been established at http://www.ncbi.nlm.nih.gov/books/bv.fcgi?rid=mono_001.TOC&depth=2. According to the nomenclature, there are seven subfamilies of ABC transporters, annotated A–G (see Table 4–1).

P-gp was the first member of this superfamily to be cloned; it was eventually placed in the B subfamily and is now named ABCB1. An alternate name for P-gp is *multiresistant drug 1* (MDR1). Of the nearly 50 known human ABC transporters, several genetic variations of ABC transporters have been associated with specific diseases, such as cystic fibrosis, adrenoleukodystrophy, Tangier disease, and Dubin–Johnson syndrome (see Table 4–1). Like ABCB1,

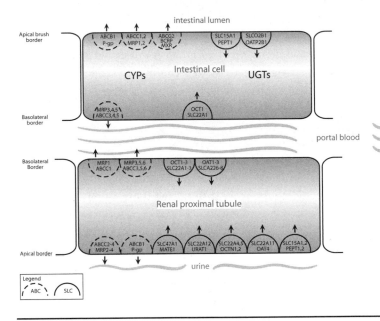

FIGURE 4–1.   Major transporters in humans.

Transporters are identified by ABC (ATP-binding cassette) or SLC (solute-linked carrier) nomenclature, preceded or followed by the transporter's previous name or alias.

other ABC members have been associated with oncological or antibiotic resistance [e.g., ABCC1-6 (MRP1-6), ABCC10 (MRP7), and ABCG2 (BCRP or MXR)]. Members of this superfamily transport endogenous compounds (e.g., sterols, lipids, large polypeptides, metabolic products), and at least 10 members from 3 different subfamilies have been associated with drug transport (see Table 4–2).

ABCB1 as a Drug Transporter

ABCB1 is the best characterized of the ABC drug transporters. Although it was known for many years that ABCB1 transported endogenous substances such as steroids or cytokines (Conrad et al. 2001), only recently has a pharmacokinetic role in drug interactions been characterized for these transporters.

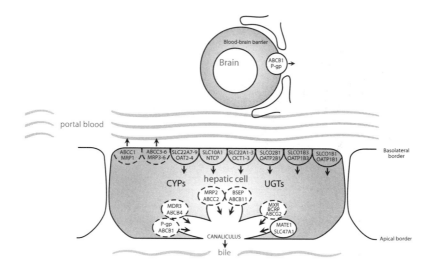

FIGURE 4–1. *(Continued)*

CYPs = cytochrome P450 enzymes; UGTs = uridine 5'-diphosphate glucuronosyltransferases.

ABCB1 is present on the apical surface of villus tips of enterocytes in the intestine in the colon, gonads, renal proximal tubules, placenta, and biliary system, as well as in the capillary endothelial cells of the blood-brain barrier and in other structures. ABCB1 can transport certain hydrophobic substances, including neutral and cationic organic compounds, across cells into the gut, into bile, into urine, and out of the gonads and the brain. The gonads and the brain can be considered "drug-free zones," and ABCB1s are postulated to play a protective role in the body by excluding drugs from these areas (Wilkinson 2001).

Drugs can be ABCB1 substrates, inhibitors, and/or inducers (see Table 4–3), and the same drugs may also involve other transporters or enzymes. A drug may not involve ABCB1 in any way, but some drugs are substrates and/or inhibitors or inducers of both ABCB1 and 3A4.

TABLE 4–1.   Human ABC transporter superfamily

Subfamily	Some functions	Example Transporter[a]	Example Associations
ABCA (12 members)	Transport cholesterol, vitamin A	ABCA1 (CERP) ABCA3 (ABCC)	Tangier disease Surfactant deficiency of newborn
ABCB (11 members)	Transport drugs, bile	ABCB1 (P-gp)	Multidrug resistance, transporters
ABCC (13 members)	Organic anion transport; toxin secretion	ABCC2 (MRP2) ABCC7 (CFTR)	Dubin-Johnson syndrome Cystic fibrosis
ABCD (4 members)	Peroxisomal transport, long-chain fatty acids	ABCD1 (ALD)	X-linked adrenoleukodystrophy
ABCE (1 member)	Immunity		
ABCF (3 members)	Unknown		
ABCG (6 members)	Transport lipids, bile, cholesterol	ABCG8 (GBD4)	Sitosterolemia

[a]Example transporters are identified by ABC (ATP-binding cassette) nomenclature with previous name shown in parentheses.

The *drug efflux-metabolism alliance* is a term coined to describe the joint ABCB1 and 3A4 functioning, particularly in the intestine and liver, for drugs that are substrates of both ABCB1 and 3A4 (Benet and Cummins 2001). In the intestine (see Figure 4–1), because ABCB1 is embedded in the apical border, drugs first come in contact with ABCB1 before becoming available to 3A4. ABCB1 repeatedly recycles drugs across the intestinal lumen to make them available to 3A4 at intracellular concentrations of drug within a range that 3A4 can manage. ABCB1 also effluxes the metabolic degradation products produced by 3A4 into the gut (Christians et al. 2005). 3A4 metabolites then pass through the basolateral border to the portal vein (see Figure 4–1). In the liver, drugs first pass through the basolateral border of the hepatocyte via an array of ABC and SLC transporters and then meet 3A4, other P450s, and phase II conjugation systems to undergo metabolism and conjugation (see Figure 4–1). There is substantial evidence that 3A4 and ABCB1 are coregulated by nuclear receptors in both the intestine and in the liver, but this regulation does not appear to occur between organs (Christians et al. 2005). In the kidney, ABCB1 transporters exist on the apical membrane and "bounce" drugs out into the urine. At the luminal surface of the blood-brain barrier, ABCB1 transporters protect the brain by effluxing a variety of substrates.

ABCB1-Based Drug Interactions

Intestine and Kidney

If ABCB1 function is inhibited or induced, changes in absorption or excretion can occur. Table 4–3 lists some drugs that are known to be *in vitro* ABCB1 substrates, inhibitors, or inducers. For example, quinidine has been known to increase digoxin levels. For many years, the mechanism of this interaction was undetermined, but it was known that P450 enzymes were not involved, since digoxin is largely unmetabolized. Quinidine has been shown to inhibit ABCB1 activity in the kidney, and digoxin is a known ABCB1 substrate. During coadministration, digoxin excretion is inhibited by quinidine's ABCB1 inhibition (Fromm et al. 1999; Su and Huang 1996). Other drugs that increase digoxin concentrations by ABCB1 inhibition either in the intestine or in the kidney include ritonavir (Ding et al. 2004), mifepristone (Woodland et al. 2003), clarithromycin (Tanaka et al 2003; Verschraagen et al. 1999; Wakasugi

TABLE 4–2.   Some ABC drug transporters

Name[a]	Site	Substrates	Inhibitors	Inducers
ABCB1 (P-gp)	I,L,K,B	Broad specificity See Table 4–3	Cyclosporine Erythrcmycin Ketoconazole Quinidine For others, see Table 4–3	Rifampin St. John's wort For others, see Table 4–3
ABCB4 (MDR3)	L	Digoxin Vinblastine	Cyclosporine Verapamil	
ABCB11 (BSEP)	L	Vinblastine	Rifampin	
ABCC1 (MRP1)	WS	Adefovir Daunorubicin Doxorubicin Indinavir Methotrexate Vincristine	Cyclosporine Indomethacin Probenecid Sulfinpyrazone	Chlorambucil Epirubicin
ABCC2 (MRP2)	I,L,K	Beta-lactam antibiotics Etoposide Indinavir Indinavir Methotrexate Pravastatin Probenecid Vinblastine Vincristine	Cyclosporine Furosemide Indomethacin Probenecid	Carbamazepine

TABLE 4–2. Some ABC drug transporters *(continued)*

Name[a]	Site	Substrates	Inhibitors	Inducers
ABCC3 (MRP3)	I,L,K	Etoposide Methotrexate Teniposide Vincristine		
ABCC4 (MRP4)	I,K	Methotrexate	Dipyridamole Probenecid	
ABCC5 (MRP5)	WS	Mercaptopurine Methotrexate	Probenecid Sildenafil Sulfinpyrazone	
ABCC6 (MRP6)	L,K	Cisplatin Daunorubicin Mercaptopurine		Retinoids
ABCG2 (BCRP, MXR)	I,L,K	Daunorubicin Doxorubicin Etoposide Mitoxantrone Pitavastatin Prazosin Topotecan Rosuvastatin Zidovudine	Cyclosporine Imatinib Reserpine Ritonavir Saquinavir Tamoxifen	

Note. ABC=ATP-binding cassette; I=intestine; L=liver; K=kidney; B=blood-brain barrier; WS=widespread.
[a]Transporters are identified by ABC nomenclature with previous name shown in parentheses.

TABLE 4–3.   Some ABCB1 nonsubstrates, substrates, inhibitors, and inducers shown by in vitro studies

Nonsubstrates	Substrates	Inhibitors	Inducers
Alfentanil	Aldosterone	Amiodarone	Dexamethasone
Amantadine	Amitriptyline	Amitriptyline	Doxorubicin
Chlorpheniramine	Amoxicillin	Atorvastatin	?Nefazodone (chronic)
Clozapine	Amprenavir	Bromocriptine	Phenobarbital
Fentanyl	Carbamazepine	Chloroquine	Prazosin
Fluconazole	Chloroquine	Chlorpromazine	Rifampin
Flunitrazepam	Cimetidine	Cimetidine	Ritonavir (chronic)
Fluoxetine	Ciprofloxacin	Clarithromycin	St. John's wort
Haloperidol	Colchicine	Cyclosporine	Trazodone
Itraconazole	Corticosteroids	Cyproheptadine	?Venlafaxine
Ketoconazole	Cyclosporine	Desipramine	
Lidocaine	Daunorubicin	Diltiazem	
Methotrexate	Digitoxin	Erythromycin	
Midazolam	Digoxin	Felodipine	
Mirtazapine	Diltiazem	Fentanyl	
Sumatriptan	Docetaxel	Fluphenazine	
Yohimbine	Doxorubicin	Garlic	
	Enoxacin	Grapefruit juice	
	Epirubicin	Green tea (catechins)	
	Erythromycin	Haloperidol	

TABLE 4–3. Some ABCB1 nonsubstrates, substrates, inhibitors, and inducers shown by in vitro studies *(continued)*

Nonsubstrates	Substrates	Inhibitors	Inducers
	Estradiol	Hydrocortisone	
	Fexofenadine	Hydroxyzine	
	Glyburide	Imipramine	
	Indinavir	Itraconazole	
	Irinotecan	Ketoconazole	
	Lansoprazole	Lansoprazole	
	Loperamide	Lidocaine	
	Losartan	Lovastatin	
	Lovastatin	Maprotiline	
	Mibefradil	Methadone	
	Morphine	Mibefradil	
	Nelfinavir	Midazolam	
	Nortriptyline	Nefazodone (acute)	
	Ondansetron	Nelfinavir	
	Paclitaxel	Omeprazole	
	Phenytoin	Orange juice (Seville)	
	Quetiapine	Pantoprazole	
	Quinidine	Phenothiazines	
	Ranitidine	Pimozide	
	Rifampin	Piperine	
	Risperidone	Probenecid	

TABLE 4–3.   Some ABCB1 nonsubstrates, substrates, inhibitors, and inducers shown by in vitro studies (*continued*)

Nonsubstrates	Substrates	Inhibitors	Inducers
	Ritonavir	Progesterone	
	Saquinavir	Propafenone	
	Tacrolimus	Propranolol	
	Talinolol	Quinidine	
	Teniposide	Ritonavir (initial)	
	Terfenadine	Saquinavir	
	Verapamil	Simvastatin	
	Vinblastine	Spironolactone	
	Vincristine	Tamoxifen	
		Terfenadine	
		Testosterone	
		Trifluoperazine	
		Valspodar	
		Verapamil	
		Vinblastine	
		Vitamin E	

et al. 1998), propafenone (Woodland et al. 1997), and itraconazole (Jalava et al. 1997).

In contrast, rifampin has been shown to decrease digoxin concentrations by inducing ABCB1 in the intestinal tract (Greiner et al. 1999; Larsen et al. 2007). Another ABCB1 inducer, St. John's wort, can also increase intestinal ABCB1 transport 1.5-fold (Durr et al. 2000). When high-dose St. John's wort is administered with digoxin, the area under the plasma concentration-time curve (AUC) of digoxin can be significantly reduced (Mueller et al. 2004). St John's wort has also been shown to reduce the AUC of talinolol, another ABCB1 substrate, by as much as 31% via induction of intestinal ABCB1 [Schwarz et al. 2007].

Blood-Brain Barrier

ABCB1 and other transporters [e.g., ABCG2 (BCRP) and ABCC2-4 (MRP2-4)] regulate the concentrations of various chemicals in the central nervous system (CNS) via transport, or "pumping out," of these chemicals. Most psychotropic drugs, however, are not substrates of these transporters and therefore achieve adequate CNS concentrations. This finding makes intuitive sense given that psychotropics require certain concentrations within the CNS to be effective. Similarly, all first-generation antihistamines cause sedation, unlike second-generation antihistamines. Second-generation antihistamines are ABCB1 substrates, unlike the nonsubstrate first-generation. As a substrate, second-generation antihistamines are regulated out of the brain and are thus less sedating (Chishty et al. 2001).

Loperamide (Imodium A-D), an over-the-counter opiate antidiarrheal, is an ABCB1 substrate. Under normal circumstances, loperamide is transported out of the brain by ABCB1 and therefore has no central opiate effects. When loperamide is given with quinidine, which is an ABCB1 inhibitor, loperamide concentrations in the brain increase, and signs of respiratory depression can ensue (Sadeque et al. 2000). Unlike in the intestine and the kidney, drug clearance is not altered, and there is no increase in systemic blood concentrations of the offending drug. The only clues to the existence of an interaction are CNS-related symptoms.

Many oncological drugs are ABCB1 substrates and are excluded from the brain at the blood-brain barrier. In cases of primary or secondary brain tumors, if an ABCB1 inhibitor could be found that would block the extrusion of the oncological drug from the blood-brain barrier without causing toxicity, the efficacy of such drugs might be improved. Several generations of ABCB inhibitors have been evaluated with modest success. A recent study of paclitaxel with elacridar, a third-generation ABCB1 inhibitor, has shown promise (Marchetti et al. 2007). A similar targeted strategy for improving drug efficacy could be useful for a wide range of disorders requiring specific or increased CNS drug levels.

ABCB1-Based and 3A4-Based Drug Interactions

While many drugs are both ABCB1 and 3A4 substrates, many are also ABCB1 and 3A4 inhibitors or inducers (Wacher et al. 1995). These "double inhibitors" (e.g., ketoconazole, erythromycin) combined with double substrates are very likely to produce clinically significant drug-drug interactions. Examples of clinically relevant DDIs involving double inhibitors and double substrates include the following: inhibitor ritonavir, substrate saquinavir; inhibitor erythromycin, substrate cyclosporine; inhibitor verapamil, substrate tacrolimus (Marchetti et al. 2007). Examples of a double inducer and double substrates are the following: inducer: St. John's wort, substrate: tacrolimus; and inducer: St. John's wort, substrate: indinavir (Marchetti et al. 2007).

Current Status of ABCB1-based Drug Interactions

Clearly, more study is needed to fully elucidate the role of ABCB1 in drug interactions. Since intestinal ABCB1 kinetics are "saturable," ABCB1 can play a part in drug interactions only when the therapeutic concentration of the substrate drug is low (e.g., digoxin, fexofenadine, talinolol [Lin 2003]).

Many factors can alter ABCB1 function and influence ABCB1-based interactions. Genetic differences in ABCB1 are known to exist. More than 100 variations have been identified, and most are single-nucleotide polymorphisms (SNPs) (Kimchi-Sarfaty et al. 2007; Marchetti et al. 2007). A polymorphism at exon 26 (C3435) has been shown to influence the level of intestinal ABCB1 and the concentration of digoxin. Variant C3435TT shows decreased ABCB1 and increased digoxin levels (Lin and Yamakazi 2003). This genotype has also been shown to be a risk factor for the side effect of orthostatic hypotension when given the ABCB1 substrate nortriptyline (Roberts et al. 2002). There also appear to be racial differences in occurrence of this polymorphism (Kim et al. 2001; Schaeffeler et al. 2001). Women may have significantly lower hepatic ABCB1 levels than men, which may account for more efficient metabolism of certain drugs by women (Cummins et al. 2002; Meibohm et al. 2002). Herbal supplements (e.g., St. John's wort) and even vitamins—certain formulations of vitamin E are ABCB1 inhibitors—can influence ABCB1s. Certain foods can alter ABCB1 function; for example, black pepper (Bhardwaj et al. 2002), grapefruit juice, and Seville orange juice (Di Marco et al. 2002) are ABCB1 inhibitors. Hormones such as levothyroxine (Siegmund et al. 2002) may affect ABCB1 transport as well.

Because ABCB1-based drug interactions are a relatively new concept, well-substantiated reports of clinical interactions are relatively few and sometimes conflicting. Contradictory findings in studies may be due to different methods in measuring ABCB1 as well as the difficulties in singling out ABCB1 effects from those of other transporters and of P450 CYPs, particularly the 3A subfamily (Fischer et al. 2005). Although *in vitro* models and intestinal biopsy of ABCB1 have been available, the development of mice lacking one or both *MDR1* alleles (knockout mice) has provided more direct evidence of ABCB1-based drug interactions (Marchetti et al. 2007).

SLC Transporters

The superfamily of SLC transporters numbers more than 300 members in 47 families. Although all superfamily members are functionally involved in the transport of solutes, individual families have no structural or ontological relation to each other. The SLC superfamily concept is a human construct. Members of each family have at least a 20% homology of their amino acids. None use ATP as an energy source, but they transport solutes by coupling with another solute transport or using favorable gradients. A relatively comprehensive list of SLC transporters (without the newest addition, SLC47) can be found at http://www.bioparadigms.org/slc/menu.asp.

Some SLC families known to be involved in drug transport include 1) proton oligopeptide transporters (SLC15); 2) organic anion transporting polpeptides (SLC21, known as SLCO); 3) organic cation, anion, and zwitterion transporters (SLC22); and 4) multidrug and toxin extrusion transporters (SLC47). Table 4–4 provides an overview of some individual transporters in these families. The anatomic distribution and representative substrates and inhibitors have been determined largely by *in vitro* methods. These transporters are organ-specific and involve only a single membrane in the organ, and they may transport only anionic or cationic drugs. Both ABC and SLC transporters can be involved at each membrane (effluxing and influxing) and, depending on the organ, phase I and phase II may also make a contribution to a drug's metabolism in the intestine and liver (see Figure 4–1). Only a modest amount of information on documented *in vivo* SLCO-based DDIs exists at this time.

SLCO1B1

SLCO1B1 (OATP1B1) is an influx transporter situated on the basolateral side, the portal vein side, of the liver. SLCO1B1 (OATP1B1) has 80% homology with SLCO1B3 (OATP1B3) and shares some substrate overlap (Smith et al. 2005). In addition to exogenous compounds, these two transporters handle endogenous compounds: bilirubin and its UGT conjugates, bile acids, hormones, and others (Smith et al. 2005).

Inhibition of hepatic uptake by this transporter leads to increased plasma levels of SLCO1B1 (OATP1B1) substrates. As can be seen in Figure 4–1, a second transporter is responsible for transporting a drug into the biliry canaliculus. As an example, pravastatin has been shown to be a SLCO1B1 (OATP1B1) and other SLCO substrate at the basolateral side (influxing) and an ABCC2 (MRP2) substrate at the apical side (effluxing) (Nakagomi-Hagihara et al. 2007). Although phase I and phase II enzymes and conjugates are situated in the hepatic parenchyma between the membranes, for pravastatin, metabolism through the 3A4 and UGT system is minimal. Gemfibrozil inhibits the influx transporter SLCO1B1 (OATP1) but not the biliary effluxer ABCC2 (MRP2) (Nakagomi-Hagihara et al. 2007). Inhibition of SLCO1B1 (OATP1B1) is the mechanism of action responsible for the 202% increase in pravastatin's AUC when combining gemfibrozil and pravastatin (Goosen et al. 2007; Pravachol 2007).

Interactions between the statins and gemfibrozil have been clinically investigated over the last few years. All of the statins are SLCO1B1 substrates, according to *in vitro* evidence. Increases in the plasma levels of statins may be associated with an increased risk of rhabdomyolisis. This risk factor is independent from the additive risk for this adverse effect from combining these drugs, which in itself may cause rhabdomyolisis individually. Most of the statins' AUCs are increased by about 200%–300% when gemfibrozil is added, but the AUCs of cerivistatin and its lactone are increased more than 400%, because in addition to gemfibrozil's inhibitory effect on SLCO1B1 (OATP1B1), cerivistatin is a 2C8 substrate and gemfibrozil is a 2C8 inhibitor (Backman et al. 2002). Cerivastatin was removed from the U.S. market in 2001.

All statins are SLCO1B1 (OATP1B1) substrates, and the open acid forms (cerivistatin, fluvastatin, and atorvastatin) may be especially likely to have hepatic transporters play a role in their clearance. Surprisingly, however, the addition of gemfibrozil to fluvastatin and atorvastatin, unlike the addition of cerivistatin, does not lead to substantial increases in the AUC of either (Goosen et al. 2007; Pravachol 2007). The reason for this unexpected result has not yet been elucidated.

The finding that many drugs are potent SLCO1B1 (OATP1B1) inhibitors has fueled the clinical evaluation of other potentially serious DDIs. Gemfibrozil has been shown to increase the plasma concentration of repaglinide more than 20-fold since the latter is both a 2C8 and a SLCO1B1 (OATP1B1) substrate (Niemi et al. 2003). Cyclosporine is also a potent SLCO1B1 (OATP1B1) and 3A4 inhibitor and is involved in many transporter-based DDIs of the substrates listed in Table 4–4. As an example, repaglinide, a 2C9, 3A4 and SLCO1B1 (OATP1B1) substrate, has its maximum concentration (C_{max}) increased more than 175% with the addition of cyclosporine (Kajosaari et al. 2005).

There has been at least one surprising SLCO1B1 (OATP1B1) inhibitor. Bosentan is metabolized by 2C9 and 3A4, and it is vulnerable to 2C9 and 3A4 inhibitors and inducers. However, it has been shown that its concentration is increased by rifampin (6-fold) (Tracleer 2007). Because rifampin is a 2C9 and 3A4 inducer, this puzzling outcome has only lately been solved by the finding that bosentan is a SLCO1B1 (OATP1B1) and SLCO1B3 (OATP1B3) substrate and rifampin a SLCO1B (OATP) inhibitor (Treiber et al. 2007). Through a similar mechanism of inhibiting SLCO1B1 (OATP1B1), rifampin also increases atorvastatin acid and its lactone metabolites (Lau et al. 2007).

TABLE 4–4.   Some SLC drug transporters

SLC	Aliases	Organ	Substrate drugs	Inhibitor drugs
SLC10A1	NTCP	I,L,K	Rosuvastatin	Cyclosporine
SLC15A1	PEPT1	I, (K)	Some dipeptides and tripeptides; some peptidomimetic drugs Amoxicillin Ampicillin Captopril Cefaclor Cephalexin Enalapril Lisinopril Midodrine Temocapril Valacyclovir Valganciclovir	Glibenclamide Repaglinide Tolbutamide
SLC15A2	PEPT2	K	Some dipeptides and tripeptides; some peptidomimetic drugs Amoxicillin Ampicillin Captopril Cefaclor Cefadroxil Valacyclovir	Amoxicillin Cefaclor Glibenclamide Tolbutamide

TABLE 4–4. Some SLC drug transporters *(continued)*

SLC	Aliases	Organ	Substrate drugs	Inhibitor drugs
SLC21	OAT1B1	L	Atorvastatin	Atorvastatin
SLCO1B1	OATP-C		Atrasentan	Clarithromycin
	OATP2		Bosentan	Cyclosporine
	LST-1		Cerivastatin	Gemfibrozil
			Fexofenadine	Indinavir
			Fluvastatin	Lovastatin
			Methotrexate	Pioglitazone
			Nateglinide	Pravastatin
			Olmesartan	Rifampin
			Penicillin G	Ritonavir
			Pitavastatin	Rosiglitazone
			Pravastatin	Saquinavir
			Repaglinide	Simvastatin
			Rifampin	
			Rosuvastatin	
			Simvastatin	
			SN-38	
			Thyroxine	
			Valsartan	

TABLE 4–4. Some SLC drug transporters *(continued)*

SLC	Aliases	Organ	Substrate drugs	Inhibitor drugs
SLC21 SLCO1B3	OATP1B3 OATP-8	L	Bosentan Digoxin Fexofenadine Fluvastatin Methotrexate Olmesartan Paclitaxel Pitavastatin Pravastatin Rifampin Rosuvastatin Telmisartan	Cyclosporine Pravastatin Rifampin
SLC21 SLCO1A2	OATP1A2	L		
SLC21 SLCO2B1	OATP-B OAT-RP2 Oatp9 Moat1	W/S	Fexofenadine Fluvastatin Penicillin G Pravastatin Rosuvastatin	

TABLE 4–4. Some SLC drug transporters *(continued)*

SLC	Aliases	Organ	Substrate drugs	Inhibitor drugs
SLC22A1	OCT-1 hOCT-1	I,L	Acyclovir Amantadine Desipramine Famotidine Ganciclovir	Disopyramide Midazolam Phenformin Quinidine Quinine Ritonavir Verapamil
SLC22A2	OCT-2 hOCT2	K	Amantadine Cimetidine Cisplatin Memantine Oxaliplatin	Cimetidine Quinine
SLC22A3	OCT-3 hOCT-3 EMT hEMT	K,L	Cimetidine Cisplatin	Desipramine Prazosin
SLC22A4	OCTN-1 hOCTN1	K	Quinidine Verapamil	Clonidine

TABLE 4–4. Some SLC drug transporters *(continued)*

SLC	Aliases	Organ	Substrate drugs	Inhibitor drugs
SLC22A5	OCTN-2 hOCTN-2 CT1 SCD	I,K,L	Quinidine Verapamil	Cimetidine Clonidine
SLC22A6	OAT-1 hOAT-1 PAHT NKT	K	Acyclovir Adefovir Cidofovir Methotrexate Mycophenolic acid Tenofovir Zidovudine	Amoxicillin Cefaclor Cefadroxil Cefamandole Cefazolin Ethacrynic acid Furosemide Pravastatin Probenecid
SLC22A7	OAT-2 hOAT-2 NLT	K,L	Allopurincl Erythromycin Fluorouracil Paclitaxel Tetracycline Zidovudine	Pravastatin

TABLE 4–4. Some SLC drug transporters *(continued)*

SLC	Aliases	Organ	Substrate drugs	Inhibitor drugs
SLC22A8	OAT-3 hOAT-3	K	Cimetidine Fexofenadine Methotrexate Mycophenolic acid Penicillin G Pravastatin Rosuvastatin Tetracycline Zidovudine	Cefadroxil Cefamandole Cefazolin Indomethacin Phenylbutazone Probenecid Salicylate
SLC47A1	MATE-1 FLJ10847	K,L	Cimetidine Cisplatin Oxaliplatin Probenecid Procainamide Quinidine Verapamil	Cimetidine Verapamil Quinidine
SLC47A2	MATE-2 MATE2-K FLJ31196	K	Cimetidine Procainamide Quinidine Quinine Verapamil	

Note. I=intestine; K=kidney; L=liver; SLC=solute-linked carrier; WS=widespread.

A complicating factor in possible rifampin DDIs is that rifampin is itself a substrate of SLCO1B1 (OATP1B1) and changes in its intracellular concentration may change its ability to influence the pregnane X receptor and therefore the induction of phase I and phase II (Tirona et al. 2003).

Genetic Polymorphisms

SLCO1B1 (OATP1B1) has many polymorphisms. A case has been reported showing an association with *SLCO1B1*15* (OATP1B1) and SN-38 toxicity (Takane et al. 2007). Higher pravastatin levels are associated with this polymorphism, which is present in 15% of European Americans and 1% of African Americans (Ho et al. 2007). This polymorphism is also associated with a higher AUC of atorvastatin (Pasanen et al. 2007), pitavastatin (Ieiri et al. 2007), and simvastatin (Pasanen et al. 2006). Fluvastatin is not affected (Niemi et al. 2006).

It is possible that because SLCO1B1 (OATP1B1) handles bilirubin as well as exogenous compounds, alterations in concentrations of this transporter through genetic polymorphisms can contribute to drug-induced hyperbilirubinemias (Campbell et al. 2004).

Current Status of SLC Transporter-Based DDIs

The documentation of SLC transporter-based DDIs remains in its infancy. The nomenclature has been standardized in the last few years, and additional transporters are still being discovered. SLC47A1 (MATE-1), an organic cation transporter responsible for the final steps of uptake in the renal proximal tubule, was added to the nomenclature only in the last 2 years. Standardization of *in vitro* methods to identify substrates, inhibitors, and inducers of transporters is ongoing. There are problems of *in vitro* methodology when one of the drugs is both a substrate and an inhibitor of a transporter. Further, *in vitro* data may not reflect *in vivo* realities. Some of the examples of SLCO1B1 (OATP1B1)–based DDIs given above exemplify the problems of teasing out the contribution of a single transporter when drugs have multiple substrates (including CYPs, UGTs, and other transporters), and the additive drug may affect multiple P450 CYPs, UGTs, and other transporters.

In the last few years, there has been an avalanche of new information about transporters and their involvement in possible DDIs. Much of the information remains largely hidden in pharmacology journals. Clinicians should try to find reviews of transporter-based DDIs in order to stay tuned to the new developments in this fast-changing area.

References

Backman JT, Kyrklund C, Neuvonen, et al: Gemfibrozil greatly increases plasma concentrations of cerivastatin. Clin Pharmacol Ther 72:685–691, 2002

Benet LZ, Cummins CL: The drug efflux-metabolism alliance: biochemical aspects. Adv Drug Deliv Rev 50 (suppl 1):S3–S11, 2001

Bhardwaj RK, Glaeser H, Becquemont L, et al: Piperine, a major constituent of black pepper, inhibits human P-glycoprotein and CYP3A4. J Pharmacol Exp Ther 302:645–650, 2002

Campbell SD, de Morais SM, Xu JJ: Inhibition of human organic anion transporting polypeptide OATP 1B1 as a mechanism of drug-induced hyperbilirubinemia. Chem Biol Interact 150:179–187, 2004

Chishty M, Reichel A, Siva J, et al: Affinity for the P-glycoprotein efflux pump at the blood-brain barrier may explain the lack of CNS side-effects of modern antihistamines. J Drug Target 9:223–228, 2001

Christians U, Schmitz V, Haschke M: Functional interactions between P-glycoprotein and CYP3A in drug metabolism. Expert Opin Drug Metab Toxicol 1:641–654, 2005

Conrad S, Kauffmann HM, Ito K, et al: Identification of human multidrug resistance protein 1 (MRP1) mutations and characterization of a G671V substitution. J Hum Genet 46:656–663, 2001

Cummins CL, Wu CY, Benet LZ: Sex-related differences in the clearance of cytochrome P450 3A4 substrates may be caused by P-glycoprotein. Clin Pharmacol Ther 72:474–489, 2002

Decleves X, Niel E, Debray M, et al: Is P-glycoprotein (ABCB1) a phase or a phase 3 colchicine transporter depending on colchicines exposure conditions? Toxicol Appl Pharmacol 21:153–160, 2006

Di Marco MP, Edwards DJ, Wainer IW, et al: The effect of grapefruit juice and seville orange juice on the pharmacokinetics of dextromethorphan: the rule of gut CYP3A and P-glycoprotein. Life Sci 71:1149–1160, 2002

Ding R, Tayrouz Y, Riedel KD, et al: Substantial pharmacokinetic interaction between digoxin and ritonavir in healthy volunteers. Clin Pharmacol Ther 76:73–84, 2004

Durr D, Stieger B, Kullak-Ublick GA, et al: St John's wort induces intestinal P-glycoprotein/MDR1 and intestinal and hepatic CYP3A4. Clin Pharmacol Ther 68:598–604, 2000

Fischer V, Einolf HJ, Cohen D: Efflux transporters and their clinical relevance. Mini Rev Med Chem 5:183–195, 2005

Fromm MF, Kim RB, Stein CM, et al: Inhibition of P-glycoprotein-mediated drug transport: a unifying mechanism to explain the interaction between digoxin and quinidine. Circulation 99:552–557, 1999

Goosen TC, Bauman JN, Davis JA, et al: Atorvastatin glucuronidation is minimally and nonselectively inhibited by the fibrates gemfibrozil, fenofibrate, and fenofibric acid. Drug Metab Dispos 35:1315–1324, 2007

Greiner B, Eichelbaum M, Fritz P, et al: The role of intestinal P-glycoprotein in the interaction of digoxin and rifampin. J Clin Invest 104:147–153, 1999

Ho RH, Choi L, Lee W, et al: Effect of drug transporter genotypes on pravastatin disposition in European- and African-American participants. Pharmacogenet Genomics 17:647–656, 2007

Ieiri I, Suwannakul S, Maeda K, et al: SLCO1B1 (OATP1B1, an uptake transporter) and ABCG2 (BCRP, an efflux transporter) variant alleles and pharmacokinetics of pitavastatin in healthy volunteers. Clin Pharmacol Ther 82:541–547, 2007

Jalava KM, Partanen J, Neuvonen PJ: Itraconazole decreases renal clearance of digoxin. Ther Drug Monit 19:609–613, 1997

Kajosaari LI, Niemi M, Neuvonen M, et al: Cyclosporine markedly raises the plasma concentrations of repaglinide. Clin Pharmacol Ther 78:388–399, 2005

Kim RB, Leake BF, Choo EF, et al: Identification of functionally variant MDR1 alleles among European Americans and African Americans. Clin Pharmacol Ther 70:189–199, 2001

Kimchi-Sarfaty C, Marple AH, Shinar S, et al: Ethnicity-related polymorphisms and haplotypes in the human ABCB1 gene. Pharmacogenomics 8:29–39, 2007

Larsen UL, Hyldahl Olesen L: Human intestinal P-glycoprotein activity estimated by the model substrate digoxin. Scand J Clin Lab Invest 67:123–134, 2007

Lau YY, Huang Y, Frassetto L, et al: Effect of OATP1B transporter inhibition on the pharmacokinetics of atorvastatin in healthy volunteers. Clin Pharmacol Ther 81:194–204, 2007

Lin JH: Drug-drug interaction mediated by inhibition and induction of P-glycoprotein. Adv Drug Deliv Rev 55:53–81, 2003

Lin JH, Yamazaki M: Role of P-glycoprotein in pharmacokinetics: clinical implications. Clin Pharmacokinet 42:59–98, 2003

Marchetti S, Mazzanti R, Beijnen JH, et al: Concise review: clinical relevance of drug drug and herb drug interactions mediated by the ABC transporter ABCB1 (MDR1, P-glycoprotein). Oncologist 12:927–941, 2007

Meibohm B, Bererle I, Derendorf H: How important are gender differences in pharmacokinetics? Clin Pharmacokinet 41:329–342, 2002

Mueller SC, Uehleke B, Woehling H, et al: Effect of St John's wort dose and preparations on the pharmacokinetics of digoxin. Clin Pharmacol Ther 75:546–557, 2004

Nakagomi-Hagihara R, Nakai D, Tokui T, et al: Gemfibrozil and its glucuronide inhibit the hepatic uptake of pravastatin mediated by OATP1B1. Xenobiotica 37:474–486, 2007

Niemi M, Backman JT, Neuvonen M, et al: Effects of gemfibrozil, itraconazole, and their combination on the pharmacokinetics and pharmacodynamics of repaglinide: potentially hazardous interaction between gemfibrozil and repaglinide. Diabetologia 46:347–351, 2003

Niemi M, Pasanen MK, Neuvonen PJ: SLCO1B1 polymorphism and sex affect the pharmacokinetics of pravastatin but not fluvastatin. Clin Pharmacol Ther 80:356–366, 2006

Pasanen MK, Neuvonen M, Neuvonen PJ, et al: SLCO1B1polymorphism markedly affects the pharmacokinetics of simvastatin acid. Pharmacogenet Genomics 16:873–879, 2006

Pasanen MK, Fredrikson H, Neuvonen PJ, et al: Different effects of SLCO1B1 polymorphism on the pharmacokinetics of atorvastatin and rosuvastatin. Clin Pharmacol Ther 82:726–733, 2007

Pravachol (package insert). Princeton, NJ, Bristol-Myers Squibb Co., 2007

Roberts RL, Joyce PR, Mulder RT, et al: A common P-glycoprotein polymorphism is associated with nortriptyline-induced postural hypotension in patients treated for major depression. Pharmacogenomics 2:191–196, 2002

Sadeque AJ, Wandel C, He H, et al: Increased drug delivery to the brain by P-glycoprotein inhibition. Clin Pharmacol Ther 68:231–237, 2000

Schaeffeler E, Eichelbaum M, Brinkmann U, et al: Frequency of C3435T polymorphism of MDR1 gene in African people (letter). Lancet 358:383–384, 2001

Schwarz UI, Hanso H, Oertel R, et al: Induction of intestinal P-glycoprotein by St John's wort reduces the oral bioavailability of talinolol. Clin Pharmacol Ther 81:669–678, 2007

Siegmund W, Altmannsberger S, Paneitz A, et al: Effect of levothyroxine administration on intestinal P-glycoprotein expression: consequences for drug disposition. Clin Pharmacol Ther 72:256–264, 2002

Smith NF, Figg WD, Sparreboom A: Role of the liver-specific transporters OATP1B1 and OATP1B3 in governing drug elimination. Expert Opin Drug Metab Toxicol 1:429–45, 2005

Su SF, Huang JD: Inhibition of the intestinal digoxin absorption and exsorption by quinidine. Drug Metab Dispos 24:142–147, 1996

Takane H, Miyata M, Burioka N, et al: Severe toxicities after irinotecan-based chemotherapy in a patient with lung cancer: a homozygote for the SLCO1B1*15 allele. Ther Drug Monit 29:666–668, 2007

Tanaka H, Matsumoto K, Ueno K, et al: Effect of clarithromycin on steady-state digoxin concentrations. Ann Pharmacother 37:178–181, 2003

Tirona RG, Leake BF, Wolkoff AW, et al: Human organic anion transporting polypeptide-C (SLC21A6) is a major determinant of rifampin-mediated pregnane X receptor activation. J Pharmacol Exp Ther 304:223–228, 2003

Tracleer (package insert). South San Francisco, CA, Actelion Pharmaceuticals US, Inc., 2007

Treiber A, Schneiter R, Hausler S, et al: Bosentan is a substrate of human OATP1B1 and OATP1B3: inhibition of hepatic uptake as the common mechanism of its interactions with cyclosporin A, rifampicin, and sildenafil. Drug Metab Dispos 35:1400–1407, 2007

Verschraagen M, Koks CH, Schellens JH, et al: P-glycoprotein system as a determinant of drug interactions: the case of digoxin-verapamil. Pharmacol Res 40:301–306, 1999

Wacher VJ, Wu CY, Benet LZ: Overlapping substrate specificities and tissue distribution of cytochrome P450 3A and P-glycoprotein: implications for drug delivery and activity in cancer chemotherapy. Mol Carcinog 13:129–134, 1995

Wakasugi H, Yano I, Ito T, et al: Effect of clarithromycin on renal excretion of digoxin: interaction with P-glycoprotein. Clin Pharmacol Ther 64:123–128, 1998

Wilkinson GR: Pharmacokinetics: the dynamics of drug absorption, distribution, and elimination, in Goodman & Gilman's The Pharmacological Basis of Therapeutics, 10th Edition. Edited by Hardman JG, Limbird LE, Gilman AG. New York, McGraw-Hill, 2001, pp 3–29

Woodland C, Verjee Z, Giesbrecht E, et al: The digoxin-propafenone interaction: characterization of a mechanism using renal tubular cell monolayers. J Pharmacol Exp Ther 283:39–45, 1997

Woodland C, Koren G, Ito S: From bench to bedside: utilization of an in vitro model to predict potential drug-drug interactions in the kidney: the digoxin-mifepristone example. J Clin Pharmacol 43:743–750, 2003

PART II

P450 Enzymes

Jessica R. Oesterheld, M.D.
Editor

Introduction to Part II

In this section, individual P450 cytochromes are introduced and discussed. For historical reasons, 2D6 is the starter P450, followed in decreasing order of importance by the other P450s. This section precedes Part III, in which classes of drugs are reviewed. The reader therefore receives two perspectives on the P450s and drugs: by metabolic schema and by pharmacologic class. Tables of substrates, inhibitors, and inducers conclude each P450 chapter. These tables are not intended to provide exhaustive coverage, but rather to present representative drugs.

A developmental perspective on the P450s can also be instructive. The following two tables highlight what is currently known of the changes in P450s and other metabolic routes in the mother during pregnancy (Anderson 2005; Tracy et al. 2005; Tsutsumi et al. 2001), in the fetus, and in the stagger-step emergence of phase I and phase II routes during fetal development, infancy, and childhood (Oesterheld et al. 2007).

References

Anderson GD: Pregnancy-induced changes in pharmacokinetics: a mechanistic-based approach. Clin Pharmacokinet 44:989–1008, 2005

Oesterheld JR, Shader RI, Martin A: Clinical and developmental aspects of pharmacokinetics and drug interactions, in Lewis's Child and Adolescent Psychiatry: A Comprehensive Textbook. Edited by Martin A, Volkmar FR. Philadelphia, PA, Wolters Kluwer Health/Lippincott Williams & Wilkins, 2007, pp 742–753

Tracy TS, Venkataramanan R, Glover DD, et al: Temporal changes in drug metabolism (CYP1A2, CYP2D6 and CYP3A activity) during pregnancy. National Institute for Child Health and Human Development Network of Maternal-Fetal-Medicine Units. Am J Obstet Gynecol 192:633–639, 2005

Tsutsumi K, Kotegawa T, Matsuki S, et al: The effect of pregnancy on cytochrome P4501A2, xanthine oxidase, and N-acetyltransferase activities in humans. Clin Pharmacol Ther 70:121–125, 2001

TABLE 1. Changes in clearance in the mother during pregnancy

	1st trimester	2nd trimester	3rd trimester
CYP 1A2	Reduced 33%	Reduced 49%	Reduced 65%
CYP2C9	No change	No change	Increased 20%
CYP2C19		Reduced 50%	Reduced 50%
CYP2D6	Increased 26%	Increased 35%	Increased 48%
CYP3A4	Increased 35%–38%	Increased 35%–38%	Increased 35%–38%
UGT1A4	Increased 200%	Increased 200%	Increased 300%
UGT2B7			Increased
Acetylatn	Indeterminate	No change	No change
Xanthine oxidase	No change	No change	No change

TABLE 2. Developmental emergence of metabolic routes

Metabolic route	Fetus	Development in childhood
CYP1A2		Becomes active after 4th or 5th month subsequent to birth; reaches 20% of adult levels at 3–12 months, 50% at 1–9 years
CYP2C9		Becomes active in first few weeks after birth; reaches 50% of adult levels by 5 months
CYP2C19		Becomes active in first few weeks after birth; reaches 50%–75% of adult levels post puberty
CYP2D6		Becomes active in first few weeks after birth; reaches adult levels by 10 years
CYP2E1		Gradually develops over first few months after birth
CYP3A4		Appears in first few weeks after birth; may not reach adult levels until 1 year of age
CYP3A5	Present	Present in 20% of population
CYP3A7	Predominant CYP	Fades after 2 weeks subsequent to birth, then is present in 20% of population
Acetylation		Reaches adult levels by 2–4 years
SULT1A1	Predominant phase II	Unchanged
UGT1A1		Immature at birth, responsible for hyperbilirubinemia; reaches adult levels 3-6 months after birth
UGT1A4		May mature as late as 18 years
UGT2B7		Immature at birth; reaches adult levels 3-6 months after birth
UGT1A9		Develops only after 2nd year after birth
UGT2B4		Develops only after 2nd year after birth

5

2D6

Jessica R. Oesterheld, M.D.

2D6 is discussed first in these reviews of the P450 enzymes because of its historical role in the development of an understanding of drug interactions. In 1988, Vaughan noted very high serum levels of desipramine and nortriptyline when fluoxetine was concomitantly administered. More reports followed Vaughan's report on two cases, and by 1990 the P450 system was recognized as being responsible for this potentially dangerous interaction (von Ammon and Cavanaugh 1990). In 1991, Muller et al. identified 2D6 as the enzyme that fluoxetine inhibits, and Skjelbo and Brosen (1992) confirmed this finding the following year. In the early 1990s, 2D6 was the most studied of the hepatic enzymes. However, the other P450 enzymes have since caught up with 2D6 in terms of amount of research. Nevertheless, 2D6 is an important metabolic enzyme for many drugs. In addition to a review of 2D6, this chapter also provides a primer for understanding important concepts of drug interactions and related genetic issues.

What Does 2D6 Do?

Located in the endoplasmic reticulum, 2D6 is involved in phase I metabolism of endogenous and exogenous compounds. Through oxidative metabolism, it hydroxylates, demethylates, or dealkylates these compounds; the actual oxidative reaction depends on the compound. Examples of 2D6's diverse oxidative reactions include hydroxylation of tricyclic antidepressants (Sawada and Ohtani 2001), fluoxetine's N-demethylation to norfluoxetine (Fjordside et al. 1999), and N-dealkylation of metoclopramide (Desta et al. 2002). 2D6 is involved in metabolism of endogenous chemicals as well, such as central nervous system hydroxylation of progesterone (Hiroi et al. 2001). 2D6 activity does not appear to change with age (Shulman and Ozdemir 1997); however, 2D6 activity may appear to be altered because of age-associated changes in hepatic blood flow or a decrease in renal elimination of metabolites. Premenopausal women had been shown to have a slight increase in 2D6 activity compared with men (Hagg et al. 2001), but a study with a larger sample failed to support this finding (Bebia et al. 2004).

For many medications, particularly psychotropic medications, 2D6 is considered a low-capacity, high-affinity enzyme. It accounts for only a small percentage of liver P450 content but appears to oxidatively clear more than its share of exogenous compounds. As a low-capacity, high-affinity enzyme, 2D6 will preferentially metabolize drugs at lower concentrations. As the concentration of a drug increases, the metabolism spills over to 3A4 or 1A2, which are high-capacity, low-affinity enzymes (Olesen and Linnet 1997). Thus, if a drug that has several metabolic pathways but is metabolized primarily by 2D6 is given to a patient with poor 2D6 activity (see section "Does 2D6 Have Any Polymorphisms?" below), the other P450 enzymes that are high capacity, low affinity will clear the drug. The drug's clearance will be slower and less efficient, and levels will increase.

Where Does 2D6 Do Its Work?

2D6 is found in many tissues, including the brain (Hiroi et al. 2001), the prostate (Finnstrom et al. 2001), bone marrow (Hodges et al. 2000), and the heart (Thum and Borlak 2000). Although metabolism of drugs may occur at these sites, 2D6 plays its role in drug metabolism primarily in the liver. 2D6 actually makes up only 2%–4% of total P450 content in the liver (Ingelman-Sundberg 2005), the smallest percentage of major drug-metabolizing P450 enzymes in humans. Nevertheless, 2D6 is a very active primary or secondary participant in the metabolism of many drugs.

Does 2D6 Have Any Polymorphisms?

The fact that 2D6 is polymorphic has been known for many years (Vesell et al. 1971). In the 1980s, 2D6 was named in relation to its probes (drugs metabolized—often exclusively—by the enzyme, resulting in a measurable and reliable metabolite); it was called the debrisoquin/sparteine enzyme (Eichelbaum 1986). By 1990, its polymorphic character was well recognized and 2D6 was given its current name (Lennard 1990).

Clinically, the polymorphic expressions of the 2D6 alleles are important to remember. The 2D6 gene is located on chromosome 22 (National Center for Biotechnology Information 2008). Bernard et al. (2006) report that there are more than 75 human alleles of 2D6 (annotated *2D6*x*). These variants may arise from mechanisms including point mutations, amino acid changes or deletions, entire gene deletion, rearrangements or duplications. From paired 2D6 alleles, the following genotypes can be distinguished: extensive ("normal") metabolizers (EMs); intermediate metabolizers (IMs); homozygous poor metabolizers (PMs); and ultraextensive metabolizers (UEMs).

These four categories are based on 2D6 activity. PMs have 2 alleles with no 2D6 activity. UEMs have 3 or more copies of the 2D6 gene and have unusually high 2D6 activity (DeVane and Nemeroff 2002). EMs are the norm and have either one or two functional alleles with "normal" 2D6 activity. Individuals who are homozygous for a 2D6 gene with reduced activity, or who have one allele that has reduced activity and one allele with no activity, are typed as IMs. In these individuals, 2D6 activity ranges from slightly more activity than PMs to close to that of EMs.

There are ethnic differences in the distribution of EMs, PMs, IMs and UEMs (see table in Bernard et al. 2006). PMs are reported to make up 8% of the Caucasian American population—usually individuals who have the *2D6*4* or *2D6*5* allele. Although Asian populations (Thai, Chinese, and Japanese) include only 1%–2% of 2D6 PMs, they have about 60% of IMs—usually involving *2D6*10* (Bernard et al. 2006). Two other 2D6 alleles (*2D6*17* and

2D6 *29) associated with lower activity are found in some African populations and in African Americans (Bapiro et al. 2002; Cai et al. 2006). UEMs are generally rare, representing 4% of the Caucasian American population but 21% and 29% of Saudi Arabians and Ethiopians, respectively (Bernard et al. 2006).

Any drug that is ingested by a PM and primarily metabolized by 2D6 will have a delayed metabolism. The PM will accumulate the parent drug and incur the risk of enhanced side effects. The drug may be secondarily metabolized by another P450 enzyme that is higher in capacity but that has a lower affinity for the drug or substrate. Often, the alternative enzyme is 3A4. Shifting to a less efficient enzyme results in higher drug levels of the parent compound.

PMs usually require lower doses to achieve desired effects. In most clinical situations, however, patients are not identified as 2D6 PMs. When a drug dependent on 2D6 is given to a PM, the untoward effects are likely assumed to be due to an unknown drug interaction or some other idiosyncratic reaction. Treatment with the poorly tolerated (and metabolized) drug may be abandoned. However, many drugs can be given to a 2D6 PM; the clinician need only decrease the dose. Failing to anticipate that a patient may be a PM can be potentially dangerous. For example, the action of some antiarrhythmics depends on 2D6 metabolism. If a patient lacks 2D6 activity, severe toxicity may occur if an antiarrhythmic such as encainide is prescribed (Funck-Brentano et al. 1992).

Chou et al. (2000) studied 100 psychiatric inpatients in relation to their 2D6 genotypes and clinical outcomes. They found that PMs (12% of the study group) had more adverse medication effects and longer, more expensive hospital stays. Andreassen et al. (1997) found that PMs were more likely to develop tardive dyskinesia over 11 years while exposed to antipsychotics. Similarly, Ellingrod et al. (2000) showed that IMs were at higher risk of developing movement disorders when exposed to antipsychotics compared with 2D6 EMs. Finally, Schillevoort et al. (2002) demonstrated that when being treated with 2D6-dependent antipsychotic drugs (see Table 5–1), PMs were four times more likely to use antiparkinsonian medications than were EMs—and this was not the case among patients using non–2D6-dependent antipsychotics.

Kirchheiner et al. (2004) have made specific clinical recommendations for dosing 2D6 PMs for the tricyclic antidepressants that are substrates of 2D6 (50% of the average dose) and for the typical antipsychotics that are substrates of 2D6 (e.g., thioridazine, haloperidol, and others 60%; see Table 5–1).

However, if the drug needs to be activated at 2D6, it may be less effective for a 2D6 PM. As an example, the pro-drug tamoxifen, a treatment for receptor-positive breast cancer, relies on its active metabolite, endoxifen, for its oncologic activity. Although tamoxifen has a complicated metabolism, 2D6 is essential for this conversion. As compared to 2D6 EMs, PMs have lower endoxifen levels and less robust clinical outcomes (Beverage et al. 2007). Other examples of 2D6 pro-drug substrates include dolasetron and codeine. The latter must be metabolized via *O*-demethylation to morphine and then glucuronidated to morphine-6-glucuronide (Bernard et al. 2006). A clinician might repeatedly increase the dose of codeine in an attempt to achieve the desired effect in a PM, with possibly poor results—leaving the patient and the clinician frustrated.

A 2D6 UEM will have a different set of clinical problems when taking 2D6 substrates or 2D6 pro-drug substrates. If a UEM is taking average doses of a 2D6 substrate such as nortriptyline, trough blood levels will be unusually low. Increased dosing will be necessary to achieve efficacy. However, if a 2D6 pro-drug substrate such as codeine is taken, levels of the active metabolite and glucuronide (i.e., morphine and morphine-6-glucuronide) will be increased, which may lead to increased sedation and adverse effects (Kirchheiner et al. 2007). A case has been reported of a toddler who received routine doses of acetaminophen and codeine but was found to be a 2D6 UEM, resulting in brain injury (Voronov et al. 2007).

Given these study findings, it would seem prudent for clinicians to consider testing 2D6 genotype before instituting drug therapy. Before the availability of laboratory genotyping and the U.S. Food and Drug Administration (FDA) approval of the AmpliChip in 2005 (see de Leon et al. 2006 for a comprehensive review of the AmpliChip), patients were administered a probe drug that is primarily metabolized by 2D6, and then specific 2D6 metabolites were evaluated. Dextromethorphan, the over-the-counter cough remedy, is one such drug; debrisoquin and sparteine are also probe drugs. Testing 2D6 activity involves giving a patient a fixed single dose of dextromethorphan 25 mg and then measuring dextrorphan, the active metabolite specific to 2D6 (created by 2D6's *O*-demethylation of dextromethorphan), in the next 8 hrs of urine. The urinary ratio of dextromethorphan to dextrorphan is calculated, with a ratio of 0.3 or greater distinguishing the EM from the PM (Schadel et al. 1995; Vanel et al. 2007). An even simpler (and definitely low tech) method

TABLE 5–1.   Some drugs metabolized by 2D6

Antidepressants	Antipsychotics	Other drugs	

Tricyclic antidepressants[1]

Amitriptyline
Clomipramine
Desipramine[2]
Doxepin
Imipramine
Nortriptyline[2]
Trimipramine

Other antidepressants

Duloxetine[3]
Fluoxetine[3]
Fluvoxamine[3]
Maprotiline[3]
Mirtazapine[3]
Minaprine
Moclobemide
Paroxetine
Sertraline[3]
Venlafaxine[3]

Aripiprazole[3]
Chlorpromazine[3]
Clozapine[4]
Fluphenazine[3]
Haloperidol[3]
Perphenazine[3]
Quetiapine[3,4]
Risperidone[3]
Thioridazine[3]
Zuclopenthixol

Other psychotropics

Amphetamines
Atomoxetine

Analgesics

Codeine[5]
Hydrocodone[3]
Methadone[4]
Oxycodone[3]
Tramadol[3,6]

Cardiovascular drugs

Alprenolol[2]
Bufuralol[3,8]
Carvedilol[3]
Diltiazem[9]
Encainide[2]
Flecainide[3]
Metoprolol[2,8]
Mexiletine[3]
Nebivolol[2]
Propafenone[3]
Propranolol[7]
Timolol[2]

Miscellaneous drugs[3]

Benztropine[2]
Cevimeline[3]
Cinacalcet[3]
Chlorpheniramine
Delavirdine[2]
Debrisoquin[3,8]
Dexfenfluramine[3]
Dextromethorphan[3,8]
Diphenhydramine[3]
Donepezil[3]
Indoramin
Loratadine[3]
Metoclopramide[2]
Ondansetron[3]
Perhexilene[3]
Tacrine[3]
Tamoxifen[3]
Tropisetron

TABLE 5–1. Some drugs metabolized by 2D6 *(continued)*

Antidepressants	Antipsychotics	Other drugs

[1]Tricyclic antidepressants (TCAs) use several enzymes for metabolism. The secondary tricyclics are preferentially metabolized by 2D6, the tertiary tricyclics by 3A4. TCAs are also oxidatively metabolized by 1A2 and 2C19.

[2]Oxidatively metabolized primarily by 2D6.

[3]Metabolized by other P450 enzymes, transporter or conjugated by phase II.

[4]2D6 is a minor pathway.

[5]O-Demethylated to morphine by 2D6, a minor pathway.

[6]Metabolized to a more active pain-relieving compound, M1.

[7]Beta-blockers are partly or primarily metabolized by 2D6.

[8]Used as a probe for 2D6 activity.

[9]Metabolites use 2D6.

to determine whether a patient is a 2D6 PM is to ask the patient whether he or she has ever had a bad reaction to dextromethorphan. At the high tech end, the AmpliChip identifies 27 2D6 alleles and allows characterization of metabolizer status (AmpliChip 2006). One limitation of the AmpliChip is that it does not quantify the number of copies of 2D6 genes for UEMs (de Leon et al. 2006).

What Drugs Significantly Inhibit 2D6 Activity?

An FDA Web site provides a list of strong, moderate, and weak inhibitors (U.S. Food and Drug Administration 2006). Inhibitors are categorized by increase in area under the plasma concentration-time curve (AUC) or percentage decrease in clearance of substrates in clinical studies with known substrates. Fluoxetine, paroxetine, and quinidine are strong inhibitors; duloxetine and terbinafine are moderate inhibitors; and amiodarone and sertraline are weak inhibitors (see Table 5–2). Unfortunately, this precision in categorizing the strength of 2D6 inhibition is not yet available for most known 2D6 inhibitors.

Selective Serotonin Reuptake Inhibitors

For further discussion of selective serotonin reuptake inhibitors (SSRIs), see the section on antidepressants in Chapter 19 ("Psychiatry").

Fluoxetine

Fluoxetine was the first drug discovered to have clinically relevant P450 inhibition (Vaughan 1988). Fluoxetine is a 50/50 racemic mix. Both the R and S forms are substrates of 2D6 and several other P450 enzymes and are metabolized to the active and racemic-mix metabolites R/S-norfluoxetine (von Moltke et al. 1997).

Norfluoxetine has an extremely long half-life, resulting in a dosing schedule as infrequent as once a week. This long half-life also means that if fluoxetine is added to a regimen including a 2D6-dependent drug, then inhibition of the second drug's metabolism will continue to be impaired for 4–8 weeks after fluoxetine therapy is discontinued.

Fluoxetine and its metabolite, norfluoxetine, are very potent inhibitors of 2D6, with Ki values between 0.22 and 1.48 µM (Stevens and Wrighton

TABLE 5–2.   Some inhibitors of 2D6

Antidepressants	Antihistamines	Antipsychotics	Other inhibitors
Amitriptyline	Chlorpheniramine[1]	Chlorpromazine	Amiodarone
Bupropion	Clemastine[1]	Clozapine	Celecoxib
Citalopram	Diphenhydramine[1]	Fluphenazine	Chloroquine
Clomipramine	Hydroxyzine	Haloperidol[3]	Cimetidine
Desipramine	Tripelennamine	Perphenazine	Cinacalcet
Doxepin		Pimozide[1]	Halofantrine
Duloxetine		Risperidone	Hydroxychloroquine
Escitalopram		Thioridazine	Lopinavir/Ritonavir
Fluoxetine		Trifluperidol	Methadone
Fluvoxamine			Methylphenidate?
Imipramine			Metoclopramide
Moclobemide			Nicardipine
Norfluoxetine			Propafenone
Paroxetine			Propoxyphene
Sertraline[2]			Quinidine/Quinine
Venlafaxine			Ritonavir
			Terbinafine
			Ticlopidine[1]

Note.   Names of potent inhibitors are in **bold** type.
[1]*In vitro* evidence exists only for the *potential* for potent inhibition of 2D6.
[2]Sertraline's inhibition seems to be dose specific, with higher doses resulting in more potent inhibition than lower doses.
[3]Inhibition from metabolite.

1993). *In vivo* studies have shown fluoxetine to significantly inhibit the metabolism of many 2D6 substrates (e.g., atomoxetine, desipramine, dextromethorphan, risperidone, imipramine, and thioridazine). Case studies demonstrate that 2D6 inhibition by fluoxetine and norfluoxetine results in increased levels or side effects of numerous drugs (e.g., benztropine, clomipramine, dextromethorphan, fluphenazine, haloperidol, nortriptyline, perphenazine). Many of these drugs have narrow therapeutic indexes and/or margins of safety. For example, fluoxetine's inhibition of thioridazine and mesoridazine metabolism can result in prolonged QTc interval and potentially in torsades de pointes. Therefore, both thioridazine and mesoridazine are contraindicated during fluoxetine therapy—and a lengthy washout period for fluoxetine is indicated before administration of type 1c antiarrythmics that are 2D6 substrates can be initiated (Armstrong and Cozza 2001).

Paroxetine

Paroxetine is as potent an inhibitor of 2D6 as fluoxetine, having a Ki of 2.0 μM (von Moltke et al. 1995). Unlike fluoxetine, paroxetine does not have a long half-life (21 hours), nor does it have active metabolites. Paroxetine's oxidative metabolism is rendered mainly by 2D6, with subsequent phase II conjugation. Because of both inhibition and metabolism via 2D6, paroxetine shows nonlinear kinetics. The manufacturer recommends caution when it is used with drugs metabolized by 2D6, including certain antidepressants (e.g., nortriptyline, amitriptyline, imipramine, and desipramine), phenothiazines, and type IC antiarrhythmics (e.g., propafenone, flecainide, and encainide). As with fluoxetine, paroxetine is contraindicated with thioridazine and mesoridazine therapy.

Sertraline, Citalopram, Escitalopram, and Fluvoxamine

All of these agents are weak 2D6 inhibitors (Sandson et al. 2005). Clinically significant inhibition of 2D6 by sertraline may be dose related (Sroule et al. 1997). For example, 50 mg of sertraline (a low dose) increased nortriptyline (dependent on 2D6 for clearance) levels only 2% in 14 elderly depressed patients (Solai et al. 1997). Interestingly, in 7 patients given sertraline at a dosage of 100–150 mg/day, nortriptyline levels increased an average of 40%—with a range of 12%–239%. It appears that higher doses of sertraline (>150 mg/day) can lead to more clinically significant 2D6 inhibition. An *in vivo* study comparing escitalopram and sertraline showed no statistically significant difference in 2D6 inhibition when either was coadministered with metoprolol (Preskorn et al. 2007).

Other Antidepressants

Bupropion

Bupropion is metabolized predominantly at 2B6 (see Chapter 11, "2A6, 2B6, and 2C8") and is a 2D6 inhibitor. A study of 15 healthy volunteers with normal 2D6 who were given 150 mg of Wellbutrin SR twice a day showed significant increases in maximum drug concentration, area under the curve, and half-life of a single dose of desipramine (50 mg) (Wellbutrin 2007). The manufacturer recommends that treatment with drugs that are predominantly metabolized by 2D6 and have a narrow therapeutic window (nortriptyline, desipramine, imipramine, beta blockers, type IC antiarrhythmics, and the potentially arrhythmogenic antipsychotics haloperidol, risperidone, and thioridazine) be initiated at lower doses in patients receiving bupropion.

Duloxetine and Venlafaxine

A serotonin and noradrenaline reuptake blocking antidepressant, duloxetine is a substrate of 2D6 and 1A2 and a moderate inhibitor of 2D6. As a substrate of 2D6, it is vulnerable to more potent 2D6 inhibitors (e.g., paroxetine increased the concentration of duloxetine about 60%; Skinner et al. 2003). Duloxetine has been shown to increase the AUC of a single dose of desipramine 3-fold (Cymbalta 2007) and metoprolol almost 2-fold (Preskorn et al. 2007). Similarly, duloxetine was shown to be a more potent 2D6 inhibitor when compared with escitalopram and sertraline (100 mg). As with other 2D6 inhibitors, caution should be used when initiating therapy combining duloxetine with narrow therapeutic indexed 2D6 substrates and type 1C antiarrhythmics. Contrary to duloxetine, venlafaxine is a relatively weak inhibitor of 2D6 (Effexor 2008).

Other Medications

Cimetidine

Cimetidine appears to be a potent and indiscriminate inhibitor of several enzymes, including 2D6, 3A4, and 1A2 as well as some transporters (Khamdang et al. 2004; Martinez et al. 1999). Multiple case reports indicate that cimeti dine increases tricyclic levels, and one study showed an increase in imipramine's half-life from 10.8 to 22.7 hours (Wells et al. 1986). Cimetidine also increases serum paroxetine levels by 50% (Greb et al. 1989), likely via 2D6 inhibition.

Quinidine/Quinine

Quinidine is a commonly used antimalarial drug and class IA antiarrhythmic agent that is metabolized predominantly via 3A4. Quinidine is the *d*-isomer of quinine and a very potent inhibitor of 2D6, with a reported Ki of 0.053 μM (von Moltke et al. 1994). This inhibition can increase the maximum concentration (C_{max}) of atomoxetine 2- to 3-fold (Strattera 2008). Quinine is also an inhibitor of 2D6. The manufacturer advises caution during coadministration of quinine with drugs dependent on 2D6 for clearance (Qualaquin 2007).

Ritonavir

Ritonavir is a human immunodeficiency virus (HIV) protease inhibitor and a potent 2D6 inhibitor with a Ki of 0.16 μM (von Moltke et al. 1998a). Ritonavir is also considered a "pan-inhibitor," with significant inhibition of 3A4, 2C9, and 2C19 (but not 1A2 or 2E1 [von Moltke et al. 1998b]), as well as many transporters. The manufacturer of Norvir (ritonavir) indicates that a single 100-mg dose of desipramine coadministered with ritonavir increases desipramine's area under the curve by 145% (Norvir 2007). Because of ritonavir's 2D6 inhibition, administration of ritonavir and most antiarrhythmics is contraindicated. Caution should also be taken when prescribing ritonavir in conjunction with tricyclics or antipsychotics (see section on antiretrovirals in Chapter 14, "Infectious Diseases").

Terbinafine

Terbinafine is an antifungal agent and inhibitor of 2D6 with an *in vitro* Ki ranging from 22 to 44 μM (Abdel-Rahman et al. 1999). *In vivo*, terbinafine appears to be a more potent 2D6 inhibitor than the Ki values would predict. A clinical study demonstrated that after a single dose of paroxetine, terbinafine increased paroxetine's C_{max} 1.9 fold (Yasui-Furukori et al. 2007). The coadministration of desipramine and terbinafine resulted in a near doubling of desipramine's C_{max} (Madani et al. 2002). There have also been several cases of severe tricyclic toxicity due to terbinafine given in conjunction with imipramine (Teitelbaum and Pearson 2001) and nortriptyline (van der Kuy and Hooymans 1998) (see section on antifungals in Chapter 14, "Infectious Diseases").

Other Possible Potent Inhibitors

Ticlopidine is an antiplatelet agent known to be a potent 2C19 inhibitor (see Chapter 9, "2C19"). *In vitro* testing with human liver microsomes has demonstrated that ticlopidine may also be a fairly potent inhibitor of 2D6, having a Ki of 3.4 μM–4.4 μM (Ko et al. 2000; Turpeinen et al. 2004). Pimozide is metabolized primarily via 3A4 and to a lesser extent via 1A2. *In vitro* tests with human liver microsomes revealed pimozide may also potently inhibit 2D6, having a Ki of approximately 1.0 μM (Desta et al. 1998).

Are There Drugs That Induce 2D6 Activity?

Research has not yet shown whether 2D6 is inducible. However, there is evidence that 2D6 is induced during pregnancy (Tracy et al. 2005; Wadelius et al. 1997). There is also *in vitro* evidence of rifampin's induction of CYP2D6 in human enterocytes (Glaeser et al. 2005) and, perhaps, *in vivo* evidence of rifampin induction of sparteine (Eichelbaum et al. 1986). These studies suggest there may be a mechanism for 2D6 induction, although that mechanism remains unclear.

References

Abdel-Rahman SM, Marcucci K, Boge T, et al: Potent inhibition of cytochrome P-450 2D6-mediated dextromethorphan O-demethylation by terbinafine. Drug Metab Dispos 27:770–775, 1999

AmpliChip (product information). Indianapolis, IN, Roche Diagnostics, 2006. Available at http://www.amplichip.us. Accessed July 14, 2008.

Andreassen OA, MacEwan T, Gulbrandsen AK, et al: Non-functional CYP2D6 alleles and risk for neuroleptic-induced movement disorders in schizophrenic patients. Psychopharmacology (Berl) 131:174–179, 1997

Armstrong SC, Cozza KL: Consultation-liaison psychiatry drug-drug interactions update. Psychosomatics 42:157–159, 2001

Bapiro TE, Hasler JA, Ridderstrom M, et al: The molecular and enzyme kinetic basis for the diminished activity of the cytochrome P450 2D6.17 (CYP2D6.17) variant: potential implications for CYP2D6 phenotyping studies and the clinical use of CYP2D6 substrate drugs in some African populations. Biochem Pharmacol 64:1387–1398, 2002

Bebia Z, Buch SC, Wilson JW, et al: Bioequivalence revisited: influence of age and sex on CYP enzymes. Clin Pharmacol Ther 76:618–627, 2004

Bernard S, Neville KA, Nguyen AT, et al: Interethnic differences in genetic polymorphisms of CYP2D6 in the U.S. population: clinical implications. Oncologist 11:126–135, 2006

Beverage JN, Sissung TM, Sion AM, et al: CYP2D6 polymorphisms and the impact on tamoxifen therapy. J Pharm Sci 96:2224–2231, 2007

Cai WM, Nikoloff DM, Pan RM, et al: CYP2D6 genetic variation in healthy adults and psychiatric African-American subjects: implications for clinical practice and genetic testing. Pharmacogenomics J 6:343–350, 2006

Chou WH, Yan FX, de Leon J, et al: Extension of a pilot study: impact from the cytochrome P450 2D6 polymorphism on outcome and costs associated with severe mental illness. J Clin Psychopharmacol 20:246–251, 2000

Cymbalta (package insert). Indianapolis, IN, Eli Lilly and Company, 2007

de Leon J, Susce MT, Murray-Carmichael E: The AmpliChip CYP450 genotyping test: integrating a new clinical tool. Mol Diagn Ther 10:135–151, 2006

Desta Z, Kerbusch T, Soukhova N, et al: Identification and characterization of human cytochrome P450 isoforms interacting with pimozide. J Pharmacol Exp Ther 285:428–437, 1998

Desta Z, Wu GM, Morocho AM, et al: The gastroprokinetic and antiemetic drug metoclopramide is a substrate and inhibitor for cytochrome P450 2D6. Drug Metab Dispos 30:336–343, 2002

DeVane CL, Nemeroff CB: 2002 guide to psychotropic drug interactions. Primary Psychiatry 9:28–57, 2002

Effexor (package insert). Philadelphia, PA, Wyeth Pharmaceuticals Inc., 2008

Eichelbaum M: Polymorphic oxidation of debrisoquine and sparteine. Prog Clin Biol Res 214:157–167, 1986

Eichelbaum M, Mineshita S, Ohnhaus EE, et al: The influence of enzyme induction on polymorphic sparteine oxidation. Br J Clin Pharmacol 22:49–53, 1986

Ellingrod VL, Schultz SK, Arndt S: Association between cytochrome P4502D6 (CYP2D6) genotype, antipsychotic exposure, and abnormal involuntary movement scale (AIMS) score. Psychiatr Genet 10:9–11, 2000

Finnstrom N, Bjelfman C, Soderstrom TG, et al: Detection of cytochrome P450 mRNA transcripts in prostate samples by RT-PCR. Eur J Invest 31:880–886, 2001

Fjordside L, Jeppesen U, Eap CB, et al: The stereoselective metabolism of fluoxetine in poor and extensive metabolizers of sparteine. Pharmacogenetics 9:55–60, 1999

Funck-Brentano C, Thomas G, Jacqz-Aigrain E, et al: Polymorphism of dextromethorphan metabolism: relationships between phenotype, genotype and response to the administration of encainide in humans. J Pharmacol Exp Ther 263:780–786, 1992

Glaeser H, Drescher S, Eichelbaum M, et al: Influence of rifampicin on the expression and function of human intestinal cytochrome P450 enzymes. Br J Clin Pharmacol 59:199–206, 2005

Greb WH, Buscher G, Dierdorf HD, et al: The effect of liver enzyme inhibition by cimetidine and enzyme induction by phenobarbitone on the pharmacokinetics of paroxetine. Acta Psychiatr Scand Suppl 350:95–98, 1989

Hagg S, Spigset O, Dahlqvist R: Influence of gender and oral contraceptives on CYP2D6 and CYP2C19 activity in healthy volunteers. Br J Pharmacol 51:169–173, 2001

Hiroi T, Kishimoto W, Chow T, et al: Progesterone oxidation by cytochrome P450 2D isoforms in the brain. Endocrinology 142:3901–3908, 2001

Hodges VM, Molloy GY, Wickramasinghe SN: Demonstration of mRNA for five species of cytochrome P450 in human bone marrow, bone marrow-derived macrophages and human haemopoietic cell lines. Br J Haematol 108:151–156, 2000

Ingelman-Sundberg M: Genetic polymorphisms of cytochrome P450 2D6 (CYP2D6): clinical consequences, evolutionary aspects and functional diversity. Pharmacogenomics 5:6–13, 2005

Khamdang S, Takeda M, Shimoda M, et al: Interactions of human- and rat-organic anion transporters with pravastatin and cimetidine. J Pharmacol Sci 94:197–202, 2004

Kirchheiner J, Nickchen K, Bauer M, et al: Pharmacogenetics of antidepressants and antipsychotics: the contribution of allelic variations to the phenotype of drug response. Mol Psychiatry 9:442–473, 2004

Kirchheiner J, Schmidt H, Tzvetkov M, et al: Pharmacokinetics of codeine and its metabolite morphine in ultra-rapid metabolizers due to CYP2D6 duplication. Pharmacogenomics J 7:257–265, 2007

Ko JW, Desta Z, Soukhova NV, et al: In vitro inhibition of the cytochrome P450 (CYP) system by the antiplatelet drug ticlopidine: potent effect on CYP2C19 and CYP2D6. Br J Clin Pharmacol 49:343–351, 2000

Lennard MS: Genetic polymorphism of sparteine/debrisoquine oxidation: a reappraisal. Pharmacol Toxicol 67:273–283, 1990

Madani S, Barilla D, Cramer J, et al: Effect of terbinafine on the pharmacokinetics and pharmacodynamics of desipramine in healthy volunteers identified as cytochrome P450 2D6 (CYP2D6) extensive metabolizers. J Clin Pharmacol 42:1211–1218, 2002

Martinez C, Albet C, Agundez JA, et al: Comparative in vitro and in vivo inhibition of cytochrome P450 CYP1A2, CYP2D6, and CYP3A by H2 receptor antagonists. Clin Pharmacol Ther 65:369–376, 1999

Muller N, Brockmoller J, Roots I: Extremely long plasma half-life of amitriptyline in a woman with the cytochrome P450IID6 29/29-kilobase wild-type allele: a slowly reversible interaction with fluoxetine. Ther Drug Monit 13:533–536, 1991

National Center for Biotechnology Information: CYP2D6 cytochrome P450, family 2, subfamily D, polypeptide 6 [Homo sapiens]. Pub Med Entrez Gene, 2008. Available at http://www.ncbi.nlm.nih.gov/sites/entrez?db=gene&cmd=retrieve&list_ulds=1565. Accessed July 14, 2008.

Norvir (package insert). North Chicago, IL, Abbott Laboratories. Revised May 2007. Available at http://www.norvir.com. Accessed July 27, 2007.

Olesen OV, Linnet K: Hydroxylation and demethylation of the tricyclic antidepressant nortriptyline by cDNA-expressed human cytochrome P450 isozymes. Drug Metab Dispos 25:740–744, 1997

Preskorn SH, Greenblatt DJ, Flockhart D, et al: Comparison of duloxetine, escitalopram, and sertraline effects on cytochrome P450 2D6 function in healthy volunteers. J Clin Psychopharmacol 27:28–34, 2007

Qualaquin (package insert). Philadelphia, PA, AR Scientific, Inc., 2007

Sandson NB, Armstrong SC, Cozza KL: An overview of psychotropic drug-drug interactions. Psychosomatics 46:464–494, 2005

Sawada Y, Ohtani H: [Pharmacokinetics and drug interactions of antidepressant agents] (in Japanese). Nippon Rinsho 59:1539–1545, 2001

Schadel M, Wu D, Otton SV, et al: Pharmacokinetics of dextromethorphan and metabolites in humans: influence of the CYP2D6 phenotype and quinidine inhibition. J Clin Psychopharmacol 15:263–269, 1995

Schillevoort I, de Boer A, van der Weide J, et al: Antipsychotic-induced extrapyramidal syndromes and cytochrome P450 2D6 genotype: a case controlled study. Pharmacogenetics 12:235–240, 2002

Shulman RW, Ozdemir V: Psychotropic medications and cytochrome P450 2D6: pharmacologic considerations in the elderly. Can J Psychiatry 42 (suppl 1):4S–9S, 1997

Skinner MH, Kuan HY, Pan A, et al: Duloxetine is both an inhibitor and a substrate of cytochromes P4502D6 in healthy volunteers. Clin Pharmacol Ther 73:170–177, 2003

Skjelbo E, Brosen K: Inhibitors of imipramine metabolism by human liver microsomes. Br J Clin Pharmacol 34:256–261, 1992

Solai LK, Mulsant BH, Pollock BG, et al: Effect of sertraline on plasma nortriptyline levels in depressed elderly. J Clin Psychiatry 58:440–443, 1997

Sroule BA, Otton SV, Cheung SW, et al: CYP2D6 inhibition in patients treated with sertraline. J Clin Psychopharmacol 17:102–106, 1997

Stevens JC, Wrighton SA: Interaction of the enantiomers of fluoxetine and norfluoxetine with the human liver cytochromes P450. J Pharmacol Exp Ther 266:964–971, 1993

Strattera (package insert). Indianapolis, IN, Eli Lilly and Company, 2008

Teitelbaum ML, Pearson VE: Imipramine toxicity and terbinafine (letter). Am J Psychiatry 158:2086, 2001

Thum T, Borlak J: Gene expression in distinct regions of the heart. Lancet 355:979–983, 2000

Tracy TS, Venkataramanan R, Glover DD, et al: Temporal changes in drug metabolism (CYP1A2, CYP2D6 and CYP3A Activity) during pregnancy. Am J Obstet Gynecol 192:633–639, 2005

Turpeinen M, Nieminen R, Juntunen T, et al: Selective inhibition of CYP2B6-catalyzed bupropion hydroxylation in human liver microsomes in vitro. Drug Metab Dispos 32:626–631, 2004

U.S. Food and Drug Administration: Drug development and drug interactions: table of substrates, inhibitors and inducers. Updated October 11, 2006. Available at http://www.fda.gov/cder/drug/drugInteractions/tableSubstrates.htm. Accessed July 27, 2007.

van der Kuy PH, Hooymans PM: Nortriptyline intoxication induced by terfenadine. BMJ 316:441, 1998

Vanel P, Talon JM, Haffen E, et al: Pharmacogenetics and drug therapy in psychiatry: the role of the CYP2D6 polymorphism. Curr Pharm Des 13:241–250, 2007

Vaughan DA: Interaction of fluoxetine with tricyclic antidepressants (letter). Am J Psychiatry 145:1478, 1988

Vesell ES, Passananti GT, Greene FE, et al: Genetic control of drug levels and of the induction of drug-metabolizing enzymes in man: individual variability in the extent of allopurinol and nortriptyline inhibition of drug metabolism. Ann N Y Acad Sci 179:752–773, 1971

von Ammon K, Cavanaugh S: Drug-drug interactions of fluoxetine with tricyclics. Psychosomatics 31:273–276, 1990

von Bahr C, Steiner E, Koike Y, et al: Time course of enzyme induction in humans: effect of pentobarbital on nortriptyline metabolism. Clin Pharmacol Ther 64:18–26, 1998 von Moltke LL, Greenblatt DJ, Cotreau-Bibbo MM, et al: Inhibition of desipramine hydroxylation in vitro by serotonin-reuptake-inhibitor antidepressants, and by quinidine and ketoconazole: a model system to predict drug interactions in vivo. J Pharmacol Exp Ther 268:1278–1283, 1994

von Moltke LL, Greenblatt DJ, Court MH: Inhibition of alprazolam and desipramine hydroxylation in vitro by paroxetine and fluvoxamine: comparison with other selective serotonin reuptake inhibitor antidepressants. J Clin Psychopharmacol 15:125–131, 1995

von Moltke LL, Greenblatt DJ, Duan SX, et al: Human cytochromes mediating N-demethylation of fluoxetine in vitro. Psychopharmacology (Berl) 132:402–407, 1997

von Moltke LL, Greenblatt DJ, Duan SX, et al: Inhibition of desipramine hydroxylation (cytochrome P450-2D6) in vitro by quinidine and by viral protease inhibitors: relation to drug interactions in vivo. J Pharm Sci 87:1184–1189, 1998a

von Moltke LL, Greenblatt DJ, Grassi JM, et al: Protease inhibitors as inhibitors of human cytochromes P450: high risk associated with ritonavir. J Clin Pharmacol 38:106–111, 1998b

Voronov P, Przybylo HJ, Jagannathan N: Apnea in a child after oral codeine: a genetic variant—an ultra-rapid metabolizer. Paediatr Anaesth 17:684–687, 2007

Wadelius M, Darj E, Frenne G, et al: Induction of CYP2D6 in pregnancy. Clin Pharmacol Ther 62:400–407, 1997

Wellbutrin (package insert). Research Triangle Park, NC, GlaxoSmithKline, 2007. Available at http://www.wellbutrin.com. Accessed July 14, 2007.

Wells BG, Pieper JA, Self TH: The effect of ranitidine and cimetidine on imipramine disposition. Eur J Pharmacol 31:285–290, 1986

Yasui-Furukori N, Saito M, Inoue Y, et al: Terbinafine increases the plasma concentration of paroxetine after a single oral administration of paroxetine in healthy subjects. Eur J Clin Pharmacol 63:51–56, 2007

Study Cases

Case 1

A 37-year-old Caucasian man who had tested positive for HIV was severely depressed and began nortriptyline therapy. He improved clinically at a dosage of 75 mg/day, and the serum tricyclic level was 87 ng/mL (therapeutic range, 50–150 ng/mL). His infectious disease physician prescribed ritonavir and saquinavir. The psychiatrist was notified about the new medication and told the patient to obtain a tricyclic level measurement 5–7 days after initiation of treatment with ritonavir and saquinavir. The patient did so and returned complaining of increasing depressive symptoms. His serum tricyclic level was 203 ng/mL.

Comment

2D6 (and 3A4) inhibition: The psychiatrist and infectious disease physician correctly anticipated that ritonavir and saquinavir might increase tricyclic levels. Indeed, the tricyclic level increased beyond the therapeutic range, which may have a negative impact on the nortriptyline therapy. The presumed mechanism for this interaction is ritonavir's potent inhibition of several P450 enzymes, including 2D6. Nortriptyline is metabolized primarily via 2D6, with minor 3A4 as a secondary route.

References

Tseng AL, Foisy MM: Management of drug interactions in patients with HIV. Ann Pharmacother 31:1040–1058, 1997

Venkatakrishnan K, von Moltke LL, Greenblatt DJ: Nortriptyline E-10-hydroxylation in vitro is mediated by human CYP2D6 (high affinity) and CYP3A4 (low affinity): implications for interactions with enzyme-inducing drugs. J Clin Pharmacol 39:567–577, 1999

Case 2

A 17-year-old Caucasian boy with bipolar disorder, a seizure disorder, and mild mental retardation was taking haloperidol (5 mg/day), valproic acid (1,750 mg/day), benztropine (4 mg/day), gabapentin (300 mg/day), and buspirone (10 mg/day). Paroxetine therapy (20 mg/day) was initiated because of depressive symptoms. On the eighth day of paroxetine therapy, the patient reported that he had a dry mouth and nausea. He also appeared disoriented, disorganized, forgetful, and lethargic and had slurred speech. Vital signs were unremarkable, and he exhibited no extrapyramidal symptoms. Pupils were dilated and sluggish to respond. Treatment with paroxetine, haloperidol, benztropine, and valproic acid was discontinued. Serum levels were as follows: valproic acid, 119 μg/mL (previous levels, 103–105 μg/mL); haloperidol, 2.9 ng/mL (therapeutic range, 5–12 ng/mL); and gabapentin, 4.1 ng/mL (therapeutic range, 4–16 ng/mL). His serum benztropine level was 35.9 ng/mL.

Benztropine levels greater than 25 ng/mL are considered toxic. Two days after discontinuation of the medications, the patient reported feeling better. He resumed treatment with haloperidol and valproic acid without incident.

Comment

2D6 inhibition: Benztropine is an older drug whose metabolism has not been well characterized. Seven cases in the literature indicate that SSRIs may interact with benztropine because of inhibition of benztropine metabolism. Because fluoxetine, paroxetine, and sertraline inhibit 2D6, it is hypothesized that benztropine is at least in part metabolized by 2D6. We suggest that in this case, paroxetine's 2D6 inhibition increased serum benztropine levels, causing a mild anticholinergic delirium.

References

Armstrong SC, Schweitzer SM: Delirium associated with paroxetine and benztropine combination. Am J Psychiatry 154:581 582, 1997

Byerly MJ, Christensen RC, Evans OL: Delirium associated with a combination of sertraline, haloperidol and benztropine. Am J Psychiatry 153:965–966, 1996

Roth A, Akyol S, Nelson JC: Delirium associated with a combination of a neuroleptic, an SSRI and benztropine. J Clin Psychiatry 55:492–495, 1994

Case 3

A 29-year-old man with major depression was treated with paroxetine 40 mg/day. Because of a partial response to treatment, desipramine 100 mg/day was added. The patient developed signs of delirium. The trough blood level of desipramine was greater than 400 ng/mL.

Comment

2D6 inhibition: Paroxetine inhibits 2D6, which desipramine uses for its clearance. Although tricyclics have alternative pathways for clearance, significant elevation in plasma levels is common. Similarly, in a PM at 2D6, one would expect lower doses of desipramine to achieve therapeutic levels while "normal" doses might result in toxicities.

References

Alderman J, Preskorn SH, Greenblatt DJ, et al: Desipramine pharmacokinetics when coadministered with paroxetine or sertraline in extensive metabolizers. J Clin Psychopharmacol 17:284–291, 1997

Brosen K, Hansen JG, Nielsen KK, et al: Inhibition by paroxetine of desipramine metabolism in extensive but not in poor metabolizers of sparteine. Eur J Clin Pharmacol 44:349–355, 1993

Case 4

A 17-year-old boy with schizophrenia experienced severe extrapyramidal side effects when prescribed risperidone 4 mg/day. Genotyping indicated he was a PM at 2D6, raising his serum levels of risperidone and 9-hydroxyrisperidone.

Comment

PM at 2D6: Often clinicians are more concerned about severe adverse drug reactions in a PM at 2D6, but in this case, determining the patient's genotype helped explain his sensitivity to risperidone. The clinician then chose an antipsychotic not predominantly cleared by 2D6. Olanzapine, quetiapine, and ziprasidone are three atypical antipsychotics that are not dependent on 2D6 for metabolism.

References

Bork JA, Rogers T, Wedlund PJ, et al: A pilot study on risperidone metabolism: the roles of cytochromes P450 2D6 and 3A. J Clin Psychiatry 60:469–476, 1999

Kohnke MD, Griese EU, Stosser D, et al: Cytochrome P450 2D6 deficiency and its clinical relevance in a patient treated with risperidone. Pharmacopsychiatry 35:116–118, 2002

6

3A4

Jessica R. Oesterheld, M.D.

3A4 is the workhorse of the P450 system, accounting for 30% of P450 activity in the liver and 70% of P450 activity in the small intestine (DeVane and Nemeroff 2002; Zhang et al. 1999). 3A4 not only accounts for a large portion of human P450 activity but also performs the bulk of oxidative metabolism of drugs. This makes 3A4 arguably the most important P450 enzyme to understand. 3A4 may account for more than half of all drug oxidation in the human liver (Guengerich 1999). Until the mid-1990s, the literature referenced 3A3 or 3A3/4. It is now understood, however, that 3A3 is not a separate P450 enzyme but a transcript variant of 3A4 (for a review, see National Center for Biotechnology Information 2008).

What Does 3A4 Do?

3A4 performs phase I metabolism of endogenous and exogenous compounds, including hydroxylation, demethylation, and dealkylation of substrates (see Table 6–1). The presence of multiple enzymatic sites on the 3A4 molecule likely explains the capacity of 3A4 to affect so many compounds through multiple enzymatic steps. 3A4 is also an important enzyme in the metabolism of endogenous steroids, cholesterol, and lipids, primarily through hydroxylation. 3A4 appears to vary in activity by age and gender: children have more 3A4 activity than adults (Strolin Benedetti et al. 2005), and women have up to 25% higher 3A4 activity than men, although this may vary over the menstrual cycle (Zhu et al. 2003).

Does 3A4 Have Any Polymorphisms?

Although almost 50 genetic variations of 3A4 have been found, they appear to have no functional consequence for 3A4 metabolic activity. 3A4 represents one variation of the 3A subfamily in humans. The two other major human 3A enzymes are 3A5 and 3A7. All three are coded on chromosome 7. These enzymes are not polymorphisms, but rather genetic variations or "isozymes" of the P450 subfamily. The 3A region on chromosome 7 (7q21.1–22.1) codes for a superfamily of genes (perhaps up to 850 genes) that are all related to the 3A subfamily. It remains unclear what these genes encode. Some genes are actually pseudogenes that code for promoter regions. Variations in promoter regions may alter the rate of transcription to messenger RNA.

3A7 is expressed in the fetus, where it plays an important role in fetal steroid metabolism. During the first 12 months after birth 3A7 diminishes in quantity, and it is replaced gradually by 3A4 (Oesterheld 1998). 3A7 continues to be present in 10%–20% of the adult population, including 3A7 high expressers, in whom this isoform contributes about 9%–36% of total 3A subfamily activity (Wojnowski and Kamdem 2006). 3A7 also resides extrahepatically in a variety of tissues, including the endometrium, adrenal gland, and prostate.

3A5 is the most common member of the CYP3A subfamily in nonhepatic tissues including the small intestine, colon, kidney, and adrenal gland. 3A5 may be the predominant 3A gene in individuals who show lower levels of 3A4. 3A5 may also play a more important role in the intestine and the kidney than 3A4 in individuals with normal amounts of 3A4, although the evidence remains unclear. Like 3A7, 3A5 is present in approximately 20% of the population ("3A5 high expressers"). In an individual with 3A5 expression, 3A5 is likely 10%–20% of the total P450 cytochrome protein (Daly 2006).

Despite the apparent genetic complexity of the 3A subfamily, all humans have considerable 3A activity. Given the breadth of molecules metabolized at 3A4, an essential lack of 3A4 activity would make little teleological sense for a human population. Current understanding suggests that an allelic pattern of poor metabolism at 3A4 would probably be inconsistent with life.

3A isozymes share many substrates but can show variable affinities for the same molecule. 3A4 frequently has the highest affinity for a given substrate, except in a few cases where 3A5 (e.g., tacrolimus) or 3A7 (e.g., retinoic acid drugs) shows higher affinity.

For 3A5 "high expressers," higher doses of drugs such as tacrolimus are required to achieve therapeutic concentrations than for individuals without 3A5 expression. However, less than half of the variation in tacrolimus dose requirement can be explained by 3A5 genetic status. The remaining variation can be accounted for by the total amount of available 3A4, co-medication, and other health factors (Haufroid et al. 2004). As a result of this variation, Haufroid et al. (2006) recommend doubling the dose of tacrolimus in high expressers of 3A5 immediately after renal transplant.

Although the amino acid sequence of 3A5 is 83% homologous to 3A4, it is unclear how differences in gene expression change drug metabolism. Some *in vitro* evidence shows 3A5 is less inducible than 3A4. Theoretically, 3A5 "low expressers" could become phenocopies of high expressers after induction (Burk et al. 2004). 3A4 and 3A5 also vary in their level of response to 3A inhibition, with *in vitro* studies showing a lower response in 3A5 (Soars et al. 2006).

The complexity of the 3A subfamily gene region indicates a wide variation in 3A enzymatic activity. Clinical observations reveal that expression of 3A varies as widely as 10- to 30-fold from individual to individual (Ketter et al. 1995).

In light of these factors, I refer in this chapter to CYP3A or 3A, whereas in other chapters, CYP3A4 or 3A4 may be specified. In most settings the difference will not give rise to significantly different clinical results, although the reader is cautioned to consider the possibility of various 3A subfamily members when evaluating possible drug-drug interactions.

Where Does 3A Do Its Work?

Small intestine enterocytes have 3A enzymes as well as P-glycoprotein (ABCB1) and other transporters (Doherty and Charman 2002). Gut lumen

TABLE 6–1.   Some drugs metabolized by 3A

Antidepressants

Amitriptyline[1]
Citalopram[2]
Clomipramine[1]
Doxepin[1]
Fluoxetine[2]
Imipramine[1]
Mirtazapine[4]
Nefazodone
Reboxetine
Sertraline[2]
Trazodone[3]
Trimipramine[1]
Venlafaxine[6]

Antipsychotics

Aripiprazole[6]
Chlorpromazine[2]
Clozapine[2]
Haloperidol[2]
Perphenazine[2]
Pimozide[5]
Quetiapine[2]
Risperidone[6]
Ziprasidone[2]

Sedative-hypnotics

Benzodiazepines
Clonazepam[2]
Diazepam[2]
Flunitrazepam[2]
Nitrazepam[6]

Triazolobenzodiazepines
Alprazolam[2]
Estazolam
Midazolam[11]
Triazolam[11]

Other sedative-hypnotics
Eszopiclone[2]
Zaleplon[2,7]
Zolpidem[2]
Zopiclone[2]

Psychotropic drugs, other

Buspirone[11]
Donepezil[2]
Galantamine[2]

Other drugs

Analgesics
Alfentanil
Buprenorphine[2]
Codeine (10%, *N*-demethylated)
Fentanyl
Hydrocodone[6]
Meperidine[2]
Methadone[7]
Propoxyphene[8]
Sufentanil

Antiarrhythmics[5]
Amiodarone[2,7]
Lidocaine[2]
Mexiletine[2]
Propafenone[2]
Quinidine[2]

Antiepileptics
Carbamazepine[2]
Ethosuximide[2]
Felbamate[2]
Tiagabine[2]
Zonisamide[2]

TABLE 6–1. Some drugs metabolized by 3A (*continued*)

Other drugs (*continued*)	Antineoplastics (*continued*)	Calcium channel blockers
Antifungals	Etoposide[2]	Amlodipine
Fluconazole[2]	Ifosfamide[2]	Diltiazem[2]
Itraconazole[2]	Paclitaxel[2]	Felodipine[11]
Ketoconazole[2]	Tamoxifen[2]	Lercanidipine
Miconazole	Teniposide[2]	Nicardipine
Voriconazole[2]	Trofosfamide[2]	Nifedipine[11]
Antihistamines	Vinblastine[2]	Nimodipine
Astemizole[9]	Vincristine[2]	Nitrendipine
Desloratidine[2]	Vindesine[2]	Verapamil[2]
Ebastine[2]	Vinorelbine[2]	*HMG-CoA reductase inhibitors (statins)*
Loratadine[7]	*Antiparkinsonian drugs*	Atorvastatin
Terfenadine[9,11]	Bromocriptine[2]	Cerivastatin[9]
Antimalarials	Pergolide[2,9]	Lovastatin[2,11]
Chloroquine[2]	Tolcapone[2]	Simvastatin[2,11]
Halofantrine[2]	*Antiprogesterone agents*	*Macrolide/ketolide antibiotics*
Primaquine[2]	Lilopristone	Clarithromycin
Antineoplastics	Mifepristone	Dirithromycin
Busulfan	Onapristone	Erythromycin
Cyclophosphamide[2]	Toremifene[2]	Telithromycin[2]
Daunorubicin	*Antirejection drugs*	Troleandomycin[9]
Docetaxel[2]	Cyclosporine[2]	
Doxorubicin[2]	Sirolimus (Rapamune)[2]	
	Tacrolimus	

TABLE 6–1.   Some drugs metabolized by 3A (continued)

Other drugs (continued)

Nonnucleoside reverse transcriptase inhibitors
Delavirdine[2]
Efavirenz[2]
Nevirapine[2]

Protease inhibitors (antivirals)
Amprenavir[2]
Atazanavir[2]
Indinavir[2]
Lopinavir[2]
Nelfinavir[2]
Ritonavir[2]
Saquinavir[2]

Proton pump inhibitors
Esomeprazole[10]
Lansoprazole[10]
Omeprazole[10]
Pantoprazole[10]
Rabeprazole[2]

Steroids
Budesonide
Cortisol
Dexamethasone[2]
Estradiol
Ethinyl estradiol[2]
Fluticasone[2]
Gestodene
Hydrocortisone
Methylprednisolone[2]
Prednisone[2]
Progesterone[2]
Testosterone[2,11]

Triptans
Almotriptan[2]
Eletriptan[2,11]

Miscellaneous drugs
Bosentan[2]
Cevimeline[2]
Cilostazol[2]
Cisapride[2,9]

Miscellaneous drugs (continued)
Colchicine[2]
Conivaptan
Cyclobenzaprine[2]
Dextromethorphan[11]
Diclofenac[2]
Ergots
Granisetron
Levomethadyl[6,9]
Meloxicam[2]
Montelukast[2]
Nateglinide[2]
Ondansetron[2]
Pioglitazone[2]
Ranolazine[2]
Sibutramine
Sildenafil[2,11]
Vardenafil[2]
Vesnarinone

TABLE 6–1. Some drugs metabolized by 3A *(continued)*

Note. HMG-CoA=hydroxymethylglutaryl–coenzyme A.

[1]Tertiary tricyclics are metabolized preferentially by 3A4 but are also metabolized by 1A2, 2C19, 2D6, and uridine 5'-diphosphate glucuronosyltransferases.

[2]Also significantly metabolized by other P450 and/or phase II enzymes or transporters.

[3]Metabolized by 3A4 to *m*-chlorophenylpiperazine.

[4]Also metabolized by 2D6 and 1A2.

[5]Potentially toxic to the cardiac conduction system at high levels and therefore should not be used with potent inhibitors of 3A4.

[6]Also metabolized by 2D6.

[7]Known intestinal 3A activity.

[8]3A4 activates to analgesic norpropoxyphene.

[9]No longer available in the United States.

[10]Also metabolized by 2C19.

[11]Reaction can be a probe for 3A4 activity, either *in vitro* or *in vivo.*

3A metabolizes ingested drugs. ABCB1 repeatedly recycles drugs through the intestinal lumen and maximizes CYP3A metabolism and "bounces" the products of 3A metabolism back into the intestinal lumen preventing absorption (Christians et al. 2005). This way, 3A and ABCB1 in the mucosal cells serve to keep the body safe from noxious exogenous compounds. ABCB1 and CYP3A are both regulated by nuclear receptors: pregnane X receptor (PXR) and others (Christians et al. 2005) (See Chapter 4, "Transporters").

Hepatic 3A has significant metabolic activity and is often thought of as a high-capacity/low-affinity enzyme, as compared with other P450 enzymes—particularly 2D6, 2C9, and 2C19, which are low-capacity/high-affinity enzymes. 3A may function as a "sink," where many drugs go to be metabolized at relatively high levels. At low levels, the low-capacity/high-affinity enzymes perform more of the enzymatic activity. If 3A is inhibited, metabolic activity may spill over to other enzymes; however, because these enzymes are often low capacity, serum substrate levels may increase, leading to potential toxicity (Olesen and Linnet 1997).

3A metabolism can occur preferentially via intestinal 3A, hepatic 3A, or both intestinal and hepatic 3A. Similarly, drugs that are inhibitors or inducers of 3A may act preferentially at one or both sites. For example, rifaximin is an intestinal 3A inducer that has no effect on ethinyl estradiol, which is metabolized primarily via hepatic 3A4 (Trapnell et al. 2007). Unfortunately, this information is frequently unknown.

Are There Drugs That Inhibit 3A Activity?

Yes. Many drugs and drug classes moderately or potently inhibit 3A (see Table 6–2). However, although a drug may be known to be a potent 3A inhibitor *in vitro*, there may be insufficient clinical data to confirm it is an inhibitor *in vivo,* as *in vitro* studies of inhibition potency may not accurately predict *in vivo* results. The U.S. Food and Drug Administration ([FDA] 2006) has created a Web page that categorizes some known *in vivo* 3A inhibitors into three categories: strong, moderate, and weak. Strong inhibitors increase the area under the plasma concentration-time curve (AUC) of a 3A substrate equal or more than 5-fold, while moderate are between 2- to 5-fold, and weak between 1.25 to 2-fold. Strong 3A inhibitors include atazanavir, clarithromycin, indinavir, itraconazole, ketoconazole, nefazodone, nelfinavir, ritonavir (initially), saquinavir, and telithromycin. Moderate 3A inhibitors include amprenavir, aprepitant (initially), diltiazem, erythromycin, fluconazole, fosamprenavir, grapefruit juice, and verapamil. Cimetidine is a weak 3A inhibitor. The actual potency of *in vivo* 3A inhibition is not always known, because clinical studies with known 3A substrates such as midazolam have not always been performed.

Many drugs that inhibit 3A less potently still have the potential for increasing levels of 3A substrate drugs. Despite the notoriety of selective serotonin reuptake inhibitors (SSRIs) as P450 enzyme inhibitors, only nefazodone, a non-SSRI antidepressant, is a potent inhibitor of 3A (Owen and Nemeroff 1998).

Another moderate 3A inhibitor is grapefruit juice. The chemical flavonoid in grapefruit is a known moderate inhibitor of 3A (Fuhr 1998; He et al. 1998) (see "Grapefruit Juice" section below). Flavonoid inhibits gut wall 3A (Wang et al. 2001). Ingesting an intestinal 3A substrate after consuming grapefruit juice will likely result in higher levels of substrate than expected. Grapefruit juice can also inhibit certain transporters such as OATP1A2 (Bailey et al. 2007) and other phase I reactions to include certain esterases (Li et al. 2007). Consumption of large amounts of grapefruit juice can also affect hepatic 3A (Veronese et al. 2003), as can obesity (Kotlyar and Carson 1999).

With so many drugs metabolized by 3A, one would expect serious consequences from 3A inhibition. One would also expect drugs that are highly dependent on 3A4 activity and have a narrow therapeutic window or margin of safety to be most at risk. The FDA Web site cited above lists the following drugs with this risk that are available in the United States: alfentanil, cyclosporine, diergotamine, ergotamine, fentanyl, quinidine, sirolimus, tacrolimus, and pimozide. For example, two deaths in 1997 were associated with concomitant use of clarithromycin and pimozide (Desta et al. 1999). In these cases clarithromycin, a commonly used macrolide and potent 3A inhibitor, was added to pimozide, which is metabolized via 3A. The resulting high serum levels of pimozide likely caused the resultant prolongation of the QTc interval. Pimozide is now contraindicated with *any* azole antifungal, macrolide antibiotic, or protease inhibitor (Orap 1999).

Some 3A inhibitors, such as azole antifungal agents, macrolide antibiotics, and quinolone antibiotics, are medications commonly prescribed by physi-

TABLE 6–2.   Some inhibitors of 3A4

Antidepressants
Fluvoxamine[1]
Nefazodone

Antimicrobials

Antibiotics, other
Chloramphenicol
Ciprofloxacin[2]
Norfloxacin[3]
Quinupristin/Dalfopristin

Azole antifungals
Fluconazole[6]
Itraconazole
Ketoconazole[1]
Miconazole[1]
Voriconazole[6]

Macrolide and ketolide antibiotics
Clarithromycin[7]
Erythromycin
Telithromycin
Troleandomycin

Nonnucleoside reverse transcriptase inhibitors
Delavirdine[1]
Efavirenz[1]

Protease inhibitors
Amprenavir (short term)[5]
Atazanavir[8]
Indinavir[1]
Nelfinavir[1]
Ritonavir (short term)[1]
Saquinavir[10]

Antipsychotics
Pimozide[10]

Other inhibitors
Amiodarone[1]
Anastrozole[1]
Androstenedione
Aprepitant (short term)
Cimetidine[1]
Conivaptan
Cyclosporine

Danazol
Diltiazem
Gestodene[5]
Grapefruit juice[1]
Imatinib[1]
Mibefradil[9]
Micafungin
Omeprazole?[6]
Oral contraceptives[1]
Propoxyphene[1]
Ranolazine[10]
Tamoxifen[4]
Verapamil[1]

TABLE 6–2. Some inhibitors of 3A4 *(continued)*

Note. Names of potent inhibitors are in bold type.
[1] Also inhibits other P450 enzymes.
[2] Also a potent inhibitor of 1A2.
[3] Also an inhibitor of 1A2.
[4] Also an inhibitor of 2C9.
[5] Also an inhibitor of 2C19.
[6] Also an inhibitor of 2C9 and 2C19.
[7] Also inhibitor of 3A5.
[8] Also and inhibitor of 2C8.
[9] No longer available in the United States.
[10] Also an inhibitor of 2D6.

cians who may be unfamiliar with or unable to obtain an individual's medication regimen (e.g. emergency departments). We recommend that primary care physicians and psychiatrists ask their patients to contact them after any new medication is prescribed. This way, the regular prescribing clinician can determine whether extra caution or monitoring is necessary. For example, the levels of clozapine or any tertiary tricyclic may be increased to toxic ranges when widely used antibiotics such as clarithromycin are administered for respiratory tract infections or when itraconazole is administered for fungal infections.

Inhibition of 3A4: Historical Issues

In the late 1990s, several 3A-related drug-drug interactions resulted in severe morbidity or mortality and removal of the drugs from the U.S. market by manufacturers. These cases are reminders of the potentially serious nature of 3A inhibition.

Nonsedating Antihistamines

In 1997, terfenadine was voluntarily withdrawn from the U.S. market by its manufacturer. Astemizole was withdrawn in 1999. Both terfenadine and astemizole, histamine type 1 (H_1) receptor antagonists, are pro-drugs requiring 3A metabolism to their active compound. The parent compounds, however, were toxic to the cardiac conduction system. Significantly increased levels resulted in QTc prolongation and the potential for supraventricular tachycardia and/or torsades de pointes. These interactions occurred when terfenadine and astemizole were coadministered with 3A inhibitors. Terfenadine is no longer available, having been replaced by the nontoxic active metabolite fexofenadine. Fexofenadine is excreted unchanged by the kidneys, and at high serum levels it is not associated with cardiac toxicity (see "Allergy Drugs" in Chapter 13, "Internal Medicine").

Mibefradil

Mibefradil, a calcium channel blocker, was voluntarily removed from the U.S. market in 1998 because of severe drug-drug interactions. Mibefradil potently inhibits both 3A4 and 2D6 and mildly inhibits 1A2. Despite warnings regarding potential interactions, physicians continued to prescribe mibefradil. Just before the manufacturer's withdrawal, the FDA published a Talk Paper (U.S. Food and Drug Administration 1998) that included a long list of drugs that are potentially dangerous when administered with mibefradil. Cardiac shock and death occurred in a number of cases when mibefradil was administered with other cardiotropic drugs such as calcium channel blockers and beta-blockers metabolized via 3A (Krum and McNeil 1998; Mullins et al. 1998). Many other calcium channel blockers without potent 3A inhibition being available, the manufacturer removed mibefradil from the market.

Cisapride

Cisapride was removed from the U.S. market in 2000, but it is available through a limited-access program. Metabolized via 3A, cisapride can have serious cardiac effects similar to those of terfenadine and astemizole. A total of 341 cases of cardiac arrhythmias have been reported, including 80 fatalities (Propulsid [drug warning] 2000). Eighty-five percent of these interactions involved 3A inhibitors or other drugs that could cause arrhythmia (U.S. Food and Drug Administration 2000). Before full understanding of this interaction, a study of low-dose cisapride to reduce SSRI-induced nausea was conducted and did not reveal any adverse cardiac events (Bergeron and Plier 1994).

Inhibition of 3A: Current Issues

Dresser et al. (2000) wrote a thorough review of the consequences of 3A inhibition and cautioned against coadministering potent 3A inhibitors and drugs with narrow safety margins. For example, drugs that can increase the QTc interval (pimozide) or cause excessive sedation (triazolobenzodiazepines), hypotension (calcium channel blockers), or rhabdomyolysis (hydroxymethylglutaryl–coenyzme A [HMG-CoA] reductase inhibitors) have potentially severe interactions with 3A inhibitors. In this section I describe some issues regarding 3A inhibition that a primary care physician or psychiatrist may encounter.

Triazolobenzodiazepines

Triazolobenzodiazepines (alprazolam, midazolam, estazolam, and triazolam) are dependent on 3A4 for metabolism. Other benzodiazepines are metabolized through multiple pathways, including via phase II enzymes. Caution is warranted when these drugs are used with any potent 3A4 inhibitors, including nefazodone, clarithromycin, and ketoconazole. A 50% reduction in the dose of alprazolam and a 75% reduction in the dose of triazolam are recommended when these are coadministered with nefazodone. Case reports and small controlled studies have indicated enhanced side effects (including severe sedation and delirium) when triazolobenzodiazepines are used with inhibitors of 3A, such as ritonavir (von Moltke et al. 1998), erythromycin (Tokinaga et al. 1996), and diltiazem (Kosuge et al. 1997).

Nonbenzodiazepine Hypnotics

The nonbenzodiazepine hypnotics zolpidem, eszopiclone, and zaleplon are metabolized via 3A4. Enhanced effects are likely to occur when these drugs are administered with 3A inhibitors. Greenblatt et al. (1998) reported impairment in motor or cognitive skills in five healthy volunteers who had increased zolpidem half-life and levels due to 3A inhibition by ketoconazole. Similarly, coadministration of ritonavir and zolpidem is not recommended because of the potential for enhanced central nervous system (CNS) effects (Norvir 2007). Voriconazole also increases the peak plasma concentration of zolpidem and prolongs half-life (Saari et al. 2007). There are limited case reports of other 3A4 inhibitors enhancing the CNS effects of these drugs. The wide safety margin of nonbenzodiazepine hypnotics may explain the lack of severe side effects despite potent 3A inhibition. Many patients and physicians may consider such reactions common and expected side effects rather than a drug-drug interaction.

Selective Serotonin Reuptake Inhibitors

Fluoxetine's active metabolite, norfluoxetine, is considered a more potent inhibitor of 3A than is fluoxetine. Reports of fluoxetine's role in enhancing 3A substrates are inconsistent. Nevertheless, there are case reports of fluoxetine increasing the half-life and serum levels of 3A substrates, including alprazolam (Lasher et al. 1991), cyclosporine (Horton and Bonser 1995), midazolam (von Moltke et al. 1996), nifedipine (Azaz-Livshits and Danenberg 1997), and tertiary tricyclics.

Buspirone

Buspirone is metabolized mainly by 3A. Enhanced effects of the drug (chiefly sedation) have been reported with coadministration of buspirone and either erythromycin or itraconazole, resulting in buspirone levels as high as 5 to 13 times expected (Kivisto et al. 1997). Administration of buspirone and other 3A4 inhibitors may have similar results.

Grapefruit Juice

There is ample evidence that furanocoumarins, present in grapefruit juice, inhibit 3A (Edwards et al. 1996). Furanocoumarins preferentially inhibit intestinal 3A, but with increasing consumption (3 double-strength glasses) will also inhibit hepatic 3A. Although effects can last up to 3 days after grapefruit juice ingestion, effects are especially predictable when grapefruit juice is consumed along with the oral dose of a 3A substrate. Furanocoumarin inhibition of 3A depends on both the type of grapefruit juice (white greater than pink) and the amount of available 3A. As little as 6–8 ounces of grapefruit juice taken with a high first-pass substrate of CYP3A can substantially alter pharmacokinetics (Grapefruit-drug interactions 2005).

Grapefruit interacts with many drugs, including cisapride (Desta et al. 2001), cyclosporine, tacrolimus, sirolimus, HMG-CoA reductase inhibitors (simvastatin, atorvastatin, and lovastatin) (Kane and Lipsky 2000), triazolam (Hukkinsen 1995), amiodarone (Libersa et al. 2000), repaglinide and nateglinide (Scheen 2007), and many others (Mertens-Talcott et al. 2006). Although earlier reports indicate grapefruit juice increases clozapine levels, the most informed study indicates the effect is not clinically relevant (Lane et al. 2001). Grapefruit juice may also worsen side effects of oral contraceptives by increasing peak levels of ethinyl estradiol by as much as 137% (Weber et al. 1996). For a more complete list of drugs that interact with grapefruit, see the Web site "Grapefruit-Drug Interactions" (2005).

Drug Sparing and Augmentation

The use of inexpensive 3A inhibitors, such as grapefruit juice or ketoconazole, as sparing agents to reduce costs for expensive drugs such as cyclosporine or antiretrovirals remains controversial. The strategy is not advisable for several reasons. First, the concentration of the furanocoumarin 3A4 inhibitor in grapefruit juice is not constant even within the same brand. Second, if the timing of grapefruit juice ingestion is not precise, the "spared" drug's serum drug levels can fluctuate significantly from dose to dose.

Keogh et al. (1995) reported that concomitant use of ketoconazole and cyclosporine reduced the dose of the latter drug by 80% and reduced the amount spent on that drug per patient by an average of $5,200 in the first year after cardiac transplantation. A subsequent report indicated that long-term use of ketoconazole may have risks, including a reduction of bone mineralization (Moore et al. 1996).

Psychiatrists have engaged in enhancement of drug effects for some time. Cimetidine had been used to augment the effects of clozapine, a 1A2 substrate, until the potential toxicity was recognized (Szymanski et al. 1991). Cimetidine, an inhibitor of 2D6, 3A, and 1A2 (Rendic 1999), can increase clozapine levels. Some clinicians have also used SSRIs to augment clozapine's effects in an effort to decrease the amount spent on clozapine. Maintaining safe serum levels of clozapine with these drug combinations is difficult and likely outweighs any greater cost savings. One series indicated a 43% increase in serum clozapine or norclozapine levels when administered with fluoxetine, paroxetine, or sertraline (Centorrino et al. 1996). Fluvoxamine, a 1A2 and 3A4 inhibitor, has the highest likelihood of interaction with clozapine among the SSRIs. Wetzel et al. (1998) demonstrated a 3-fold increase in clozapine or norclozapine serum level and half-life with the addition of just 50 mg of fluvoxamine per day. Coadministration of fluvoxamine and clozapine has been found, in a well-monitored setting, to be both safe and economically beneficial in schizophrenic patients requiring clozapine (Lu et al. 2000). However, we believe that this strategy has a potential for disaster should fluvoxamine be discontinued or the dose increased without close monitoring (which is probably impractical).

Are There Drugs That Induce 3A Activity?

Yes. Through receptors such as the pregnane X receptor (PXR) and the constitutively activated receptor (CAR), a number of compounds can induce 3A (see Table 6–3). The best-known inducer of 3A is carbamazepine. For years it was known that carbamazepine, a 3A4 substrate, induces its own metabolism. Neurologists and psychiatrists recognized that after initiation of carbamazepine therapy, a dose increase would be necessary within 4–8 weeks to maintain stable carbamazepine levels. The mechanism is now known to be 3A4 induction by carbamazepine (Levy 1995). Carbamazepine also induces phase II conjugation enzymes (Ketter et al. 1999).

TABLE 6–3.   Some inducers of 3A4

Antiepileptics	Other inducers	
Carbamazepine[1]	Aprepitant (long term)	Prednisone
Felbamate	Bosanten	Rifabutin
Oxcarbazepine	Dexamethasone	Rifampin[1]
Phenobarbital[1]	Efavirenz?	Rifapentine[1]
Phenytoin[1]	Griseofulvin	Rifaximin[3]
Primidone	Methylprednisolone	Ritonavir and ritonavir-boosted drugs
Topiramate (>200mg/d)	Modafinil[3]	(long term)
	Nafcillin	St. John's wort (long term)[3]
	Nelfinavir	Troglitazone[2]
	Nevirapine	

[1]"Pan-inducers"—also induce most other P450 enzymes.
[2]Removed from the U.S. market.
[3]Known intestinal 3A activity.

Oxcarbazepine also induces 3A, potentially leading to adverse outcomes such as ineffective oral contraceptive therapy (Fattore et al. 1999). Studies have shown that the induction by carbamazepine is 46% greater than by oxcarbazepine (Andreasen et al. 2007).

Other antiepileptics are also 3A inducers. These include phenytoin, phenobarbital, primidone (Anderson 1998), and topiramate. Adding topiramate to valproic acid in patients taking 35 µg of ethinyl estradiol (EE) oral contraceptives resulted in decreases in plasma levels of EE by 18%–30%, depending on the topiramate dose (dose range: 200–800mg/day [Topamax 2007]). A clinical study using less than 200 mg/day of topiramate in conjunction with 35 µg of ethinyl estradiol showed no such effect (Doose et al. 2003).

The antitubercular drugs rifampin and rifabutin (Strayhorn et al. 1997), nevirapine (Barry et al. 1997; Tseng and Foisy 1997), and troglitazone (Caspi 1997) induce 3A4. Dexamethasone and prednisone likewise are 3A4 inducers (Pichard et al. 1992). Caution should be taken when coadministering any of these drugs with 3A substrates, as toxicities may result.

Modafinil, a CNS stimulant, can induce its own metabolism at high doses via induction of 3A4 (Provigil 2007). A case report of a 41-year-old woman taking cyclosporine showed cyclosporine levels were reduced by 50% when modafinil 200 mg/day was administered (Provigil 2007).

The herbal supplement St. John's wort, in forms that have high hyperforin content, induces 3A (Mueller et al. 2006) as well as ABCB1 (Hennessy et al. 2002). Use of the supplement in conjunction with cyclosporine has led to decreased cyclosporine levels and, ultimately, transplant rejection in certain cases (Barone et al. 2000; Karliova et al. 2000; Ruschitzka et al. 2000). St. John's wort has also been shown to decrease digoxin levels, presumably through ABCB1 induction (Johne et al. 1999). Individuals should be cautioned regarding the ingestion of St. John's wort in conjunction with 3A substrates because of potential reductions in plasma levels of the 3A substrate.

Although initially a potent inhibitor of 3A, ritonavir becomes an *inducer* of 3A metabolism after several weeks. The mechanism for this change is unclear but may involve a feedback loop response to chronic exposure, as ritonavir has a high affinity for 3A4. Ritonavir can also increase the metabolism of meperidine, resulting in increased formation of the neurotoxic metabolite normeperidine (Piscitelli et al. 2000). Due to 3A induction, ethinyl estradiol levels may decrease when this agent is administered with ritonavir, leading to potentially ineffective therapy (Ouellet et al. 1998).

Similar to ritonavir, aprepitant is initially a dose-related 3A inhibitor but becomes a 3A inducer over time. Oral midazolam combined with aprepitant can result in a 2- to 3-fold increase in the AUC of midazolam, leading to increased side effects during 3A inhibition by aprepitant. In contrast, long-term use in conjunction with ethinyl estradiol can lead to as much as an 18% AUC decrease in ethinyl estradiol (Emend 2008).

Bosentan is both a substrate and an inducer of 2C9 and 3A. Coadministration with ethinyl estradiol leads to decreases in ethinyl estradiol levels (Tracleer 2007). Sildenafil and bosentan, both used in pulmonary arterial hypertension, can result in as much as a 63% decrease in sildenafil levels via 3A induction by bosentan. This combination also results in a 50% increase in bosentan levels through an unknown mechanism (Tracleer 2007). The addition of bosentan to cyclosporine at steady state necessitates a dose increase of as much as 35% to maintain previous levels while bosentan levels were 30-fold higher than expected, likely due to transporter inhibition by cyclosporine (Treiber et al. 2004).

In all, the number of potential interactions involving 3A can seem quite large and very daunting. Despite this appearance, a general understanding of 3A induction and inhibition can be achieved with the help of an easily available reference guide (such as this text) maintained in the clinical setting. A periodic review of any individual's medication regimen with close attention to 3A substrates, inhibitors, and inducers may help avoid potential complications, especially when an individual is on a polypharmacy regimen.

References

Anderson GD: A mechanistic approach to antiepileptic drug interactions. Ann Pharmacother 32:554–563, 1998

Andreasen AH, Brosen K, Damkier P: A comparative pharmacokinetic study in healthy volunteers of the effect of carbamazepine and oxcarbazepine on CYP3A4. Epilepsia 48:490–496, 2007

Azaz-Livshits TL, Danenberg HD: Tachycardia, orthostatic hypotension and profound weakness due to concomitant use of fluoxetine and nifedipine. Pharmacopsychiatry 30:274–275, 1997

Bailey DG, Dresser GK, Leake BF, et al. Naringin is a major and selective clinical inhibitor of organic anion-transporting polypeptide 1A2 (OATP1A2) in grapefruit juice. Clin Pharmacol Ther 81:495–502, 2007

Barone GW, Gurley BJ, Ketel BL, et al: Drug interaction between St John's wort and cyclosporine. Ann Pharmacother 34:1013–1016, 2000

Barry M, Gibbons S, Mulchay F: Protease inhibitors in patients with HIV disease: clinically important pharmacokinetic considerations. Clin Pharmacokinet 32:194–209, 1997

Bergeron R, Plier P: Cisapride for the treatment of nausea produced by selective serotonin inhibitors. Am J Psychiatry 151:1084–1086, 1994

Burk O, Koch I, Raucy J, et al: The induction of cytochrome P450 3A5 (CYP3A5) in the human liver and intestine is mediated by the xenobiotic sensors pregnane X receptor (PXR) and constitutively activated receptor (CAR). J Biol Chem 279:38379–38385, 2004

Caspi A: Troglitazone. Pharmacy and Therapeutics 22:198–205, 1997

Centorrino F, Baldessarini RJ, Frankenburg FR, et al: Serum levels of clozapine and norclozapine in patients treated with selective serotonin reuptake inhibitors. Am J Psychiatry 153:820–822, 1996

Christians U, Schmitz V, Haschke M: Functional interactions between P-glycoprotein and CYP3A in drug metabolism. Expert Opin Drug Metab Toxicol 1:641–654, 2005

Daly AK: Significance of the minor cytochrome P450 3A isoforms. Clin Pharmacokinet 45:13–31, 2006

Desta Z, Kerbusch T, Flockhart DA: Effect of clarithromycin on the pharmacokinetics and pharmacodynamics of pimozide in healthy poor and extensive metabolizers of cytochrome P450 2D6 (CYP2D6). Clin Pharmacol Ther 65:10–20, 1999

Desta Z, Kivisto KT, Lilja JJ, et al: Stereoselective pharmacokinetics of cisapride in healthy volunteers and the effect of repeated administration of grapefruit juice. Br J Clin Pharmacol 52:399-407, 2001

DeVane CL, Nemeroff CB: 2002 guide to psychotropic drug interactions. Prim Psychiatry 9:28–57, 2002

Doherty MM, Charman WN: The mucosa of the small intestine: how clinically relevant as an organ of drug metabolism? Clin Pharmacokinet 41:235–253, 2002

Doose DR, Wang SS, Padmanabhan M, et al: Effect of topiramate or carbamazepine on the pharmacokinetics of an oral contraceptive containing norethindrone and ethinyl estradiol in healthy obese and nonobese female subjects. Epilepsia 44:540–549, 2003

Dresser GK, Spence JD, Bailey DG: Pharmacokinetic-pharmacodynamic consequences and clinical relevance of cytochrome P450 3A4 inhibition. Clin Pharmacokinet 38:41–57, 2000

Edwards DJ, Bellevue FH 3rd, Woster PM: Identification of 6',7'-dihydroxybergamottin, a cytochrome P450 inhibitor, in grapefruit juice. Drug Metab Dispos 24:1287–1290, 1996

Emend (package insert). Whitehouse Station, NJ, Merck & Co., Inc., 2008

Fattore C, Cipolla G, Gatti G, et al: Induction of ethinylestradiol and levonorgestrel metabolism by oxcarbazepine in healthy women. Epilepsia 40:783–787, 1999

Fuhr U: Drug interactions with grapefruit juice: extent, probable mechanism and clinical relevance. Drug Saf 18:251–272, 1998

Grapefruit-drug interactions (Web site). 2005. Available at http://www.powernetdesign.com/grapefruit/. Accessed July 10, 2007.

Greenblatt DJ, von Moltke LL, Harmatz JS, et al: Kinetic and dynamic interaction study of zolpidem with ketoconazole, itraconazole, and fluconazole. Clin Pharmacol Ther 64:661–671, 1998

Guengerich FP: Cytochrome P-450 3A4: regulation and the role in drug metabolism. Annu Rev Pharmacol Toxicol 39:1–17, 1999

Haufroid V, Mourad M, Van Kerckhove V, et al: The effect of CYP3A5 and MDR1 (ABCB1) polymorphisms on cyclosporine and tacrolimus dose requirements and trough blood levels in stable renal transplant patients. Pharmacogenetics 14:147–154, 2004

Haufroid V, Wallemacq P, Van Kerckhove V, et al: CYP3A5 and ABCB1 polymorphisms and tacrolimus pharmacokinetics in renal transplant candidates: guidelines from an experimental study. Am J Transplant 6:2706–2713, 2006

He K, Iyer KR, Hayes RN, et al: Inactivation of cytochrome P450 3A4 by bergamottin, a component of grapefruit juice. Chem Res Toxicol 11:252–259, 1998

Hennessy M, Kelleher D, Spiers JP, et al: St John's wort increases expression of P-glycoprotein: implications for drug interactions. Br J Clin Pharmacol 53:75–82, 2002

Horton RC, Bonser RS: Interaction between cyclosporin and fluoxetine. BMJ 311:422, 1995

Hukkinsen SK: Plasma concentrations of triazolam are increased by concomitant ingestion of grapefruit juice. Clin Pharmacol Ther 58:127–131, 1995

Johne A, Brockmoller J, Bauer S, et al: Pharmacokinetic interaction of digoxin with an herbal extract from St John's wort (*Hypericum perforatum*). Clin Pharmacol Ther 66:338–345, 1999

Kane GC, Lipsky JJ: Drug-grapefruit juice interactions. Mayo Clin Proc 75:933–942, 2000

Karliova M, Treichel U, Malago M, et al: Interaction of *Hypericum perforatum* (St John's wort) with cyclosporin A metabolism in a patient after liver transplantation. J Hepatol 33:853–855, 2000

Keogh A, Spratt P, McCosker C, et al: Ketoconazole to reduce the need for cyclosporine after cardiac transplantation. N Engl J Med 333:628–633, 1995

Ketter TA, Flockhart DA, Post RM, et al: The emerging role of cytochrome P450 3A in psychopharmacology. J Clin Psychopharmacol 15:387–398, 1995

Ketter TA, Frye MA, Cora-Locatelli G, et al: Metabolism and excretion of mood stabilizers and new anticonvulsants. Cell Mol Neurobiol 19:511–532, 1999

Kivisto KT, Lamberg TS, Kantola T, et al: Plasma buspirone concentrations are greatly increased by erythromycin and itraconazole. Clin Pharmacol Ther 62:348–354, 1997

Kosuge K, Nishimoto M, Kimura M, et al: Enhanced effect of triazolam with diltiazem. Br J Clin Pharmacol 43:367–372, 1997

Kotlyar M, Carson SW: Effects of obesity on the cytochrome P450 enzyme system. Int J Clin Pharmacol Ther 37:8–19, 1999

Krum H, McNeil JJ: The short life and rapid death of a novel antihypertensive and antianginal agent. Med J Aust 169:408–409, 1998

Lane H-Y, Jann MW, Chang Y-C, et al: Repeated ingestion of grapefruit juice does not alter clozapine's steady-state plasma levels, effectiveness, and tolerability. J Clin Psychiatry 62:812–817, 2001

Lasher TA, Fleishaker JC, Steenwyk RC, et al: Pharmacokinetic pharmacodynamic evaluation of the combined administration of alprazolam and fluoxetine. Psychopharmacology (Berl) 104:323–327, 1991

Levy RH: Cytochrome P450 isoenzymes and antiepileptic drugs. Epilepsia 36 (suppl 5):S8–S13, 1995

Li P, Callery PS, Gan LS, et al: Esterase inhibition by grapefruit juice flavonoids leading to a new drug interaction. Drug Metab Dispos 35:1203–1208, 2007

Libersa CC, Brique SA, Motte KB, et al: Dramatic inhibition of amiodarone metabolism induced by grapefruit juice. Br J Clin Pharmacol 49:373–378, 2000

Lu ML, Lane HY, Chen KP, et al: Fluvoxamine reduces the clozapine dosage needed in refractory schizophrenic patients. J Clin Psychiatry 61:594–599, 2000

Mertens-Talcott SU, Zadezensky I, De Castro WV, et al: Grapefruit-drug interactions: can interactions with drugs be avoided? J Clin Pharmacol 46:1390–416, 2006

Moore LW, Alloway RR, Acchiardo SR, et al: Clinical observations of metabolic changes occurring in renal transplant recipients receiving ketoconazole. Transplantation 61:537–541, 1996

Mueller SC, Majcher-Peszynska J, Uehleke B, et al: The extent of induction of CYP3A by St. John's wort varies among products and is linked to hyperforin dose. Eur J Clin Pharmacol 62:29–36, 2006

Mullins ME, Horowitz BZ, Linden DH, et al: Life-threatening interaction of mibefradil and beta-blockers with dihydropyridine calcium channel blockers. JAMA 280:157–158, 1998

National Center for Biotechnology Information: CYP3A4 cytochrome P450, family 3, subfamily A, polypeptide 4 [Homo sapiens]. PubMed Entrez Gene, 2008. Available at http://www.ncbi.nlm.nih.gov/sites/entrez?Db=gene&Cmd=Show-DetailView&TermToSearch=1576&ordinalpos=1&itool=EntrezSystem2.PEntrez.Gene.Gene_ResultsPanel.Gene_RVDocSum. Accessed July 14, 2008.

Nefazodone hydrochloride (patient information sheet). U.S. Food and Drug Administration Center for Drug Evaluation and Research, 2005. Available at http://www.fda.gov/cder/drug/infopage/nefazodone/default.htm. Accessed July 14, 2008.

Norvir (package insert). Chicago, IL, Abbott Laboratories, 2007. Available at http://www.norvir.com. Accessed July 1, 2007.

Oesterheld JR: A review of developmental aspects of cytochrome P450. J Child Adolesc Psychopharmacol 8:161–174, 1998

Olesen OV, Linnet K: Metabolism of the tricyclic antidepressant amitriptyline by cDNA-expressed human cytochrome P450 enzymes. Pharmacology 55:235–243, 1997

Orap (package insert). Sellersville, PA, Gate Pharmaceuticals, 1999. Available at http://www.gatepharma.com/ORAP/orap.html. Accessed July 10, 2007.

Ouellet D, Hsu A, Qian J, et al: Effect of ritonavir on the pharmacokinetics of ethinyl oestradiol in healthy female volunteers. Br J Clin Pharmacol 46:111–116, 1998

Owen JR, Nemeroff CB: New antidepressants and the cytochrome P450 system: focus on venlafaxine, nefazodone, and mirtazapine. Depress Anxiety 7 (suppl 1):24–32, 1998

Pichard L, Fabre I, Daujat M, et al: Effect of corticosteroids on the expression of cytochromes P450 and on cyclosporin A oxidase activity in primary cultures of human hepatocytes. Mol Pharmacol 41:1047–1055, 1992

Piscitelli SC, Kress DR, Bertz RJ, et al: The effect of ritonavir on the pharmacokinetics of meperidine and normeperidine. Pharmacotherapy 20:549–553, 2000

Propulsid (drug warning). Titusville, NJ, Janssen Pharmaceutica, April 12, 2000. Available at http://www.fda.gov/MedWatch/safety/2000/propul.htm. Accessed July 14, 2008.

Provigil (package insert). West Chester, PA, Cephalon, Inc., 1999, 2007. Available at http://www.provigil.com. Accessed July 11, 2007.

Rendic S: Drug interactions of H2-receptor antagonists involving cytochrome P450 (CYPs) enzymes: from the laboratory to the clinic. Croat Med J 40:357–367, 1999

Ruschitzka F, Meier PJ, Turina M, et al: Acute heart transplant rejection due to Saint John's wort (letter). Lancet 355:548–549, 2000

Saari TI, Laine K, Leino K, et al: Effect of voriconazole on the pharmacokinetics and pharmacodynamics of zolpidem in healthy subjects. Br J Clin Pharmacol 63:116–120, 2007

Scheen AJ: Drug-drug and food-drug pharmacokinetic interactions with new insulinotropic agents repaglinide and nateglinide. Clin Pharmacokinet 46:93–108, 2007

Soars MG, Grime K, Riley RJ: Comparative analysis of substrate and inhibitor interactions with CYP3A4 and CYP3A5. Xenobiotica 36:287–299, 2006

Strayhorn VA, Baciewicz AM, Self TH: Update on rifampin drug interactions, III. Arch Intern Med 157:2453–2458, 1997

Strolin Benedetti M, Whomsley R, Baltes EL: Differences in absorption, distribution, metabolism and excretion of xenobiotics between the paediatric and adult populations. Expert Opin Drug Metab Toxicol 1:447–471, 2005

Szymanski S, Lieberman JA, Picou D, et al: A case report of cimetidine-induced clozapine toxicity. J Clin Psychiatry 52:21–22, 1991

Tokinaga N, Kondo T, Kaneko S, et al: Hallucinations after a therapeutic dose of benzodiazepine hypnotics with co-administration of erythromycin. Psychiatry Clin Neurosci 50:337–339, 1996

Topamax (package insert). Titusville, NJ, Ortho-McNeil Neurologics, Inc., 2007

Tracleer (package insert). South San Francisco, CA, Actelion Pharmaceuticals US, Inc., 2007

Trapnell CB, Connolly M, Pentikis H, et al: Absence of effect of oral rifaximin on the pharmacokinetics of ethinyl estradiol/norgestimate in healthy females. Ann Pharmacother 41:222–228, 2007

Treiber A, Schneiter R, Delahaye S, et al: Inhibition of organic anion transporting polypeptide-mediated hepatic uptake is the major determinant in the pharmacokinetic interaction between bosentan and cyclosporin A in the rat. J Pharmacol Exp Ther 308:1121–1129, 2004

Tseng AL, Foisy MM: Management of drug interactions in patients with HIV. Ann Pharmacother 31:1040–1058, 1997

U.S. Food and Drug Administration: Report of pimozide and macrolide antibiotic interaction. FDA Medical Bulletin 26:3, 1996

U.S. Food and Drug Administration: Roche Laboratories announces withdrawal of Posicor from the market (FDA Talk Paper T98–33). Rockville, MD, National Press Office, June 8, 1998

U.S. Food and Drug Administration: Janssen Pharmaceutica stops marketing cisapride in the US (FDA Talk Paper T00–14). Rockville, MD, National Press Office, March 23, 2000

U.S. Food and Drug Administration: Drug development and drug interactions: table of substrates, inhibitors and inducers. 2006. Available at http://www.fda.gov/cder/drug/drugInteractions/tableSubstrates.htm. Accessed July 11, 2007.

Veronese ML, Gillen LP, Burke JP, et al: Exposure-dependent inhibition of intestinal and hepatic CYP3A4 in vivo by grapefruit juice. J Clin Pharmacol 43:831–839, 2003

von Moltke LL, Greenblatt DJ, Schmider J, et al: Midazolam hydroxylation by human liver microsomes in vitro: inhibition by fluoxetine, norfluoxetine, and by the azole antifungal agents. J Clin Pharmacol 36;783–791, 1996

von Moltke LL, Greenblatt DJ, Grassi JM, et al: Protease inhibitors as inhibitors of human cytochrome P450: high risk associated with ritonavir. J Clin Pharmacol 38:106–111, 1998

Wang EJ, Casciano CN, Clement RP, et al: Inhibition of P-glycoprotein transport function by grapefruit juice psoralen. Pharm Res 18:432–438, 2001

Weber A, Jager R, Borner A, et al: Can grapefruit juice influence ethinylestradiol bio-availability? Contraception 53:41–47, 1996

Wetzel H, Anghelescu I, Szegedi A, et al: Pharmacokinetic interactions of clozapine with selective serotonin reuptake inhibitors: differential effects of fluvoxamine and paroxetine in a prospective study. J Clin Psychopharmacol 18:2–9, 1998

Wojnowski L, Kamdem LK: Clinical implications of CYP3A polymorphisms. Expert Opin Drug Metab Toxicol 2:171–182, 2006

Zhang QY, Dunbar D, Ostrooska A, et al: Characterization of human small intestinal cytochromes P-450. Drug Metab Dispos 27:804–809, 1999

Zhu B, Liu ZQ, Chen GL, et al: The distribution and gender difference of CYP3A activity in Chinese subjects. Br J Clin Pharmacol 55:264–269, 2003

Study Cases

Case 1

A 38-year-old Caucasian man who tested positive for the human immunode-ficiency virus (HIV) and has several chronic problems, including allergic rhin-itis, degenerative joint disease, and gastroesophageal reflux, began nelfinavir and nevirapine therapy. He subsequently became depressed, and treatment with doxepin was begun. Over time, the doxepin dose was titrated to 300 mg/

day, with few side effects but with questionable efficacy despite good compliance. Because of efficacy concerns, serum tricyclic levels were measured and showed low levels, ranging from 35 to 50 ng/mL.

Comment

3A4 induction: Nelfinavir has only mild inhibitory effects on 3A4 and 1A2. However, nevirapine induces 3A4. Although the metabolism of doxepin has not been studied as well as the metabolisms of other tertiary tricyclics (amitriptyline and imipramine), doxepin is likely metabolized via 3A4 with minor contribution by other enzymes. In this case, nevirapine likely induced doxepin metabolism, causing low tricyclic levels despite a relatively high dose of doxepin.

References

Lemoine A, Gautier JC, Azoulay D, et al: Major pathway of imipramine metabolism is catalyzed by cytochromes P-450 1A2 and P-450 3A4 in human liver. Mol Pharmacol 43:827–832, 1993

Tseng AL, Foisy MM: Management of drug interactions in patients with HIV. Ann Pharmacother 31:1040–1058, 1997

Venkatakrishnan K, Greenblatt DJ, von Moltke LL, et al: Five distinct human cytochromes mediate amitriptyline N-demethylation in vitro: dominance of CYP 2C19 and 3A4. J Clin Pharmacol 38:112–121, 1998

Case 2

A 35-year-old HIV-positive African American male had been taking fluoxetine (20 mg/day) for several years and ritonavir for several months without problems. The patient tried his mother's zolpidem (10 mg) because of sleeplessness. He slept for 14 hours and had a "hangover" the next day.

Comment

3A inhibition: Zolpidem is metabolized via 3A. Ritonavir (potent) and fluoxetine both inhibit 3A. It is likely that 3A inhibition caused a delay in zolpidem clearance, leading to oversedation.

References

Tseng AL, Foisy MM: Management of drug interactions in patients with HIV. Ann Pharmacother 31:1040–1058, 1997

von Moltke LL, Greenblatt DJ, Granda BW, et al: Zolpidem metabolism in vitro: responsible cytochromes, chemical inhibitors, and *in vivo* correlations. Br J Clin Pharmacol 48:89–97, 1999

Case 3

A 45-year-old Caucasian man with chronic schizophrenia was being treated with haloperidol decanoate at a stable dose of 200 mg/month. He developed a seizure disorder and was prescribed phenytoin by his neurologist. Within 2 months, the patient's psychotic symptoms worsened and he required hospitalization. Although baseline haloperidol levels had not been obtained before initiation of phenytoin therapy, a haloperidol level obtained in the hospital was low (less than 2 ng/mL) during therapy with both haloperidol and phenytoin.

Comment

3A4 inhibition: Haloperidol is metabolized via several P450 enzymes, including 3A4, as well as via phase II enzymes. Phenytoin induces 3A4. It is likely that over several weeks, phenytoin induced 3A4 and decreased serum haloperidol levels, leading to a recurrence of psychotic symptoms.

References

Kudo S, Ishizaki T: Pharmacokinetics of haloperidol: an update. Clin Pharmacokinet 37:435–456, 1999

Linnoila M, Viukari M, Vaisanen K: Effect of anticonvulsants on plasma haloperidol and thioridazine levels. Am J Psychiatry 137:819–821,1980

Case 4

A teenage boy with Tourette's syndrome was treated with pimozide 2 mg/day by his psychiatrist. Subsequently, a primary care physician treated the patient for pharyngitis with clarithromycin. Twenty-four hours after initiating clarithromycin therapy, the patient developed heart palpitations.

Comment

3A inhibition: Clarithromycin is a potent inhibitor of 3A. Pimozide requires 3A4 for clearance, and toxic cardiac side effects may occur when pimozide levels are increased. Deaths have been reported with coadministration of pimozide and erythromycin or clarithromycin.

References

Desta Z, Kerbusch T, Flockhart DA: Effect of clarithromycin on the pharmacokinetics and pharmacodynamics of pimozide in healthy poor and extensive metabolizers of cytochrome P450 2D6 (CYP2D6). Clin Pharmacol Ther 65:10–20, 1999

U.S. Food and Drug Administration: Report of pimozide and macrolide antibiotic interaction. FDA Med Bull 26:3, 1996

7

1A2

Miia R. Turpeinen, M.D., Ph.D.

What Does 1A2 Do?

1A2 is primarily involved in phase I metabolism of xenobiotics. Its actions, like those of other P450 enzymes, include oxidative, peroxidative, and reductive metabolism of compounds (see Table 7–1). 1A2 is a major enzyme in the metabolism of a number of important chemicals (Brosen 1995). Besides detoxification of compounds, 1A2 is often responsible for metabolic activation of polycyclic aromatic hydrocarbons found in cigarette smoke and aromatic amines, and thus it has been linked to chemical carcinogenesis (Boobis et al. 1994; Ioannides and Lewis 2004; Yamazaki et al. 2000). 1A2 is also the main enzyme involved in the oxidative metabolism of melatonin and the methylxanthines: caffeine and theophylline. These drugs are also frequently used as probe drugs for 1A2 (Faber et al. 2005; Hartter et al. 2001).

TABLE 7–1.   Some drugs metabolized by 1A2

Antidepressants	Antipsychotics	Other drugs	
Amitriptyline[1,2]	Chlorpromazine[3]	Anagrelide	Phenacetin
Clomipramine[1,3]	Clozapine[5]	Caffeine	Ramelteon
Duloxetine[3]	Fluphenazine[3]	Cinacalcet[3]	Riluzole
Fluvoxamine[3]	Olanzapine[6]	Cyclobenzaprine[3]	Ropinirole
Imipramine[1,3]	Perphenazine[3]	Dacarbazine[3]	Ropivacaine
Mirtazapine[3]	Thioridazine[3]	Erlotinib[3]	Tacrine
	Thiothixene[3]	Flutamide[3]	Theophylline[3]
	Trifluoperazine[3]	Frovatriptan	Tizanidine
		Lidocaine	R-Warfarin[8]
		Melatonin	Zolmitriptan
		Mexiletine	
		Naproxen	
		Ondansetron	

Note.   Names of drugs are in **bold** type if there is evidence that in normal human use, the drugs are potent inhibitors.
[1]Tertiary tricyclic antidepressants (TCAs) are demethylated by 1A2 and 3A4. 2D6, 2C9, and 2C19 also metabolize tertiary TCAs at a minor extent.
[2]*N*-Demethylation may be preferentially done by 2C19.
[3]Metabolized by other P450 enzymes as well.
[4]Also metabolized by 3A4.
[5]Demethylated by 1A2 to norclozapine. Clozapine is metabolized to clozapine-*N*-oxide by 3A4 and, to a lesser extent, 2D6 and others.
[6]Metabolized 30%–40% by 1A2 and some by 2D6, glucuronidated by the glucuronosyltransferase (UGT) 1A4.
[7]1A2 is a minor route of metabolism.
[8]Weaker pharmacological isomer of racemic warfarin (see Chapter 9, "2C9").

Where Is 1A2?

1A2, the only hepatic member of the CYP1 family, is expressed almost exclusively in the liver. The gene for 1A2 is located on chromosome 15 (National Center for Biotechnology Information 2008). 1A2 accounts for about 15% of the total cytochromes present in the liver. There is significant intra- and interindividual variation in the metabolic activity of 1A2. This variability is related to enzyme induction and inhibition via concomitant medications or dietary and environmental factors.

Does 1A2 Have Any Polymorphisms?

Yes. More than twenty different single nucleotide polymorphisms (SNPs) of 1A2 have been reported, although most of them have been found to be quite rare (Aitchison et al. 2000; Nakajima et al. 1994; Sachse et al. 1999). Certain SNPs have been associated with altered probe drug metabolism (Eap et al. 2004; Han et al. 2002; Saito et al. 2005), adverse drug reactions (Tiwari et al. 2005), and certain disease states (Chen et al. 2006; Cornelis et al. 2006; Moonen et al. 2005). The *1A2*1F* genotype is known to be associated with increased inducibility of 1A2, whereas the *1A2*1C* genotype is associated with lower inducibility of 1A2 (Aitchison et al. 2000; Sachse et al. 1999). Thus, these polymorphisms can affect the likelihood and intensity of drug interactions involving a 1A2 inducer. *1A2*1F* and *1A2*1C* are also associated with a susceptibility to develop tardive dyskinesia with administration of typical antipsychotics (Basile et al. 2000; Tiwari et al. 2005).

The presence of *1A2*F* may increase the risk of developing prolonged QTc interval in patients treated with high doses of antipsychotics (Tay et al. 2007), although this does not hold true for thioridazine or mesoridazine (Dorado et al. 2007).

Several studies have shown a relationship between caffeine metabolic ratios and steady-state plasma concentrations of olanzapine and clozapine. 1A2 phenotyping with caffeine has therefore been suggested as a convenient tool to distinguish between nonresponse and noncompliance in individualizing the doses of olanzapine and clozapine (Ozdemir et al. 2001; Carrillo et al. 2003; Faber et al. 2005; Shirley et al. 2003). Studies evaluating the benefit of routine 1A2 phenotyping for clinical practice are currently lacking. Until further research is completed, genetic phenotyping of 1A2 should not be considered for routine practice.

In some studies, individuals with decreased 1A2 activity, as determined by the probe drug caffeine, who ingested large amounts of coffee had an increased risk of nonfatal myocardial infarction (Cornelis et al. 2004, 2006). These associations, however, remain controversial.

Variation in 1A2 activity can also be due to the effect of a variety of external factors (Aklillu et al. 2003; Butler et al. 1992; Eap et al. 2004; Nordmark et al. 2002). These are discussed in more detail in the sections below on 1A2 inhibition and induction.

Are There Drugs That Inhibit 1A2 Activity?

Yes. There are several potent and clinically relevant inhibitors of 1A2. The clinical significance of 1A2 inhibition may be notable if the inhibitor is co-administered with a drug relying primarily on 1A2 for clearance. In addition to the previously mentioned probe drugs, caffeine, melatonin, and theophylline (Faber et al. 2005; Hartter et al. 2001; Miners and Birkett 1996), such drugs include clozapine (Eiermann et al. 1997), cyclobenzaprine (Wang et al. 1996), flutamide (Shet et al. 1997), frovatriptan (Buchan et al. 2002), mirtazapine (Stormer et al. 2000), naproxen (Miners et al. 1996), ropivacaine (Arlander et al. 1998), riluzole (Sanderink et al. 1997), tacrine (Madden et al. 1995), tizanidine (Granfors et al. 2004a), and zolmitriptan (Wild et al. 1999). When any of these drugs is used concomitantly with an inhibitor of 1A2, plasma levels of the former may increase and lead to enhanced effects and even to toxicity. Depending on the nature and therapeutic window of the drug, the inhibition can be insignificant (caffeine, melatonin) or have potentially hazardous, life-threatening consequences (clozapine, theophylline, tizanidine). The potency of the inhibitor also plays a role (see Table 7–2). Fluvoxamine, a potent inhibitor of 1A2, has been shown to raise serum clozapine concentrations up to 10-fold (Hiemke et al. 1994; Koponen et al. 1996), whereas ciprofloxacin, a less potent inhibitor, increases clozapine concentration only by one-third (Raaska and Neuvonen 2000).

TABLE 7–2.   Some inhibitors of 1A2

Fluoroquinolone antibiotics	Selective serotonin reuptake inhibitors		Other drugs
Ciprofloxacin	Fluvoxamine	Acyclovir	Oral contraceptives
Enoxacin		Amiodarone	Perphenazine
Lomefloxacin		Anastrozole	Phenacetin
Norfloxacin		Caffeine	**Propafenone**
Ofloxacin		Cimetidine	Ropinirole
		Famotidine	Tacrine
		Flutamide[1]	Ticlopidine
		Grapefruit juice	Tocainide
		Lidocaine	Verapamil
		Mexiletine	Zileuton
		Moclobemide	

Note.   Names of drugs are in **bold** type if there is evidence that in normal human use, the drugs are potent inhibitors.
[1]The primary metabolite of flutamide is a potent inhibitor of 1A2.

Fluvoxamine

Fluvoxamine, unlike the other selective serotonin reuptake inhibitors, is a potent inhibitor of 1A2 (Preskorn 1997; Richelson 1997). In addition to 1A2, fluvoxamine is a known potent inhibitor of 2C9 and 2C19 and moderate inhibitor of 3A4 and 2D6 (Brosen et al. 1993; Rasmussen et al. 1997). Administration of fluvoxamine in conjunction with haloperidol, clozapine, imipramine, tizanidine, and theophylline results in an elevated plasma level of the substrate due to enzymatic inhibition by fluvoxamine (Brosen 1995; Granfors et al. 2004c; Hiemke et al. 1994). In the case of clozapine, fluvoxamine should likely be avoided because fluvoxamine inhibits all major metabolic pathways of clozapine. As little as 50 mg a day of fluvoxamine can result in significantly elevated clozapine levels and result in clozapine toxicity (Wetzel et al. 1998).

Amitriptyline and imipramine, tertiary tricyclic antidepressants (TCAs), are metabolized via 1A2 and 3A4 (Madsen et al. 1997; Venkatakrishnan et al. 2001). Inhibition of 1A2 selectively can result in increases in TCA levels, and inhibition of both 1A2 and 3A4 by drugs such as fluvoxamine have an even greater likelihood of causing TCA toxicity.

Fluoroquinolone Antibiotics

Nearly all fluoroquinolones are known to inhibit 1A2 to some extent. Fluoroquinolones that potently inhibit 1A2 include ciprofloxacin, enoxacin, lomefloxacin, and ofloxacin (Fuhr et al. 1990, 1992, 1993). Several case reports and controlled clinical studies have shown the potent inhibitory effect of 1A2 by ciprofloxacin, along with adverse effects from increased plasma levels of clozapine (Markowitz et al. 1997; Raaska and Neuvonen 2000), tizanidine (Granfors 2004b; Momo et al. 2004), and theophylline (Holden 1988; Thomson et al. 1987).

Oral Contraceptives

Estrogen-containing oral contraceptives and hormone replacement therapy moderately inhibit 1A2 and can decrease the clearance of 1A2 substrates such as caffeine, clozapine, chlorpromazine, and tacrine (Abernethy and Todd 1985; Laine et al. 1999; Pollock et al. 1999). This inhibition, though moderate, can result in toxicities for drugs with narrow therapeutic windows (see Chapter 12, "Gynecology: Oral Contraceptives").

Are There Drugs That Induce 1A2 Activity?

Yes. 1A2 is regulated by the aryl hydrocarbon receptor and can be induced by several drugs, dietary components, and environmental factors (Hankinson 1995; Nebert et al. 2004) (see Table 7–3).

Smoking

Polycyclic aromatic hydrocarbons present in cigarette smoke (including marijuana) are one of the best-known inducers of 1A2 (Kalow and Tang 1991; Vistisen et al. 1991, 1992). The clinical relevance of this effect has been studied extensively regarding smoking during antipsychotic therapy. 1A2 induction by smoking increases the metabolic clearance of 1A2 substrates such as clozapine and olanzapine, necessitating higher doses to achieve effective therapy (Bozikas et al. 2004; Conley and Kelly 2007; de Leon 2004; Desai et al. 2001; Dratcu et al. 2007; van der Weide et al. 2003; Zevin and Benowitz 1999). Given the number of schizophrenia patients who smoke, this interaction has a high likelihood of affecting therapy (de Leon and Diaz 2005).

Smoking cessation results in a decrease in 1A2 activity over the course of several weeks due to the removal of induction. During smoking cessation, the plasma level of any 1A2 substrate at a steady state will likely increase. Monitoring serum levels of drugs with narrow therapeutic indices such as clozapine and theophylline during these transition periods is strongly recommended. Individuals should be cautioned against alterations in smoking habits while using 1A2 substrates such as clozapine or theophylline, except under the close monitoring of a physician.

Carbamazepine

Carbamazepine, an antiepileptic, is a well-known inducer of a wide spectrum of P450s, including 1A2. Carbamazepine has been shown to significantly decrease the plasma concentrations of risperidone and its pharmacologically active metabolite 9-hydroxyrisperidone, both 1A2 substrates. Carbamazepine may also decrease the plasma concentrations of olanzapine, clozapine, ziprasidone, haloperidol, zuclopenthixol, flupenthixol, and possibly chlorpromazine and fluphenazine secondary to 1A2 induction (Besag and Berry 2006; Perucca 2006; Spina and de Leon 2007).

TABLE 7–3. Some inducers of 1A2

Drugs	Foods	Other inducers
Carbamazepine	Broccoli	Chronic hepatitis
Esomeprazole	Brussels sprouts	Chronic smoking
Griseofulvin	Cabbage	PAHs
Lansoprazole	Cauliflower	PCBs
Moricizine	Charbroiled foods	TCDD
Omeprazole		
Rifampin		
Ritonavir		

Note. PAHs=polycyclic aromatic hydrocarbons; PCBs=polychlorinated biphenyls; TCDD= 2,3,7,8-tetrachlorodibenzo-*p*-dioxin.

Foods

Certain dietary compounds are also inducers of 1A2. Daily consumption of cruciferous vegetables like brussels sprouts, broccoli, and cabbage (Jefferson 1998; Jefferson and Griest 1996; Lampe et al. 2000; Vistisen et al. 1992), as well as charcoal-grilled foods (Boobis et al. 1996; Larsen and Brosen 2005), may result in significant 1A2 induction. Heavy exercise may also induce 1A2 activity to the same magnitude as heavy smoking, although the mechanism remains unclear (Vistisen et al. 1991, 1992).

References

Abernethy DR, Todd EL: Impairment of caffeine clearance by chronic use of low-dose oestrogen-containing oral contraceptives. Eur J Clin Pharmacol 28:425–428, 1985

Aitchison KJ, Gonzalez FJ, Quattrochi LC, et al: Identification of novel polymorphisms in the 5′ flanking region of CYP1A2, characterization of interethnic variability, and investigation of their functional significance. Pharmacogenetics 10:695–704, 2000

Aklillu E, Carrillo JA, Makonnen E: Genetic polymorphism of CYP1A2 in Ethiopians affecting induction and expression: characterization of novel haplotypes with single-nucleotide polymorphisms in intron 1. Mol Pharmacol 64:659–669, 2003

Arlander E, Ekstrom G, Alm C, et al: Metabolism of ropivacaine in humans is mediated by CYP1A2 and to a minor extent by CYP3A4: an interaction study with fluvoxamine and ketoconazole as in vivo inhibitors. Clin Pharmacol Ther 64:484–491, 1998

Basile VS, Ozdemir V, Maseelis M, et al: A functional polymorphism of the cytochrome P450 1A2 (CYP1A2) gene: association with tardive dyskinesia in schizophrenia. Mol Psychiatry 5:410–417, 2000

Besag FM, Berry D: Interactions between antiepileptic and antipsychotic drugs. Drug Saf 29:95–118, 2006

Boobis AR, Lynch AM, Murray S, et al: CYP1A2-catalyzed conversion of dietary heterocyclic amines to their proximate carcinogens is their major route of metabolism in humans. Cancer Res 54:89–94, 1994

Boobis AR, Gooderham NJ, Edwards RJ: Enzymic and interindividual differences in the human metabolism of heterocyclic amines. Arch Toxicol Suppl 18:286–302, 1996

Bozikas VP, Papakosta M, Niopas I, et al: Smoking impact on CYP1A2 activity in a group of patients with schizophrenia. Eur Neuropsychopharmacol 14:39–44, 2004

Brosen K: Drug interactions and the cytochrome P450 system: the role of cytochrome P450 1A2. Clin Pharmacokinet 29 (suppl 1):20–25, 1995

Brosen K, Skjelbo E, Rasmussen BB, et al: Fluvoxamine is a potent inhibitor of cytochrome P4501A2. Biochem Pharmacol 45:1211–1214, 1993

Buchan P, Keywood C, Wade A, et al: Clinical pharmacokinetics of frovatriptan. Headache 42 (suppl 2):54–62, 2002

Butler MA, Iwasaki M, Guengerich FP: Human cytochrome P-450PA (P-450IA2), the phenacetin O-deethylase, is primarily responsible for the hepatic 3-demethylation of caffeine and N-oxidation of carcinogenic arylamines. Proc Natl Acad Sci U S A 86:7696–7700, 1992

Carrillo JA, Herraíz AG, Ramos SI, et al: Role of the smoking-induced cytochrome P450 (CYP)1A2 and polymorphic CYP2D6 in steady-state concenration of olanzapine. J Clin Psychopharmacol 23:119–127, 2003

Chen X, Wang H, Xie W: Association of CYP1A2 genetic polymorphisms with hepatocellular carcinoma susceptibility: a case-control study in a high-risk region of China. Pharmacogenet Genomics 16:219–227, 2006

Conley RR, Kelly DL: Drug-drug interactions associated with second-generation antipsychotics: considerations for clinicians and patients. Psychopharmacol Bull 40:77–97, 2007

Cornelis MC, El-Sohemy A, Campos H: Genetic polymorphism of CYP1A2 increases the risk of myocardial infarction. J Med Genet 41:758–762, 2004

Cornelis MC, El-Sohemy A, Kabagambe EK: Coffee, CYP1A2 genotype, and risk of myocardial infarction. JAMA 295:1135–1141, 2006

de Leon J: Atypical antipsychotic dosing: the effect of smoking and caffeine. Psychiatr Serv 55:491–493, 2004

de Leon J, Diaz FJ: A meta-analysis of worldwide studies demonstrates an association between schizophrenia and tobacco smoking behaviors. Schizophr Res 76:135–157, 2005

Desai HD, Seabolt J, Jann MW: Smoking in patients receiving psychotropic medications: a pharmacokinetic perspective. CNS Drugs 15:469–494, 2001

Dorado P, Berecz R, Penas-Lledo EM, et al: No effect of the CYP1A2*1F genotype on thioridazine, mesoridazine, sulforidazine plasma concentrations in psychiatric patients. Eur J Clin Pharmacol 63:527–528, 2007

Dratcu L, Grandison A, McKay G: Clozapine-resistant psychosis, smoking, and caffeine: managing the neglected effects of substances that our patients consume every day. Am J Ther 14:314–318, 2007

Eap CB, Bender S, Jaquenoud SE, et al: Nonresponse to clozapine and ultrarapid CYP1A2 activity: clinical data and analysis of CYP1A2 gene. J Clin Psychopharmacol 24:214–219, 2004

Eiermann B, Engel G, Johansson I, et al: The involvement of CYP1A2 and CYP3A4 in the metabolism of clozapine. Br J Clin Pharmacol 44:439–446, 1997

Faber MS, Jetter A, Fuhr U: Assessment of CYP1A2 activity in clinical practice: why, how, and when? Basic Clin Pharmacol Toxicol 97:125–134, 2005

Fuhr U, Wolff T, Harder S, et al: Quinolone inhibition of cytochrome P-450-dependent caffeine metabolism in human liver microsomes. Drug Metab Dispos 18:1005–1010, 1990

Fuhr U, Anders EM, Mahr G, et al: Inhibitory potency of quinolone antibacterial agents against cytochrome P450IA2 activity in vivo and in vitro. Antimicrob Agents Chemother 36:942–948, 1992

Fuhr U, Strobl G, Manaut F, et al: Quinolone antibacterial agents: relationship between structure and in vitro inhibition of the human cytochrome P450 isoform CYP1A2. Mol Pharmacol 43:191–199, 1993

Granfors MT, Backman JT, Laitila J, et al: Tizanidine is mainly metabolized by cytochrome p450 1A2 in vitro. Br J Clin Pharmacol 57:349–353, 2004a

Granfors MT, Backman JT, Neuvonen M, et al: Ciprofloxacin greatly increases concentrations and hypotensive effect of tizanidine by inhibiting its cytochrome P450 1A2-mediated presystemic metabolism. Clin Pharmacol Ther 76:598–606, 2004b

Granfors MT, Backman JT, Neuvonen M, et al: Fluvoxamine drastically increases concentrations and effects of tizanidine: a potentially hazardous interaction. Clin Pharmacol Ther 75:331–341, 2004c

Han XM, Ouyang DS, Chen XP, et al: Inducibility of CYP1A2 by omeprazole in vivo related to the genetic polymorphism of CYP1A2. Br J Clin Pharmacol 54:540–543, 2002

Hankinson O: The aryl hydrocarbon receptor complex. Annu Rev Pharmacol Toxicol 35:307–340, 1995

Hartter S, Ursing C, Morita S, et al: Orally given melatonin may serve as a probe drug for cytochrome P450 1A2 activity in vivo: a pilot study. Clin Pharmacol Ther 70:10–16, 2001

Hiemke C, Weigmann H, Hartter S, et al: Elevated levels of clozapine in serum after addition of fluvoxamine. J Clin Psychopharmacol 14:279–281, 1994

Holden R: Probable fatal interaction between ciprofloxacin and theophylline. BMJ 297:1, 1988

Ioannides C, Lewis DF: Cytochromes P450 in the bioactivation of chemicals. Curr Top Med Chem 4:1767–1788, 2004

Jefferson JW: Drug and diet interactions: avoiding therapeutic paralysis. J Clin Psychiatry 59 (suppl 16):31–39, 1998

Jefferson JW, Griest JH: Brussels sprouts and psychopharmacology: understanding the cytochrome P450 enzyme system. Psychiatr Clin North Am: Annual on Drug Therapy 3:205–222, 1996

Kalow W, Tang B: Caffeine as a metabolic probe: exploration of the enzyme-inducing effect of cigarette smoking. Clin Pharmacol Ther 49:44–48, 1991

Koponen H, Leinonen E, Lepola U: Fluvoxamine increases the clozapine serum levels significantly. Eur Neuropsychopharmacol 6: 69–71, 1996

Laine K, Palovaara S, Tapanainen P, et al: Plasma tacrine concentrations are significantly increased by concomitant hormone replacement therapy. Clin Pharmacol Ther 66:602–608, 1999

Lampe JW, King IB, Li S: Brassica vegetables increase and apiaceous vegetables decrease cytochrome P450 1A2 activity in humans: changes in caffeine metabolite ratios in response to controlled vegetable diets. Carcinogenesis 21:1157–1162, 2000

Larsen JT, Brosen K: Consumption of charcoal-broiled meat as an experimental tool for discerning CYP1A2-mediated drug metabolism in vivo. Basic Clin Pharmacol Toxicol 97:141–148, 2005

Madden S, Spaldin V, Park BK: Clinical pharmacokinetics of tacrine. Clin Pharmacokinet 28:449–457, 1995

Madsen H, Rasmussen BB, Brosen K: Imipramine demethylation in vivo: impact of CYP1A2, CYP2C19, and CYP3A4. Clin Pharmacol Ther 61:319–324, 1997

Markowitz JS, Gill HS, DeVane CL, et al: Fluoroquinolone inhibition of clozapine metabolism (letter). Am J Psychiatry 154:881, 1997

Miners JO, Birkett DJ: The use of caffeine as a metabolic probe for human drug metabolizing enzymes. Gen Pharmacol 27:245–249, 1996

Miners JO, Couler S, Tukey RH, et al: Cytochromes P450, 1A2, and 2C9 are responsible for the human hepatic O-demethylation of R- and S-naproxen. Biochem Pharmacol 51:1003–1008, 1996

Momo K, Doki K, Hosono H, et al: Drug interaction of tizanidine and fluvoxamine. Clin Pharmacol Ther 76:509–510, 2004

Moonen H, Engels L, Kleinjans J, et al: The CYP1A2-164A–>C polymorphism (CYP1A2*1F) is associated with the risk for colorectal adenomas in humans. Cancer Lett 229:25–31, 2005

Nakajima M, Yokoi T, Mizutani M, et al: Phenotyping of CYP1A2 in Japanese population by analysis of caffeine urinary metabolites: absence of mutation prescribing the phenotype in the CYP1A2 gene. Cancer Epidemiol Biomarkers Prev 3:413–421, 1994

National Center for Biotechnology Information: CYP1A2 cytochrome P450, family 1, subfamily A, polypeptide 2 [Homo sapiens]. Pub Med Entrez Gene, 2008. Available at http://www.ncbi.nlm.nih.gov/sites/entrez?Db=gene&Cmd=ShowDetailView&TermToSearch=1544&ordinalpos=1&itool=EntrezSystem2.PEntrez.Gene.Gene_ResultsPanel.Gene_RVDocSum. Accessed July 15, 2008.

Nebert DW, Dalton TP, Okey AB, et al: Role of aryl hydrocarbon receptor-mediated induction of the CYP1 enzymes in environmental toxicity and cancer. J Biol Chem 279:23847–23850, 2004

Nordmark A, Lundgren S, Ask B, et al: The effect of the CYP1A2 *1F mutation on CYP1A2 inducibility in pregnant women. Br J Clin Pharmacol 54:504–510, 2002

Ozdemir V, Kalow W, Okey AB, et al: Treatment resistance to clozapine in association with ultrarapid CYP1A2 activity and the C>A polymorphism in intron 1 of the CYP1A2 gene: effect of grapefruit juice and low-dose fluvoxamine. J Clin Psychopharmacol 21:603–607, 2001

Perucca E: Clinically relevant drug interactions with antiepileptic drugs. Br J Clin Pharmacol 61:246–255, 2006

Pollock BG, Wylie M, Stack JA, et al: Inhibition of caffeine metabolism by estrogen replacement therapy in postmenopausal women. J Clin Pharmacol 39:936–940, 1999

Preskorn SH: Clinically relevant pharmacology of selective serotonin reuptake inhibitors: an overview with emphasis on pharmacokinetics and effects on oxidative drug metabolism. Clin Pharmacokinet 32 (suppl 1):1–21, 1997

Raaska K, Neuvonen PJ: Ciprofloxacin increases serum clozapine and N-desmethyl-clozapine: a study in patients with schizophrenia. Eur J Clin Pharmacol 56:585–589, 2000

Rasmussen BB, Jeppesen U, Gaist D, et al: Griseofulvin and fluvoxamine interactions with the metabolism of theophylline. Ther Drug Monit 19:56–62, 1997

Richelson E: Pharmacokinetic drug interactions of new antidepressants: a review of the effects on the metabolism of other drugs. Mayo Clin Proc 72:835–847, 1997

Sachse C, Brockmoller J, Bauer S, et al: Functional significance of a C–>A polymorphism in intron 1 of the cytochrome P450 CYP1A2 gene tested with caffeine. Br J Clin Pharmacol 47:445–449, 1999

Saito Y, Hanioka N, Maekawa K, et al: Functional analysis of three CYP1A2 variants found in a Japanese population. Drug Metab Dispos 33:1905–1910, 2005

Sanderink GJ, Bournique B, Stevens J, et al: Involvement of human CYP1A isoenzymes in the metabolism and drug interactions of riluzole in vitro. J Pharmacol Exp Ther 282:1465–1472, 1997

Shet MS, McPhaul M, Fisher CW, et al: Metabolism of the antiandrogenic drug (flutamide) by human CYP1A2. Drug Metab Dispos 25:1298–1303, 1997

Shirley KL, Hon YY, Penzak SR, et al: Correlation of cytochrome P450 (CYP) 1A2 activity using caffeine phenotyping and olanzapine disposition in healthy volunteers. Neuropsychopharmacology 28:961–966, 2003

Spina E, de Leon J: Metabolic drug interactions with newer antipsychotics: a comparative review. Basic Clin Pharmacol Toxicol 100:4–22, 2007

Stormer E, von Moltke LL, Shader RI, et al: Metabolism of the antidepressant mirtazapine in vitro: contribution of cytochromes P-450 1A2, 2D6, and 3A4. Drug Metab Dispos 28:1168–1175, 2000

Tay JK, Tan CH, Chong SA, et al: Functional polymorphisms of the cytochrome P450 1A2 (CYP1A2) gene and prolonged QTc interval in schizophrenia. Prog Neuropsychopharmacol Biol Psychiatry 31:1297–1302, 2007

Thomson AH, Thomson GD, Hepburn M, et al: A clinically significant interaction between ciprofloxacin and theophylline. Eur J Clin Pharmacol 33:435–436, 1987

Tiwari AK, Deshpande SN, Rao AR, et al: Genetic susceptibility to tardive dyskinesia in chronic schizophrenia subjects, I: association of CYP1A2 gene polymorphism. Pharmacogenomics J 5:60–69, 2005

Tiwari AK, Deshpande SN, Lerer B, et al: Genetic susceptibility to Tardive Dyskinesia in chronic schizophrenia subjects, V: association of CYP1A2 1545 C>T polymorphism. Pharmacogenomics J 7:305–311, 2007

van der Weide J, Steijns LS, van Weelden MJ: The effect of smoking and cytochrome P450 CYP1A2 genetic polymorphism on clozapine clearance and dose requirement. Pharmacogenetics 13:169–172, 2003

Venkatakrishnan K, Schmider J, Harmatz JS, et al: Relative contribution of CYP3A to amitriptyline clearance in humans: in vitro and in vivo studies. J Clin Pharmacol 41:1043–1054, 2001

Vistisen K, Loft S, Poulsen HE: Cytochrome P450 IA2 activity in man measured by caffeine metabolism: effect of smoking, broccoli and exercise. Adv Exp Med Biol 283:407–411, 1991

Vistisen K, Poulsen HE, Loft S: Foreign compound metabolism capacity in man measured from metabolites of dietary caffeine. Carcinogenesis 13:1561–1568, 1992

Wang RW, Liu L, Cheng H: Identification of human liver cytochrome P450 isoforms involved in the in vitro metabolism of cyclobenzaprine. Drug Metab Dispos 24:786–791, 1996

Wetzel H, Anghelescu I, Szegedi A, et al: Pharmacokinetic interactions of clozapine with selective serotonin reuptake inhibitors: differential effects of fluvoxamine and paroxetine in a prospective study. J Clin Psychopharmacol 18:2–9, 1998

Wild MJ, McKillop D, Butters CJ: Determination of the human cytochrome P450 isoforms involved in the metabolism of zolmitriptan. Xenobiotica 29:847–857, 1999

Yamazaki H, Hatanaka N, Kizu R, et al: Bioactivation of diesel exhaust particle extracts and their major nitrated aromatic hydrocarbon components, 1-nitropyrene and dinitropyrenes, by human cytochromes P450 1A1, 1A2, and 1B1. Mutat Res 472:129–138, 2000

Zevin S, Benowitz NL: Drug interactions with tobacco smoking: an update. Clin Pharmacokinet 36:425–438, 1999

Study Cases

Case 1

A 35-year-old Caucasian woman with a history of multiple psychiatric hospitalizations had major depression and psychosis that were unresponsive to numerous treatments, including electroconvulsive therapy. Her fluvoxamine dose was titrated to 300 mg/day, with minimal benefit. Psychotic symptoms began to predominate. Her clozapine dose was titrated to 200 mg/day over several weeks. The patient began to complain of dizziness and had mild hypotension. Her serum clozapine level was 1,950 ng/mL, and a confirmatory measurement revealed a level of 2,040 ng/mL, well above normal. Fluvoxamine therapy was discontinued while clozapine levels were closely monitored. Three days after discontinuation of her fluvoxamine therapy, the clozapine level had decreased to 693 ng/mL, and on the fifth day after cessation of fluvoxamine, the level was 175 ng/mL. The aforementioned side effects disappeared.

Comment

1A2 inhibition: Fluvoxamine potently inhibits several P450 enzymes, including 1A2, 2C9, and 2C19. It also moderately inhibits 3A4. Clozapine is metabolized via 1A2 with minor contribution from 3A4 and 2D6. This case and others reported in the literature reveal the potent effect of fluvoxamine on clozapine levels. The case also illustrates how quickly inhibition can be reversed, in several days, when treatment with the offending inhibitor is discontinued, in contrast to several weeks for discontinuation of an inducer.

References

Armstrong SC, Stephans JR: Blood clozapine levels elevated by fluvoxamine: potential for side effects and lower clozapine dosage (letter). J Clin Psychiatry 58:499, 1997

Wetzel H, Anghelescu I, Szegedi A, et al: Pharmacokinetic interactions of clozapine and selective serotonin reuptake inhibitors: differential effects of fluvoxamine and paroxetine in a prospective study. J Clin Psychopharmacol 18:2–9, 1998

Case 2

A 48-year-old Caucasian man with schizophrenia was effectively treated with clozapine 500 mg/day. The patient was a longtime smoker but was able to spontaneously stop smoking 2 years after the clozapine dose was established. Several weeks after smoking cessation, the patient complained of side effects including sedation and constipation. Serum clozapine levels were found to be more than 700 ng/mL, nearly twice what the clozapine levels had been when the patient was smoking.

Comment

1A2 induction: The major P450 enzyme involved in metabolism of clozapine is 1A2, although 3A4 and 2D6 are also involved. Smoking induces 1A2. When smoking is stopped, 1A2 activity returns to normal after 3–6 weeks. Patients taking drugs such as clozapine have increased serum levels several weeks after cessation of smoking.

References

Derenne JL, Baldessarini RJ: Clozapine toxicity associated with smoking cessation: case report. Am J Ther 12:469–471, 2005

Seppala NH, Leinonen EV, Lehtonen ML, et al: Clozapine serum concentrations are lower in smoking than non-smoking schizophrenic patients. Pharmacol Toxicol 85:244–246, 1999

Case 3

After multiple trials of other antipsychotics over many years, a 64-year-old Caucasian woman with chronic undifferentiated schizophrenia responded to clozapine. Her condition was stable when she was given 125 mg twice a day. No baseline measurements of clozapine or norclozapine levels were obtained. Her primary care physician began ciprofloxacin 500 mg twice a day for a urinary tract infection. Near the end of the 10-day course of antibiotic treatment, the patient was lethargic and drooled profusely. Her psychiatrist checked serum clozapine and norclozapine levels, which were 1,043 and 432 ng/mL, respectively—indicating mild to moderate toxicity. Clozapine therapy was temporarily discontinued; it was restarted soon after completion of treatment with ciprofloxacin.

Comment

1A2 inhibition: Ciprofloxacin is a commonly used antibiotic that potently inhibits 1A2. Clozapine is preferentially metabolized via 1A2 to its active component norclozapine (*N*-desmethylclozapine). In this case, because baseline levels were not available, one cannot prove that this exact interaction occurred. However, the relatively high ratio of clozapine to norclozapine (2.4:1) suggests that 1A2 was inhibited.

References

Dumortier G, Lochu A, Zerrouk A, et al: Whole saliva and plasma levels of clozapine and desmethylclozapine. J Clin Pharm Ther 32:35–40, 1998

Raaska K, Neuvonen PJ: Ciprofloxacin increases serum clozapine and N-desmethylclozapine: a study in patients with schizophrenia. Eur J Clin Pharmacol 56:585–589, 2000

8

2C9

Jessica R. Oesterheld, M.D.

2C9 is one of a family of four 2C alleles on chromosome 10, including 2C8 (see Chapter 11), 2C18, and 2C19 (Chapter 9) (National Center for Biotechnology Information 2008). Older literature also often refers to 2C10, now considered a variant of 2C9. Of the 2C subfamily, 2C9 makes the largest contribution in terms of liver CYP content. 2C9 is responsible for the metabolism of 15% of drugs metabolized by phase I enzymes (Rettie and Jones 2005). There are more than 100 drugs metabolized by 2C9 (Kirchheiner and Brockmöller 2005), including medications such as S-warfarin, phenytoin, many nonsteroidal anti-inflammatory drugs (NSAIDs) and some oral hypoglycemic agents (see Table 8–1).

What Does 2C9 Do?

2C9, like other P450 enzymes, performs hydroxylation, demethylation, and dealkylation of endogenous and exogenous compounds. 2C9 (like 2C8) also

TABLE 8–1. Some drugs metabolized by 2C9

Angiotensin II blockers	Hypoglycemics, oral[2]	NSAIDs[3]	Other drugs
Candesartan	*Sulfonylureas*	Aceclofenac	Bosentan[1,7]
Irbesartan	Chlorpropamide	Celecoxib	Dapsone[2,3]
Losartan[1] (pro-drug)	Glimepiride	Diclofenac[1,3,7]	Fluvastatin[4,7]
	Glipizide	Flurbiprofen[3]	Mestranol[5] (pro-drug)
	Glyburide[1,7]	Ibuprofen[1,3,7]	Phenobarbital
Antidepressants	Nateglinide[1,7]	Indomethacin[3,7]	Phenytoin[3]
Fluoxetine[1]	Rosiglitazone[2]	Lornoxicam	Tamoxifen[1]
Sertraline[1]	Tolbutamide	Mefenamic acid	Tetrahydrocannabinol[1]
		Meloxicam[1]	Torsemide[7]
		Piroxicam	S-Warfarin[6]
		Tenoxicam	
		Valdecoxib[1]	

Note. NSAID=nonsteroidal anti-inflammatory drug.
[1]Metabolized by other P450 cytochromes as well.
[2]Most glitazones are metabolized by 2C8 and 3A4.
[3]Also metabolized by phase II enzymes.
[4]This is an exception among the statin drugs, most of which are partly or wholly metabolized by 3A4.
[5]Pro-drug metabolized to active ethinyl estradiol.
[6]S-Warfarin is the more active isomer of warfarin.
[7]Handled by transporters as well.

metabolizes liver arachidonic acid (Rifkind et al. 1995), 5-hydroxy-tryptamine, and linoleic acid (Rettie and Jones 2005).

Where Is 2C9?

2C9 is found in many tissues, including the kidney, testes, adrenal gland, prostate, ovary, and duodenum (Klose et al. 1999). Most of 2C9's activity appears to occur in the liver. Along with 2C19, 2C9 accounts for about 18% of liver cytochrome content (DeVane and Nemeroff 2002).

Does 2C9 Have Any Polymorphisms?

Researching different ethnic populations has shown that 2C9 is highly polymorphic. More than 16 alleles are listed on the Web site "CYP2C9 Allele Nomenclature" (1997). Kirchheiner and Brockmöller (2005) have reported that two of the non-wild-type alleles, *2C9*2* and *2C9*3*, are clinically significant. These alleles are found, respectively, in 11% and 7% of Caucasians while occurring much less frequently in African Americans and Asians (Rettie and Jones 2005). Individuals who have these alleles show a reduced ability to metabolize drugs dependent on 2C9. For example, phenytoin metabolism is reduced by as much as a third in patients with at least one *2C9*2* or *2C9*3* allele (Odani et al. 1997; van der Weide et al. 2001).

Poor metabolizers (PMs) at 2C9 have increased levels of 2C9 substrates. 2C9 PMs receiving 2C9 substrates with a narrow safety margin or therapeutic window are at risk for toxicities from elevated plasma levels of substrate. The most thoroughly described of these substrates is *S*-warfarin (the more active isomer of racemic warfarin). Individuals with the *2* and *3* polymorphisms have a lower *S*-warfarin dosing requirement, which may result in toxicity with "standard" dosing; however, lower dosing is not associated with increased bleeding if international normalized ratios (INR) are properly monitored (Joffe et al. 2004). An algorithm has been developed that utilizes genetic information regarding 2C9 and the vitamin K epoxide reductase complex subunit 1 (VKORC1) polymorphisms to estimate initial warfarin dosing. As genetic information becomes more readily available, this algorithm will serve to assist clinicians in appropriate initial dosing regimens.

Other drugs influenced by 2C9 polymorphism include nateglinide (Kirchheiner et al. 2005), piroxicam (Perini et al. 2005), celecoxib, and fluvastatin (Kirchheiner and Brockmöller 2005). As with 2D6, we suspect that most clinicians never consider the possibility that a patient may be a 2C9 PM. Unless a severe interaction occurs, most clinicians would likely attribute unusually high levels of phenytoin or hypoglycemia with tolbutamide to some other clinically unpredictable variable.

Are There Drugs That Inhibit 2C9 Activity?

Yes. According to the FDA's drug interactions Web site, *in vitro* human liver microsome studies suggest the preferred inhibitor of 2C9 is sulfaphenazole (Ki, 0.3 μM) (U.S. Food and Drug Administration 2006). Fluconazole and fluvoxamine may also be categorized as acceptable 2C9 inhibitors. Selective serotonin reuptake inhibitors other than fluvoxamine only minimally inhibit 2C9 (Hemeryck et al. 1999). *In vivo* studies have shown 2C9 inhibitors to include fluconazole and amiodarone. Other 2C9 inhibitors are listed in Table 8–2.

In the following sections, we discuss substrates that have narrow therapeutic indices or have adverse outcomes associated with 2C9 inhibition.

Warfarin

S-Warfarin is the more active isomer of the racemic mix constituting warfarin. *S*-Warfarin is metabolized primarily by 2C9, whereas *R*-warfarin is metabolized by multiple pathways, including 1A2 and 3A4 (PharmGkb 2008). Although the literature identifies hundreds of drugs that have the potential to interact pharmacokinetically and/or pharmacodynamically with warfarin, it can be difficult to differentiate the effects of 2C9 inhibition from displacement of plasma proteins or a direct pharmacodynamic effect as in the interactions with fenofibrate (Kim and Mancano 2003) and phenylbutazone (Harder and Thurmann 1996). Drugs that have been shown to increase *S*-warfarin levels through inhibition of 2C9 include fluconazole (Black et al. 1996), sulfamethoxazole (Glasheen et al. 2005), fluvastatin (Kim et al. 2006), and zafirlukast (Dekhuijzen and Koopmans 2002). Concomitant use of any 2C9 inhibitors and warfarin may result in an increased risk of bleeding due to excessive warfarin levels.

TABLE 8–2.   Some inhibitors of 2C9

SSRIs	Other inhibitors	
Fluoxetine[2]	Amiodarone[1]	Ketoconazole
Fluvoxamine	Amodiaquine	Leflunomide
Sertraline[2]	Anastrozole	Miconazole
	Cimetidine	Modafinil?
	Delavirdine?	Phenylbutazone
	Efavirenz?	Sulfamethoxazole
	Fenofibrate	**Sulfaphenazole**
	Fluconazole	Tamoxifen
	5-Fluorouracil	Teniposide
	Fluvastatin	Valproic acid
	Isoniazid	Voriconazole
		Zafirlukast

Note.   Names of potent inhibitors are in **bold** type.
SSRIs=selective serotonin reuptake inhibitors.
[1]Amiodarone, an inhibitor of 1A2 and 3A4, is an insignificant 2C9 inhibitor. However, its metabolite, desethylamiodarone, is a clinically relevant 2C9 inhibitor.
[2]Modest inhibitor.

There are numerous case reports of potent 1A2 or 3A inhibitors that also increase *S*-warfarin levels. Many of these drug interactions occur because of inhibition of *R*-warfarin metabolism leading to increased *R*-warfarin levels, which can inhibit 2C9 and thus cause elevations in *S*-warfarin. Drugs that inhibit 3A4 or 1A2 in addition to 2C9 can change both *R*- and *S*-warfarin levels. As an example, the presence of amiodarone's metabolite desethylamiodarone (Naganuma et al. 2001) in conjunction with warfarin necessitates a decrease in warfarin dosing of 25%–30% (Rettie and Jones 2005). Fluvoxamine and others drugs that inhibit several enzymes involved in warfarin metabolism may also increase warfarin levels (Limke et al. 2002).

Phenytoin

Although phenytoin metabolism is complicated and not fully understood, 2C9 and 2C19 appear to play the major roles in metabolism. Fluconazole (Blum et al. 1991; Cadle et al. 1994), amiodarone (Nolan et al. 1990), voriconazole (Purkins et al. 2003), and fluoxetine (Nelson et al. 2001; Nightingale

1994) all increase phenytoin levels through 2C9 inhibition. Fluvoxamine administration with phenytoin resulted in increases in phenytoin levels through 2C9 inhibition, according to a case report (Mamiya et al. 2001). Given phenytoin's narrow therapeutic window, careful monitoring of phenytoin level is warranted during treatment with known 2C9 inhibitors (Barry et al. 1997; Schmider et al. 1997). Finally, although no cases are currently found in the literature, the package insert for Provigil (modafinil) cautions that administration of modafinil and phenytoin may result in increased phenytoin levels through 2C9 inhibition (Provigil 1999).

The lack of case reports on 2C9 inhibition of phenytoin with administration of other drugs is curious, in view of phenytoin's continued commonplace use in clinical practice. We suspect that many cases are missed because toxic levels are not reached or the clinician is not aware of the drug-drug interaction causing the high phenytoin level and simply decreases the phenytoin dose during regular monitoring of phenytoin.

Oral Hypoglycemics

Many oral hypoglycemics of the sulfonylurea class (tolbutamide, glipizide, glyburide, and others) are metabolized primarily by 2C9. 2C9 inhibitors can increase concentrations of these oral hypoglycemics, perhaps leading to decreases in serum glucose levels. In seven nondiabetic control subjects, tolbutamide's half-life increased 3-fold, and the area under the curve increased by 66% when the drug was coadministered with ketoconazole, a known 2C9 inhibitor (Krishnaiah et al. 1994). All seven individuals experienced decreases in serum glucose levels, and five had symptoms of mild hypoglycemia. Although this type of drug-drug interaction has not been reported with other oral sulfonylureas and potent inhibitors of 2C9, we suspect that it occurs and is often not recognized by the pharmacist or clinician.

Nonsteroidal Anti-Inflammatory Drugs

Although multiple studies have indicated that many NSAIDs depend on 2C9 for their metabolism, there are very few reports of adverse outcomes in PMs at 2C9 or in patients receiving 2C9 inhibitors concomitant with NSAIDs. This may be because NSAIDs have a wide therapeutic window and also undergo conjugation via uridine 5'-diphosphate glucuronosyltransferases (UGTs).

2C9 plays only a minor role in the metabolism of sulindac, naproxen, diclofenac, and ketoprofen, but it has a major role in the metabolism of ibuprofen, indomethacin, flurbiprofen, celecoxib, lornoxicam, tenoxicam, meloxicam, and piroxicam (Rodrigues 2005). For more details on NSAID-drug interactions, see Chapter 17 ("Pain Management I: Nonnarcotic Analgesics").

Are There Drugs That Induce 2C9 Activity?

Yes. Rifampin has been shown to reduce, by induction of 2C9, serum levels of drugs including celecoxib (Jayasagar et al. 2003), warfarin (Heimark et al. 1987), tolbutamide (Zilly et al. 1977), glyburide, glipizide (Niemi et al. 2001), phenytoin (Kay et al. 1985), and nateglinide (Scheen 2007). Other drugs known to be 2C9 inducers include aprepitant with long-term usage (Depre et al 2005), carbamazepine (Herman et al. 2006), phenobarbital, St. John's wort (Chen et al. 2004), bosentan (Dingemanse and van Giersbergen 2004), and lopinavir/ritonavir (Lim et al. 2004; Yeh et al. 2006). Long-term use of ritonavir and other ritonavir-boosted HIV drugs may have a similar effect (Llibre et al. 2002). Some inducers of 2C9 are listed in Table 8–3.

TABLE 8–3.   Some inducers of 2C9

Aprepitant, long term	Rifampin
Barbiturates	Ritonavir and ritonavir-boosted
Bosentan	HIV agents, long term
Carbamazepine	St. John's wort, long term

References

Baldwin SJ, Bloomer JC, Smith GJ, et al: Ketoconazole and sulphaphenazole as the respective selective inhibitors of P4503A and 2C9. Xenobiotica 25:261–270, 1995

Barry M, Gibbons S, Back D, et al: Protease inhibitors in patients with HIV disease: clinically important pharmacokinetic considerations. Clin Pharmacokinet 32:194–209, 1997

Black DJ, Kunze KL, Wienkers LC, et al: Warfarin-fluconazole, II: a metabolically based drug interaction—in vivo studies. Drug Metab Dispos 24:422–428, 1996

Blum RA, Wilton JH, Hilligoss DM, et al: Effect of fluconazole on the disposition of phenytoin. Clin Pharmacol Ther 49:420–425, 1991

Cadle RM, Zenon GJ 3rd, Rodriguez-Barradas MC, et al: Fluconazole-induced symptomatic phenytoin toxicity. Ann Pharmacother 28:191–195, 1994

Chen Y, Ferguson SS, Negishi M, et al: Induction of human CYP2C9 by rifampicin, hyperforin, and phenobarbital is mediated by the pregnane X receptor. J Pharmacol Exp Ther 308:495–501, 2004

CYP2C9 Allele Nomenclature. 2007. Available at http://www.cypalleles.ki.se/cyp2c9.htm. Accessed July 15, 2007.

Dekhuijzen PN, Koopmans PP: Pharmacokinetic profile of zafirlukast. Clin Pharmacokinet 41:105–114, 2002

de Longueville F, Surry D, Meneses-Lorente G, et al: Gene expression profiling of drug metabolism and toxicology markers using a low-density DNA microarray. Biochem Pharmacol 64:137–149, 2002

Depre M, Van Hecken A, Oeyen M, et al: Effect of aprepitant on the pharmacokinetics and pharmacodynamics of warfarin. Eur J Clin Pharmacol 61:341–346, 2005

DeVane CL, Nemeroff CB: 2002 guide to psychotropic drug interactions. Primary Psychiatry 9:28–57, 2002

Dingemanse J, van Giersbergen PL: Clinical pharmacology of bosentan, a dual endothelin receptor antagonist. Clin Pharmacokinet 43:1089–1115, 2004

Fuhr U, Jetter A, Kirchheiner J: Appropriate phenotyping procedures for drug metabolizing enzymes and transporters in humans and their simultaneous use in the "cocktail" approach. Clin Pharmacol Ther 81:270–283, 2007

Glasheen JJ, Fugit RV, Prochazka AV: The risk of overanticoagulation with antibiotic use in outpatients on stable warfarin regimens. J Gen Intern Med 20:653–656, 2005

Harder S, Thurmann P: Clinically important drug interactions with anticoagulants: an update. Clin Pharmacokinet 30:416–444, 1996

Heimark LD, Gibaldi M, Trager WF, et al: The mechanism of the warfarin-rifampin drug interaction. Clin Pharmacol Ther 42:388–394, 1987

Hemeryck A, De Vriendt C, Belpaire FM: Inhibition of CYP2C9 by selective serotonin reuptake inhibitors: in vitro studies with tolbutamide and (S)-warfarin using human liver microsomes. Eur J Clin Pharmacol 54:947–951, 1999

Herman D, Locatelli I, Grabnar I, et al: The influence of co-treatment with carbamazepine, amiodarone and statins on warfarin metabolism and maintenance dose. Eur J Clin Pharmacol 62:291–296, 2006

Jayasagar G, Krishna Kumar M, Chandrasekhar K: Influence of rifampicin pretreatment on the pharmacokinetics of celecoxib in healthy male volunteers. Drug Metabol Drug Interact 19:287–295, 2003

Joffe HV, Xu R, Johnson FB, et al: Warfarin dosing and cytochrome P450 2C9 polymorphisms. Thromb Haemost 91:1123–1128, 2004

Kay L, Kampmann JP, Svendsen TL, et al: Influence of rifampicin and isoniazid on the kinetics of phenytoin. Br J Clin Pharmacol 20:323–326, 1985

Kim KY, Mancano MA: Fenofibrate potentiates warfarin effects. Ann Pharmacother 37:212–215, 2003

Kim MJ, Nafziger AN, Kashuba AD: Effects of fluvastatin and cigarette smoking on CYP2C9 activity measured using the probe S-warfarin. Eur J Clin Pharmacol 62:431–436, 2006

Kirchheiner J, Brockmöller J: Clinical consequences of cytochrome P450 2C9 polymorphisms. Clin Pharmacol Ther 77:1–16, 2005

Kirchheiner J, Roots I, Goldammer M, et al: Effect of genetic polymorphisms in cytochrome p450 (CYP) 2C9 and CYP2C8 on the pharmacokinetics of oral antidiabetic drugs: clinical relevance. Clin Pharmacokinet 44:1209–1225, 2005

Klose TS, Blaisdell JA, Goldstein JA: Gene structure of CYP2C8 and extrahepatic distribution of human CYP2Cs. J Biochem Mol Toxicol 13:289–295, 1999

Krishnaiah YS, Satyanarayana S, Visweswaram D: Interaction between tolbutamide and ketoconazole in healthy subjects. Br J Clin Pharmacol 37:205–207, 1994

Lebovitz HE: Differentiating members of the thiazolidinedione class: focus on safety. Diabetes Metab Res Rev 18 (suppl 2):S23–S29, 2002

Liedtke MD, Lockhart SM, Rathbun RC: Anticonvulsant and antiretroviral interactions. Ann Pharmacother 38:482–489, 2004

Lim ML, Min SS, Eron JJ, et al: Coadministration of lopinavir/ritonavir and phenytoin results in two-way drug interaction through cytochrome P-450 induction. J Acquir Immune Defic Syndr 36:1034–1040, 2004

Limke KK, Shelton AR, Elliott ES: Fluvoxamine interaction with warfarin. Ann Pharmacother 36:1890–1892, 2002

Llibre JM, Romeu J, Lopez E, et al: Severe interaction between ritonavir and acenocoumarol. Ann Pharmacother 36:621–623, 2002

Mamiya K, Kojima K, Yukawa E, et al: Phenytoin intoxication induced by fluvoxamine. Ther Drug Monit 23:75–77, 2001

Miners JO, Birkett DJ: Use of tolbutamide as a substrate probe for human hepatic cytochrome P450 2C9. Methods Enzymol 272:139–145, 1996

Miners JO, Birkett DJ: Cytochrome P4502C9: an enzyme of major importance in human drug metabolism. Br J Clin Pharmacol 45:525–538, 1998

Moody GC, Griffin SJ, Mather AN, et al: Fully automated analysis of activities catalysed by the major human liver cytochrome P450 (CYP) enzymes: assessment of human CYP inhibition potential. Xenobiotica 29:53–75, 1999

Naganuma M, Shiga T, Nishikata K, et al: Role of desethylamiodarone in the antico-agulant effect of concurrent amiodarone and warfarin therapy. J Cardiovasc Pharmacol Ther 6:363–367, 2001

National Center for Biotechnology Information: CYP2C9 cytochrome P450, family 2, subfamily C, polypeptide 9 [Homo sapiens]. PubMed Entrez Gene, 2008. Available at http://www.ncbi.nlm.nih.gov/sites/entrez?db=gene&Cmd=Show-DetailView&TermToSearch=1559&ordinalpos=1&itool=EntrezSystem2.PEntrez.Gene.Gene_ResultsPanel.Gene_RVDocSum. Accessed July 15, 2008.

Nelson MH, Birnbaum AK, Remmel RP: Inhibition of phenytoin hydroxylation in human liver microsomes by several selective serotonin re-uptake inhibitors. Epilepsy Res 44:71–82, 2001

Niemi M, Backman JT, Neuvonen M, et al: Effects of rifampin on the pharmacokinetics and pharmacodynamics of glyburide and glipizide. Clin Pharmacol Ther 69:400–406, 2001

Nightingale SL: From the Food and Drug Administration. JAMA 271:1067, 1994

Nolan PE Jr, Erstad BL, Hoyer GL, et al: Steady-state interaction between amiodarone and phenytoin in normal subjects. Am J Cardiol 65:1252–1257, 1990

Odani A, Hashimoto Y, Otsuki Y, et al: Genetic polymorphism of the CYP2C subfamily and its effect on the pharmacokinetics of phenytoin in Japanese patients with epilepsy. Clin Pharmacol Ther 62:287–292, 1997

Perini JA, Vianna-Jorge R, Brogliato, et al: Influence of CYP2C9 genotypes on the pharmacokinetics and pharmacodynamics of piroxicam. Clin Pharmacol Ther 78:362–369, 2005

PharmGkb: Warfarin pathway. 2008. Available at http://www.pharmgkb.org/do/serve?objId=PA145011113&objCls=Pathway. Accessed July 15, 2008.

Purkins L, Wood N, Ghahramani P, et al: Coadministration of voriconazole and phenytoin: pharmacokinetic interaction, safety, and toleration. Br J Clin Pharmacol 56 (suppl 1):37–44, 2003

Provigil (package insert). West Chester, PA, Cephalon, Inc., 1999

Rettie AE, Jones JP: Clinical and toxicological relevance of CYP2C9: drug-drug interactions and pharmacogenetics. Annu Rev Pharmacol Toxicol 45:477–494, 2005

Rifkind AB, Lee C, Chang TK, et al: Arachidonic acid metabolism by human cytochrome P450s 2C8, 2C9, 2E1, and 1A2: regioselective oxygenation and evidence for a role for CYP2C enzymes in arachidonic acid epoxygenation in human liver microsomes. Arch Biochem Biophys 320:380–389, 1995

Rodrigues AD: Impact of CYP2C9 genotype on pharmacokinetics: are all cyclooxygenase inhibitors the same? Drug Metab Dispos 33:1567–1575, 2005

Scheen AJ: Drug-drug and food-drug pharmacokinetic interactions with new insulinotropic agents repaglinide and nateglinide. Clin Pharmacokinet 46:93–108, 2007

Schmider J, Greenblatt DJ, von Moltke LL, et al: Inhibition of CYP2C9 by selective serotonin reuptake inhibitors in vitro: studies of phenytoin p-hydroxylation. Br J Clin Pharmacol 44:495–498, 1997

U.S. Food and Drug Administration: Drug Development and Drug Interactions: Table of Substrates, Inhibitors and Inducers. Center for Drug Evaluation and Research, 2006. Available at http://www.fda.gov/cder/drug/drugInteractions/tableSubstrates.htm. Accessed July 15, 2008.

van der Weide J, Steijns LS, van Weelden MJ, et al: The effect of genetic polymorphism of cytochrome P4502C9 on phenytoin dose requirement. Pharmacogenetics 11:287–291, 2001

Wen X, Wang JS, Backman JT, et al: Trimethoprim and sulfamethoxazole are selective inhibitors of CYP2C8 and CYP2C9, respectively. Drug Metab Dispos 30:631–635, 2002

Yeh RF, Gaver VE, Patterson KB, et al: Lopinavir/ritonavir induces the hepatic activity of cytochrome P450 enzymes CYP2C9, CYP2C19, and CYP1A2 but inhibits the hepatic and intestinal activity of CYP3A as measured by a phenotyping drug cocktail in healthy volunteers. J Acquir Immune Defic Syndr 42:52–60, 2006

Zilly W, Breimer DD, Richter E: Stimulation of drug metabolism by rifampicin in patients with cirrhosis or cholestasis measured by increased hexobarbital and tolbutamide clearance. Eur J Clin Pharmacol 11:287–293, 1977

Study Cases

Case 1

An 83-year-old man with dementia of the Alzheimer's type and behavioral disturbances was receiving 5 mg of oral glipizide twice a day for non–insulin-dependent diabetes mellitus. He was prescribed valproic acid (250 mg orally three times a day) for his behavioral problems. The drug appeared to have a positive effect, and his valproic acid level was an acceptable 51 ng/mL. However, his serum glucose level, which had been 120–140 mg/dL before initiation of valproic acid therapy, became consistently low in the morning, ranging from 48 to 60 mg/dL. The glipizide dose was reduced to compensate.

Comment

2C9 inhibition: Valproic acid inhibits 2C9 and resulted in this patient's glipizide level increasing which decreased serum glucose levels significantly. Instead of abandoning one of the drugs (both of which seemed effective), the clinician decreased the glipizide dose to compensate.

References

Wen X, Wang J-S, Kivisto KT, et al: In vitro evaluation of valproic acid as an inhibitor of human cytochrome P450 isoforms: preferential inhibition of cytochrome P450 2C9 (CYP2C9). Br J Clin Pharmacol 52:547–533, 2001

Case 2

A 40-year-old Caucasian woman was prescribed fluoxetine 20 mg/day for anxiety and depressive symptoms. She had been treated for many years with phenytoin 300 mg/day for a seizure disorder, without significant side effects. After 1 week of fluoxetine therapy, she complained of dizziness and somnolence. Serum phenytoin levels were 30 μg/mL (reference range, 10–20 μg/mL).

Comment

2C9 and 2C19 inhibition: Phenytoin is metabolized by 2C9, 2C19, and phase II conjugation enzymes. Although not a potent inhibitor of 2C9 and 2C19, fluoxetine does inhibit these enzymes and most certainly was the cause of the increase in phenytoin levels found in this woman.

References

Nelson MH, Birnbaum AK, Remmel RP: Inhibition of phenytoin hydroxylation in human liver microsomes by several selective serotonin re-uptake inhibitors. Epilepsy Res 44:71–82, 2001
Nightingale SL: From the Food and Drug Administration. JAMA 271:1067, 1994

9

2C19

Jessica R. Oesterheld, M.D.

The gene for 2C19 is located on chromosome 10 close to its related P450 enzymes—2C8, 2C9, and 2C18 (National Center for Biotechnology Information 2008). 2C19 differs from 2C9 by only 43 of 490 amino acids (Jung et al. 1998). Although some authors in older literature refer to 2C9 and 2C19 collectively as 2Cs and some similarities in their activity exist, differences with regard to genetic variability, inducers, and inhibitors necessitate referring to them individually.

What Does 2C19 Do?

2C19 is involved in phase I metabolism. Like other P450 enzymes, 2C19 performs hydroxylation, demethylation, and dealkylation of various substrates.

Where Is 2C19?

2C19 is found in the duodenum (Klose et al. 1999), but its major role is in hepatic metabolism. Together, 2C19 and 2C9 are responsible for approximately 20% of P450 activity in the liver.

Does 2C19 Have Any Polymorphisms?

Yes. There are polymorphisms that create poor metabolizers (PMs) and ultra-extensive metabolizers (UEMs). S-Mephenytoin has long been used in the laboratory as a drug probe for 2C19 activity (Wedlund et al. 1984; Wrighton et al. 1993). In fact, older literature often referred to 2C19 as the S-mephenytoin enzyme. Measuring the clearance of S-mephenytoin can help establish an individual's 2C19 activity.

Studies show that 2%–4% of Caucasians, 10%–25% of East Asians, 1% of African Americans, 4% of Mexican Americans, and 2% of populations from North Africa and the Middle East are 2C19 PMs (de Leon et al. 2006).

More than 10 defective alleles of 2C19 have been characterized (CYP2C19 Allele Nomenclature 2007).

Poor Metabolizers

The most common polymorphism, *2C19*2*, differs from the *2C19*1* of extensive metabolizers (EMs) by a nucleotide substitution that prevents enzymatic activity. This polymorphism is responsible for about 85% of 2C19 PMs. The *2C19*3* polymorphism accounts for most of the rest of the Asian PMs but is rare in Caucasians. Identification of *2C19*2* and *2C19*3* will include 99% of Asians and 84% of Caucasian PMs. Adding *2C19 *4*, **5*, and **6* will account for 92% of Caucasian PMs (Andersson 2004). The AmpliChip identifies only *2C19*1*, **2*, and **3*, accounting for 99% of Asian PMs and 84% of Caucasian PMs (AmpliChip 2006, 2007).

2C19 polymorphisms show wide intrapopulation variability. For example, the percentage of Polynesians who are PMs ranges from 38% to 79%, depending on the location within Polynesia. This variability was discovered somewhat accidentally during a study of patients from the Vanuatu Islands who were reported to have high or toxic levels of proguanil, the pro-drug antimalarial agent metabolized by 2C19 (Kaneko et al. 1999). When proguanil metabolism to its active component is inhibited, increased pro-drug can result in side effects including nausea, vomiting, and diarrhea (Funck-Brentano et al. 1997; Kaneko et al. 1999).

Chang et al. (1997) demonstrated that the antitumor agents cyclophosphamide and ifosfamide (also pro-drugs) are 4-hydroxylated to their active forms by 2C9 and 2C19. These pro-drugs are also 4-hydroxylated and activated by other CYPs (Roy et al. 1999). 2C19 PM status primarily influences cyclophosphamide clearance at doses less than 1,000 mg/m^2 (Timm et al. 2005).

Beyond influencing pro-drug activation and clearance, 2C19 phenotypic variability has other clinical implications. For example, Jiang et al. (2002) demonstrated that area under the plasma concentration-time curve (AUC) of amitriptyline (which is N-demethylated by 2C19) is significantly higher in PMs than in EMs. By contrast, the AUC of nortriptyline, metabolized primarily by 2D6, does not differ between PMs and EMs. This study, along with other similar studies (e.g., Venkatakrishnan et al. 1998), indicates that 2C19 plays a major role in amitriptyline metabolism despite the involvement of other P450 enzymes, including 1A2, 3A4, and 2C9. Because several tricyclic antidepressants (e.g., clomipramine, imipramine, trimipramine) with narrow therapeutic indices are demethylated by 2C19 (see Table 9–1), de Leon et al. (2006) recommend genetic testing of 2C19 for individuals who have poor tolerance to these drugs.

Most proton pump inhibitors (PPIs) are primarily or significantly metabolized via 2C19 (see the section"Gastrointestinal Agents" in Chapter 13, "Internal Medicine"). Aoyama et al. (1999) discovered that patients with *Helicobacter pylori* gastritis who were 2C19 PMs responded better to treatment with omeprazole, a PPI, in terms of both amelioration of symptoms and eradication of *H. pylori*. This outcome is thought to be due to the higher exposure to serum omeprazole in PMs. The EMs in the study had poorer outcomes at all doses.

Researchers have divided the *H. pylori* gastritis clinical populations into three 2C19 genetic groups, which include homozygotic EMs, heterozygotic EMs (with 1 PM allele), and PMs. This categorization can assist with specific dosage recommendations of PPIs for each group in the treatment of *H. pylori* gastritis (Furuta et al. 2004; Sugimoto et al. 2004). Desta et al. (2002) suggested that in Asian populations, which include more PMs than do Caucasian

TABLE 9–1.   Some drugs metabolized by 2C19

Antidepressants	Barbiturates	Proton pump inhibitors	Other drugs
Amitriptyline[1]	Hexobarbital	Esomeprazole[3]	Carisoprodol
Citalopram[1]	Mephobarbital[1]	Lansoprazole	Cilostazol[1]
Clomipramine[1]	Phenobarbital[1]	Omeprazole[3]	Cyclophosphamide[1] (pro-drug)
Fluoxetine[1,2]		Pantoprazole	Diazepam[1]
Imipramine[1]		Rabeprazole[4]	Flunitrazepam[2]
Moclobemide[1]			Ifosfamide[1] (pro-drug)
Sertraline[1]			Mephenytoin
Trimipramine[1]			Nelfinavir[1]
			Phenytoin[1]
			Progesterone[1]
			Proguanil[1] (pro-drug)
			Propranolol[3]
			R-warfarin[1]
			Tolbutamide
			Voriconazole[1]

[1]Also metabolized by other cytochromes.
[2]2C19 is a minor pathway.
[3]Also metabolized by 3A4.
[4]Cytochromes are a lesser pathway compared with other proton pump inhibitors.

populations, routine 2C19 polymorphic testing could save $5,000 (U.S.) per 100 patients tested for *H. pylori* infection.

According to de Leon (2007), genetic testing for 2C19 activity may prevent adverse effects in 2C19 substrates that have a narrow therapeutic index (e.g., TCAs), but for drugs that have a wide therapeutic index (e.g., PPIs), genetic testing will not prevent adverse effects. Knowing 2C19 activity may, however, provide help in improving efficacy for EMs who can benefit from increased dosing.

Determining 2C19 activity in Asian or Caucasian populations may help guide therapy of 2C19 PMs when treating with drugs such as diazepam (Klotz 2007), quazepam (Fukasawa et al. 2007) carisoprodol (Bramness et al. 2005), voriconazole (Mikus et al. 2006), citalopram and escitalopram (Yu et al. 2003), sertraline (Wang et al. 2001), and nelfinavir (Hass et al. 2005). These drugs can have increased plasma levels and side effects when administered to a 2C19 PM.

2C19 status and clearance can be used to discover if particular drugs are principally metabolized via 2C19 (see Table 9–1). As an example, although 2C19 plays a part in phenytoin metabolism, 2C9 plays a more important role in phenytoin metabolism (Mamiya et al. 1998). Investigators have noted, however, that at higher concentrations of phenytoin, 2C19 becomes more significant in phenytoin clearance (Ninomiya et al. 2000). Studies have borne out this differentiation and have resulted in the understanding that increased levels of phenytoin occur with a heterozygous EM for both 2C9 and 2C19, with a heterozygous EM for 2C9, and with a PM for 2C19 (Klotz 2007).

Ulltraextensive Metabolizers (UEMs)

A new polymorphism of 2C19, *2C19*17*, has been identified. This polymorphism shows increased 2C19 activity. *2C19*17* is estimated to be present in 18% of Swedes and Ethiopians and in only 4% of Chinese (Sim et al. 2006). The mean omeprazole AUC in homozygotic *2C19*17* individuals was shown to be 2.1-fold lower than in those homozygotic for *2C19*1* (Baldwin et al. 2008). Similar results were found for escitalopram, another 2C19 substrate (Rudberg et al. 2008). Individuals with this polymorphism are at risk for increased therapeutic failure with drugs that are 2C19 substrates.

TABLE 9–2.   Some inhibitors of 2C19

Antidepressants	Other drugs	
Fluoxetine	Artemisinin	Isoniazid
Fluvoxamine	Chloramphenicol	Modafinil
Moclobemide	Delavirdine	Omeprazole
	Efavirenz	Oxcarbazepine
	Esomeprazole	Probenecid
	Ethinyl estradiol	Ritonavir
	Felbamate	Ticlopidine
	Fluconazole	Topiramate
	Gestodene	Voriconazole
	Indomethacin	

Note.   Names of drug is in **bold** type if there is evidence of potent inhibition.

Are There Drugs That Inhibit 2C19 Activity?

Yes. The table of substrates at the Web site of the U.S. Food and Drug Administration (FDA; 2006) lists four "probe" substrates of 2C19: omeprazole, esomeprazole, lansoprazole, and pantoprazole (Blume et al. 2006). These probes can be used to test other drugs for 2C19 inhibition.

Only fluvoxamine is listed at the FDA Web site as a potent 2C19 inhibitor; it has a Ki of 0.69 µM (Rasmussen et al. 1998). Even subtherapeutic doses of fluvoxamine inhibit 2C19 (Christensen et al. 2002). Yasui-Furukori et al. (2004) found that administration of fluvoxamine with omeprazole resulted in an increase in omeprazole of 3.7-fold and 2.0-fold in homozygous EMs and heterozygous EMs, respectively.

Ticlopidine is another potent inhibitor of 2C19, with a Ki of 1.2 µM (Ko et al. 2000). Ticlopidine increases omeprazole maximum plasma concentration (C_{max}) by more than 40% (Tateishi et al. 1999) when coadministered.

Omeprazole is not only a 2C19 substrate but also a 2C19 inhibitor and has been shown to increase levels of the 2C19 substrates cilostazol (Suri and Bramer 1999) and diazepam (Andersson et al. 1990). Esomeprazole appears to inhibit 2C19 as strongly as the racemic mix of omeprazole (Blume et al. 2006). Concurrent use of esomeprazole and diazepam leads to an 81% increase in the AUC of diazepam and an increase from 43 to 86 hours in diazepam's elimination half-life (Andersson et al. 2001). Although lansoprazole is a potent 2C19 inhibitor *in vitro*, this effect does not appear to occur *in vivo* (Blume et al. 2006).

Ethinyl estradiol (EE), in oral contraceptives, moderately inhibits 2C19 and may cause phenytoin toxicity (De Leacy et al. 1979). Other documented examples of 2C19 substrate inhibition by EEs include carisoprodol, omeprazole, and proguanil. See Chapter 12 ("Gynecology: Oral Contraceptives") for a complete review of 2C19 inhibition by oral contraceptives.

The antiepileptic drugs oxcarbazepine (Lakehal et al. 2002) and felbamate (Glue et al. 1997) have been shown to inhibit 2C19 and increase levels of phenytoin. For other 2C19 inhibitors, see Table 9–2.

Are There Drugs That Induce 2C19 Activity?

Yes (see Table 9–3). Rifampin has been known to induce 2C19 as well as 2C9 (Zhou et al. 1990). St. John's wort has been shown to induce both the 2C19 hydroxylation and 3A4 sulfoxidation of omeprazole, and during omeprazole administration to homozygous 2C19 EMs, the C_{max} of omeprazole decreased by 49% (Wang et al. 2004). Ginkgo biloba can also induce 2C19 (Yin et al. 2004). For the "pan-inducers" phenobarbital and carbamazepine, there is uncertainty whether these drugs induce 2C19 *in vivo*. There is *in vitro* evidence that phenobarbital induces 2C19 (Madan et al. 2003), but *in vivo* induction is unknown (G. Anderson, personal communication, 2007). There is modest *in vivo* evidence that carbamazepine, in fact, can inhibit 2C19 (Lakehal et al. 2002).

TABLE 9–3.   Some inducers of 2C19

Ginkgo biloba
Rifampin
St. John's wort

References

AmpliChip (product information). Indianapolis, IN, Roche Diagnostics, 2007. Available at http://www.amplichip.us/documents/CYP450_P.I._US-IVD_Sept_15_2006.pdf. Accessed July 19, 2007.

AmpliChip (product monograph). Indianapolis, IN, Roche Diagnostics, 2006. Available at http://www.amplichip.us/documents/Product_Monograph.pdf. Accessed July 16, 2008.

Andersson T: Drug Metabolizing Enzymes and Pharmacogenomic Testing Workshop. Rockville, MD, September 13–14, 2004

Andersson T, Cederberg C, Edvardsson G, et al: Effect of omeprazole treatment on diazepam plasma levels in slow versus normal rapid metabolizers of omeprazole. Clin Pharmacol Ther 47:79–85, 1990

Andersson T, Hassan-Alin M, Hasselgren G, et al: Drug interaction studies with esomeprazole, the (S)-isomer of omeprazole. Clin Pharmacokinet 40:523–537, 2001

Aoyama N, Tanigawara Y, Kita T, et al: Sufficient effect of 1-week omeprazole and amoxicillin dual treatment for Helicobacter pylori eradication in cytochrome P450 2C19 poor metabolizers. J Gastroenterol 34 (suppl 11):80–83, 1999

Baldwin RM, Ohlsson S, Pedersen RS, et al: Increased omeprazole metabolism in carriers of the CYP2C19*17 allele: a pharmacokinetic study in healthy volunteers. Br J Clin Pharmacol 65:767–774, 2008

Blume H, Donath F, Warnke A, et al: Pharmacokinetic drug interaction profiles of proton pump inhibitors. Drug Saf 29:769–784, 2006

Bramness JG, Skurtveit S, Gulliksen M, et al: The CYP2C19 genotype and the use of oral contraceptives influence pharmacokinetics of carisoprodol in healthy human subjects. Eur J Clin Pharmacol 61:499–506, 2005

Chang TK, Yu L, Goldstein JA, et al: Identification of the polymorphically expressed CYP2C19 and the wild-type CYP2C9-ILE359 allele as low-Km catalysts of cyclophosphamide and ifosfamide. Pharmacogenetics 7:211–221, 1997

Christensen M, Tybring G, Mihara K, et al: Low daily 10-mg and 20-mg doses of fluvoxamine inhibit the metabolism of both caffeine (cytochrome P4501A2) and omeprazole (cytochrome P4502C19). Clin Pharmacol Ther 71:141–152, 2002

CYP2C19 Allele Nomenclature. 2007. Available at http://www.cypalleles.ki.se/cyp2c19.htm. Accessed July 18, 2008.

De Leacy EA, McLeay CD, Eadie MJ, et al: Effects of subjects' sex, and intake of tobacco, alcohol and oral contraceptives on plasma phenytoin levels. Br J Clin Pharmacol 8:33–36, 1979

de Leon J: The crucial role of the therapeutic window in understanding the clinical relevance of the poor versus the ultrarapid metabolizer phenotypes in subjects taking drugs metabolized by CYP2D6 or CYP2C19. J Clin Psychopharmacol 27:241–245, 2007 de Leon J, Armstrong SC, Cozza KL: Clinical guidelines for psychiatrists for the use of pharmacogenetic testing for CYP450 2D6 and CYP450 2C19. Psychosomatics 47:75–85, 2006

Desta Z, Zhao X, Shin JG, et al: Clinical significance of the cytochrome P450 2C19 genetic polymorphism. Clin Pharmacokinet 41:913–958, 2002

Fukasawa T, Suzuki A, Otani K: Effects of genetic polymorphism of cytochrome P450 enzymes on the pharmacokinetics of benzodiazepines. J Clin Pharm Ther 32:333–341, 2007

Funck-Brentano C, Becquemont L, Lenevu A, et al: Inhibition by omeprazole of proguanil metabolism mechanism of the interaction in vitro and prediction of in vivo results from the in vitro experiments. J Pharmacol Exp Ther 280:730–738, 1997

Furuta T, Shirai N, Sugimoto M, et al: Pharmacogenomics of proton pump inhibitors. Pharmacogenomics 5:181–202, 2004

Glue P, Banfield CR, Perhach JL, et al: Pharmacokinetic interactions with felbamate: in vitro-in vivo correlation. Clin Pharmacokinet 33:214–224, 1997

Haas DW, Smeaton LM, Shafer RW, et al: Pharmacogenetics of long-term responses to antiretroviral regimens containing efavirenz and/or nelfinavir: an Adult AIDS Clinical Trials Group Study. J Infect Dis 192:1931–1942, 2005

Jiang ZP, Shu Y, Chen XP, et al: The role of CYP2C19 in amitriptyline N-demethylation in Chinese subjects. Eur J Clin Pharmacol 58:109–113, 2002

Jung F, Griffin KJ, Song W, et al: Identification of amino acid substitutions that confer a high affinity for sulfaphenazole binding and a high catalytic efficiency for warfarin metabolism to P450 2C19. Biochemistry 37:16270–16279, 1998

Kaneko A, Lum JK, Yaviong L, et al: High and variable frequencies of CYP2C19 mutations: medical consequences of poor metabolism in Vanuatu and other Pacific islands. Pharmacogenetics 9:581–590, 1999

Klose TS, Blaisdell JA, Goldstein JA: Gene structure of CYP2C8 and extrahepatic distribution of the human CYP2Cs. J Biochem Mol Toxicol 13:289–295, 1999

Klotz U: The role of pharmacogenetics in the metabolism of antiepileptic drugs: pharmacokinetic and therapeutic implications. Clin Pharmacokinet 46:271–279, 2007

Ko JW, Desta Z, Soukhova NV, et al: In vitro inhibition of the cytochrome P450 (CYP450) system by the antiplatelet drug ticlopidine: potent effect on CYP2C19 and CYP2D6. Br J Clin Pharmacol 49:343–351, 2000

Lakehal F, Wurden CJ, Kalhorn TF, et al: Carbamazepine and oxcarbazepine decrease phenytoin metabolism through inhibition of CYP2C19. Epilepsy Res 52:79–83, 2002

Madan A, Graham RA, Carroll KM, et al: Effects of prototypical microsomal enzyme inducers on cytochrome P450 expression in cultured human hepatocytes. Drug Metab Dispos 31:421–431, 2003

Mamiya K, Ieiri I, Shimamoto J, et al: The effects of genetic polymorphisms of CYP2C9 and CYP2C19 on phenytoin metabolism in Japanese adult patients with epilepsy: studies in stereoselective hydroxylation and population pharmacokinetics. Epilepsia 39:1317–1323, 1998

Mikus G, Schowel V, Drzewinska M, et al: Potent cytochrome P450 2C19 genotype-related interaction between voriconazole and the cytochrome P450 3A4 inhibitor ritonavir. Clin Pharmacol Ther 80:126–135, 2006

National Center for Biotechnology Information: CYP2C19 cytochrome P450, family 2, subfamily C, polypeptide 19 [Homo sapiens]. PubMed Entrez Gene, 2008. Available at http://www.ncbi.nlm.nih.gov/sites/entrez?Db=gene&Cmd=ShowDetailView&TermToSearch=1557&ordinalpos=4&itool=EntrezSystem2.PEntrez.Gene.Gene_ResultsPanel.Gene_RVDocSum. Accessed July 16, 2008.

Ninomiya H, Mamiya K, Matsuo S, et al: Genetic polymorphism of the CYP2C subfamily and excessive serum phenytoin concentration with central nervous system intoxication. Ther Drug Monit 22:230–232, 2000

Rasmussen BB, Nielsen TL, Brosen K: Fluvoxamine inhibits the CYP2C19-catalysed metabolism of proguanil in vitro. Eur J Clin Pharmacol 54:735–740, 1998

Roy P, Yu LJ, Crespi CL, et al: Development of a substrate-activity based approach to identify the major human liver P-450 catalysts of cyclophosphamide and ifosfamide activation based on cDNA-expressed activities and liver microsomal P-450 profiles. Drug Metab Dispos 27:655–666, 1999

Rudberg I, Mohebi B, Hermann M, et al: Impact of the ultrarapid CYP2C19*17 allele on serum concentration of escitalopram in psychiatric patients. Clin Pharmacol Ther 83:322–327, 2008

Sim SC, Risinger C, Dahl M, et al: A common novel CYP2C19 gene variant causes ultrarapid drug metabolism relevant for the drug response to proton pump inhibitors and antidepressants. Clin Pharmacol Ther 79:103–113, 2006

Sugimoto M, Furuta T, Shirai N, et al: Different dosage regimens of rabeprazole for nocturnal gastric acid inhibition in relation to cytochrome P450 2C19 genotype status. Clin Pharmacol Ther 76:290–301, 2004

Suri A, Bramer SL: Effect of omeprazole on the metabolism of cilostazol. Clin Pharmacokinet 37 (suppl 2):53–59, 1999

Tateishi T, Kumai T, Watanabe M, et al: Ticlopidine decreases the in vivo activity of CYP2C19 as measured by omeprazole metabolism. Br J Clin Pharmacol 47:454–457, 1999

Timm R, Kaiser R, Lotsch J, et al: Association of cyclophosphamide pharmacokinetics to polymorphic cytochrome P450 2C19. Pharmacogenomics J 5:365–373, 2005

U.S. Food and Drug Administration: Drug Development and Drug Interactions: Table of Substrates, Inhibitors and Inducers. Center for Drug Evaluation and Research, 2006. Available at http://www.fda.gov/cder/drug/drugInteractions/tableSubstrates.htm. Accessed July 14, 2007.

Venkatakrishnan K, Greenblatt DJ, von Moltke LL, et al: Five distinct human cytochromes mediate amitriptyline N-demethylation in vitro: dominance of CYP 2C19 and 3A4. J Clin Pharmacol 38:112–121, 1998

Wang JH, Liu ZQ, Wang W, et al: Pharmacokinetics of sertraline in relation to genetic polymorphism of CYP2C19. Clin Pharmacol Ther 70:42–47, 2001

Wang LS, Zhou G, Zhu B, et al: St John's wort induces both cytochrome P450 3A4-catalyzed sulfoxidation and 2C19-dependent hydroxylation of omeprazole. Clin Pharmacol Ther 75:191–197, 2004

Wedlund PJ, Aslanian WS, McAllister CB, et al: Mephenytoin hydroxylation deficiency in Caucasians: frequency of a new oxidative drug metabolism polymorphism. Clin Pharmacol Ther 36:773–780, 1984

Wrighton SA, Stevens JC, Becker GW, et al: Isolation and characterization of human liver cytochrome P450 2C19: a correlation between 2C19 and S-mephenytoin 4- hydroxylation. Arch Biochem Biophys 306:240–245, 1993

Yasui-Furukori N, Takahata T, Nakagami T, et al: Different inhibitory effect of fluvoxamine on omeprazole metabolism between CYP2C19 genotypes. Br J Clin Pharmacol 57:487–494, 2004

Yin OQ, Tomlinson B, Waye MM, et al: Pharmacogenetics and herb-drug interactions: experience with Ginkgo biloba and omeprazole. Pharmacogenetics 14:841–850, 2004

Yu BN, Chen GL, He N, et al: Pharmacokinetics of citalopram in relation to genetic polymorphism of CYP2C19. Drug Metab Dispos 31:1255–1259, 2003

Zhou HH, Anthony LB, Wood AJ, et al: Induction of polymorphic 4'-hydroxylation of S-mephenytoin by rifampicin. Br J Clin Pharmacol 30:471–475, 1990

Study Case

A 42-year-old Caucasian businessman had been taking diazepam (5 or 10 mg prn) once or twice a week for several years to treat anxiety associated with his work. It generally calmed him, and he did not experience sedation or light-headedness. His physician prescribed omeprazole 40 mg/day for relief of gastroesophageal reflux. The patient reported that the next time he had taken 10 mg of diazepam, he had felt dizzy and lethargic the entire day.

Comment

2C19 inhibition: Omeprazole inhibits 2C19. Diazepam has a complicated metabolism, but 2C19 appears to be the primary enzyme involved in its oxidative demethylation. Both racemic omeprazole and S-omeprazole can inhibit diazepam metabolism.

References

Andersson T, Andren K, Cederberg C, et al: Effect of omeprazole and cimetidine on plasma diazepam levels. Eur J Clin Pharmacol 39:51–54, 1990

Andersson T, Hassan-Alin M, Hasselgren G, et al: Drug interaction studies with esomeprazole, the (S)-isomer of omeprazole. Clin Pharmacokinet 40:523–537, 2001

10

2E1

Miia R. Turpeinen, M.D., Ph.D.

What Does 2E1 Do?

2E1 is unique among P450 enzymes because of its toxicological importance. Only a few pharmaceuticals are metabolized via 2E1, but it plays an important role in the bioactivation of several industrial solvents (Raucy et al. 1993), in acetaminophen-related hepatotoxicity (Rumack 2004), as an activator of chemical carcinogenesis, and as a producer of free radicals causing cytotoxicity and tissue injury (Caro and Cederbaum 2004; Gonzalez 2005). 2E1 has also been extensively studied because of its pivotal role in ethanol metabolism (Kessova and Cederbaum 2003; Lieber 1997, 2004). Most of the known substrates of 2E1 are small hydrophobic compounds (Koop 1992; Lewis et al. 2002). Chlorzoxazone, a centrally acting skeletal muscle relaxant, is metabolized to 6-hydroxychlorzoxazone solely via 2E1, which makes chlorzoxazone a widely used probe for 2E1 activity (Lucas et al. 1999). In addition, the metabolism of the anesthetics enflurane, halothane, and others is mediated to some extent via 2E1 (Raucy et al. 1993). For other substrates of 2E1, see Table 10–1.

TABLE 10–1.   Some drugs metabolized by 2E1

Anesthetics[1]	Other drugs and chemicals
Enflurane	Acetaminophen[2]
Halothane	Aniline
Isoflurane	Benzene
Methoxyflurane	Capsaicin
Sevoflurane	Carbon tetrachloride[3]
	Chlorzoxazone[4]
	Dacarbazine[3,5]
	Ethanol[5]
	Ethylene glycol
	Ketones
	Nitrosamines[6]

[1]2E1 defluorinates these anesthetics.
[2]2E1 is normally a minor metabolic pathway. In cases of overdose or induction of 2E1, the hepatotoxic metabolite N-acetyl-p-benzoquinone imine (NAPQI) is created.
[3]Metabolism by 2E1 leads to production of a hepatotoxic metabolite.
[4]Used as a probe for 2E1 activity.
[5]Metabolized also by other hepatic and extrahepatic enzymes.
[6]2A6 is a major pathway.

Where Is 2E1?

2E1 is found almost exclusively in the liver and accounts for about 5%–7% of all hepatic cytochrome activity. *CYP2E1* is the only gene of its subfamily and is located on chromosome 10 (see the PubMed Entrez Gene review [National Center for Biotechnology Information 2008]).

Does 2E1 Have Any Polymorphisms?

Yes. A number of single nucleotide polymorphisms within the 2E1 gene have been characterized, although none of these polymorphisms is associated with a complete loss of activity (Gonzalez 2005, 2007). Thus far, no reliable predictions of clinical relevance based on 2E1 genotype and metabolic phenotype have been substantiated (Carriere et al. 1996; Ingelman-Sundberg 2004).

Extensive research regarding the association between 2E1 polymorphisms and liver, stomach, and esophageal cancer has been performed, although the correlation remains uncertain. There may also be a relationship between 2E1 polymorphisms and the risk of alcohol dependence (Harada et al. 2001) as well as alcohol-associated liver disease (Ingelman-Sundberg et al. 1994).

Unfortunately, the application of bench research to clinical practice has been difficult. Environmental factors greatly change 2E1 activity. Therefore, the importance of 2E1 polymorphisms can be difficult to measure, particularly when cancers take years to develop. Although different alleles have been found in different ethnic groups, namely Mexican Americans (Wan et al. 1998), Japanese (Sun et al. 1999), and Caucasians (Grove et al. 1998), clinical conclusions regarding 2E1 polymorphisms and cancer or alcoholism cannot be drawn at this time.

Are There Drugs That Inhibit 2E1 Activity?

Yes (see Table 10–2). Disulfiram may be the best-known and most potent inhibitor of 2E1. Disulfiram is also utilized in clinical practice in the treatment of alcohol dependence. Disulfiram, a selective 2E1 inhibitor, also inhibits alcohol and aldehyde dehydrogenase (Emery et al. 1999; Kharasch et al. 1993, 1999). In a clinical study of six healthy volunteers, concomitant use of disulfiram (a single 500-mg dose) and chlorzoxazone was found to significantly inhibit 2E1-mediated chlorzoxazone 6-hydroxylation (Kharasch et al. 1993, 1999). The disulfiram metabolite diethyl carbamate also shows inhibition of 2E1 (Guengerich et al. 1991).

TABLE 10–2.   Some inhibitors of 2E1

Acute ethanol

Acute isoniazid

Diethyl carbamate

Disulfiram

Watercress[1]

Note.   Name of drug is in **bold** type if there is evidence of clinically potent inhibition.
[1]Inhibition possibly due to phenethyl isothiocyanate.

Isoniazid (isonicotinyl hydrazine; INH) both inhibits and induces 2E1 (Zand et al. 1993). When treatment with isoniazid is initiated, the drug immediately inhibits 2E1 through competitive inhibition, resulting in increasing levels of 2E1 substrates. After several weeks of isoniazid therapy, there is evidence of 2E1 induction. 2E1 induction by isoniazid may be the underlying reason for hepatotoxicity when acetaminophen is coadministered with INH (Self et al. 1999) (see "Miscellaneous Issues" section later in this chapter for details.) 2E1 is also inhibited by acute alcohol consumption (Lieber 1997), insulin (Woodcroft et al. 2002), and, surprisingly, by a single serving of watercress (Leclercq et al. 1998).

Are There Drugs That Induce 2E1 Activity?

Yes. Several substrates of 2E1 also act as inducers of 2E1 (see Table 10–3). The mechanisms of induction appears to be very complex. The best-known inducer of 2E1 is chronically ingested ethanol, which can increase 2E1 activity 10-fold. Other known inducers of 2E1 include isoniazid, acetone, and retinoids. Certain pathophysiological conditions and environmental factors like obesity (Kotlyar and Carson 1999; Raucy et al. 1991), uncontrolled diabetes (Lieber 1997, 2004; Woodcroft et al. 2002), and smoking cigarettes, including those made with marijuana (Desai et al. 2001), are known to induce 2E1. It is unclear why uncontrolled diabetes can induce 2E1 (Lieber 1997), but it is believed that starvation and chronic ketone formation may play a role. Smoking induces 2E1 (and 1A2), probably through chronic exposure to polycyclic aromatic hydrocarbons in smoke.

Increased activity of CYP2E1 has important clinical implications. It has been associated with increased risk of cancer through an increased production of reactive metabolites and enhanced activation of a variety of carcinogens. For example, if organic compounds such as nitrosamines are metabolized to carcinogenic compounds by 2E1, then having more 2E1 activity can affect the risk of cancer.

Miscellaneous Issues

Acetaminophen is a very widely used over-the-counter antipyretic and analgesic drug. In normal circumstances, the predominant pathways of acet-

TABLE 10–3.   Some inducers of 2E1

Chronic ethanol	Retinoids
Chronic isoniazid and discontinuation	Smoking
	Starvation
Obesity	Uncontrolled diabetes

aminophen metabolism are direct formation of glucuronide and sulfate conjugates, and the role of 2E1 in metabolism is negligible. When the usual hepatic metabolism of acetaminophen is overwhelmed, either by an acetaminophen overdose or by induction of 2E1, the contribution of 2E1 to acetaminophen metabolism increases. 2E1 is the predominant enzyme that creates a hepatotoxic metabolite of acetaminophen, N-acetyl-p-benzoquinone imine (NAPQI) (Dahlin et al. 1984; Gonzalez 2007; Manyike et al. 2000). NAPQI is a chemically reactive moiety that interacts with hepatocellular proteins and other molecules and triggers a cascade of intracellular changes leading to hepatic necrosis (Nelson 1990). The small amounts of NAPQI that are usually produced are quickly conjugated with glutathione in a phase II reaction. Unfortunately, glutathione exists in relatively small quantities, and during an overdose or chronic induction, glutathione stores are rapidly depleted. The clinical response to this situation is administration of the antidote N-acetylcysteine, which competitively inhibits NAPQI at 2E1 and allows glutathione stores to regenerate (Atkuri et al. 2007; Makin and Williams 1997; McClain et al. 1999). N-Acetylcysteine prevents further formation of NAPQI and "buys time" for existing NAPQI to be conjugated by glutathione (Kozer and Koren 2001).

Acetaminophen overdose is a very common cause of drug-induced hepatotoxicity and is a leading cause of acute liver failure in the United States (Bower et al. 2007; Larson et al. 2005; Ostapowicz et al. 2002). Because of 2E1 induction, alcoholics have a higher risk of developing hepatotoxicity from even therapeutic doses of acetaminophen (Seeff et al. 1986; Thummel et al. 2000; Zimmerman and Maddrey 1995). There is also an increased risk for acetaminophen hepatitis in other conditions where glutathione levels are reduced or 2E1 is induced (e.g., in cigarette smoking, opiate consumption, malnutrition, illness-induced starvation, HIV infection, and chronic hepatitis C infection). In these conditions, acetaminophen should be used judiciously

(Barbaro et al. 1996; Lauterburg 2002). Administration of 2E1 inhibitors (such as disulfiram) to alcoholic patients might diminish the formation of NAPQI and further protect these patients from acetaminophen hepatotoxicity (Hazai et al. 2002; Manyike et al. 2000). Despite this potential, no studies have definitively shown benefit from routine use of disulfiram in these cases (Manyike et al. 2000).

References

Atkuri KR, Mantovani JJ, Herzenberg LA, et al: N-Acetylcysteine: a safe antidote for cysteine/glutathione deficiency. Curr Opin Pharmacol 7:355–359, 2007

Barbaro G, Di Lorenzo G, Soldini M, et al: Hepatic glutathione deficiency in chronic hepatitis C: quantitative evaluation in patients who are HIV positive and HIV negative and correlations with plasmatic and lymphocytic concentrations and with the activity of the liver disease. Am J Gastroenterol 91:2569–2573, 1996

Bower WA, Johns M, Margolis HS, et al: Population-based surveillance for acute liver failure. Am J Gastroenterol 102:2459–2463, 2007

Caro AA, Cederbaum AI: Oxidative stress, toxicology, and pharmacology of CYP2E1. Annu Rev Pharmacol Toxicol 44:27–42, 2004

Carriere V, Berthou F, Baird S, et al: Human cytochrome P450 2E1 (CYP2E1): from genotype to phenotype. Pharmacogenetics 6:203–211, 1996

Dahlin DC, Miwa GT, Lu AY, et al: N-acetyl-p-benzoquinone imine: a cytochrome P-450-mediated oxidation product of acetaminophen. Proc Natl Acad Sci U S A 81:1327–1331, 1984

Desai HD, Seabolt J, Jann MW: Smoking in patients receiving psychotropic medications: a pharmacokinetic perspective. CNS Drugs 15:469–494, 2001

Emery MG, Jubert C, Thummel KE, et al: Duration of cytochrome P-450 2E1 (CYP2E1) inhibition and estimation of functional CYP2E1 enzyme half-life after single-dose disulfiram administration in humans. J Pharmacol Exp Ther 291:213–219, 1999

Gonzalez FJ: Role of cytochromes P450 in chemical toxicity and oxidative stress: studies with CYP2E1. Mutat Res 569:101–110, 2005

Gonzalez FJ: The 2006 Bernard B. Brodie Award Lecture: CYP2E1. Drug Metab Dispos 35:1–8, 2007

Grove J, Brown AS, Daly AK, et al: The RsaI polymorphism of CYP2E1 and susceptibility to alcoholic liver disease in Caucasians: effect on age of presentation and dependence on alcohol dehydrogenase genotype. Pharmacogenetics 8:335–342, 1998

Guengerich FP, Kim DH, Iwasaki M: Role of human cytochrome P-450 IIE1 in the oxidation of many low molecular weight cancer suspects. Chem Res Toxicol 4:168–179, 1991

Harada S, Agarwal DP, Nomura F, et al: Metabolic and ethnic determinants of alcohol drinking habits and vulnerability to alcohol-related disorder. Alcohol Clin Exp Res. 25 (5 Suppl ISBRA):71S–75S, 2001

Hazai E, Vereczkey L, Monostory K: Reduction of toxic metabolite formation of acetaminophen. Biochem Biophys Res Commun 291:1089–1094, 2002

Ingelman-Sundberg M: Pharmacogenetics of cytochrome P450 and its applications in drug therapy: the past, present and future. Trends Pharmacol Sci 25:193–200, 2004

Ingelman-Sundberg M, Ronis MJ, Lindros KO, et al: Ethanol-inducible cytochrome P4502E1: regulation, enzymology and molecular biology. Alcohol Alcohol Suppl 2:131–139, 1994

Kessova I, Cederbaum AI: CYP2E1: biochemistry, toxicology, regulation and function in ethanol-induced liver injury. Curr Mol Med 3:509–518, 2003

Kharasch ED, Thummel KE, Mhyre J, et al: Single-dose disulfiram inhibition of chlorzoxazone metabolism: a clinical probe for P450 2E1. Clin Pharmacol Ther 53:643–650, 1993

Kharasch ED, Hankins DC, Jubert C, et al: Lack of single-dose disulfiram effects on cytochrome P-450 2C9, 2C19, 2D6, and 3A4 activities: evidence for specificity toward P-450 2E1. Drug Metab Dispos 27:717–723, 1999

Koop DR: Oxidative and reductive metabolism by cytochrome P450 2E1. FASEB J 6:724–730, 1992

Kotlyar M, Carson SW: Effects of obesity on the cytochrome P450 enzyme system. Int J Clin Pharmacol Ther 37:8–19, 1999

Kozer E, Koren G: Management of paracetamol overdose: current controversies. Drug Saf 24:503–512, 2001

Larson AM, Polson J, Fontana RJ, et al: Acetaminophen-induced acute liver failure: results of a United States multicenter, prospective study. Hepatology 42:1364–1372, 2005

Lauterburg BH: Analgesics and glutathione. Am J Ther 9:225–233, 2002

Leclercq I, Desager JP, Horsmans Y: Inhibition of chlorzoxazone metabolism, a clinical probe for CYP2E1, by a single ingestion of watercress. Clin Pharmacol Ther 64:144–149, 1998

Lewis DF, Modi S, Dickins M: Structure-activity relationship for human cytochrome P450 substrates and inhibitors. Drug Metab Rev 34:69–82, 2002

Lieber CS: Cytochrome P-4502E1: its physiological and pathological role. Physiol Rev 77:517–544, 1997

Lieber CS: The discovery of the microsomal ethanol oxidizing system and its physiologic and pathologic role. Drug Metab Rev 36:511–529, 2004

Lucas D, Ferrara R, Gonzalez E, et al: Chlorzoxazone, a selective probe for phenotyping CYP2E1 in humans. Pharmacogenetics 9:377–388, 1999

Makin AJ, Williams R: Acetaminophen-induced hepatotoxicity: predisposing factors and treatments. Adv Intern Med 42:453–483, 1997

Manyike PT, Kharasch ED, Kalhorn TF, et al: Contribution of CYP2E1 and CYP3A to acetaminophen reactive metabolite formation. Clin Pharmacol Ther 67:275–282, 2000

McClain CJ, Price S, Barve S, et al: Acetaminophen hepatotoxicity: an update. Curr Gastroenterol Rep 1:42–49, 1999

National Center for Biotechnology Information: CYP2E1 cytochrome P450, family 2, subfamily E, polypeptide 1 [Homo sapiens]. Pub Med Entrez Gene, 2008. Available at http://www.ncbi.nlm.nih.gov/sites/entrez?Db=gene&Cmd= retrieve&dopt=full_report&list_uids=1571&log$=database.PEntrez.Gene .Gene_ResultsPanel.Gene_RVDocSum. Accessed July 16, 2008.

Nelson SD: Molecular mechanisms of the hepatotoxicity caused by acetaminophen. Semin Liver Dis 10:267–278, 1990

Ostapowicz G, Fontana RJ, Schiodt FV, et al: Results of a prospective study of acute liver failure at 17 tertiary care centers in the United States. Ann Intern Med 137:947–954, 2002

Raucy JL, Lasker JM, Kraner JC, et al: Induction of cytochrome P450IIE1 in the obese overfed rat. Mol Pharmacol 39:275–280, 1991

Raucy JL, Kraner JC, Lasker JM: Bioactivation of halogenated hydrocarbons by cytochrome P4502E1. Crit Rev Toxicol 23:1–20, 1993

Rumack BH: Acetaminophen misconceptions. Hepatology 40:10–15, 2004

Seeff LB, Cuccherini BA, Zimmerman HJ, et al: Acetaminophen hepatotoxicity in alcoholics: a therapeutic misadventure. Ann Intern Med 104:399–404, 1986

Self TH, Chrisman CR, Baciewicz AM, et al: Isoniazid drug and food interactions. Am J Med Sci 317:304–311, 1999

Sun F, Tsuritani I, Honda R, et al: Association of genetic polymorphisms of alcohol-metabolizing enzymes with excessive alcohol consumption in Japanese men. Hum Genet 105:295–300, 1999

Thummel KE, Slattery JT, Ro H: Ethanol and production of the hepatotoxic metabolite of acetaminophen in healthy adults. Clin Pharmacol Ther 67:591–599, 2000

Wan YJ, Poland RE, Lin KM: Genetic polymorphism of CYP2E1, ADH2, and ALDH2 in Mexican-Americans. Genet Test 2:79–83, 1998

Woodcroft KJ, Hafner MS, Novak RF: Insulin signaling in the transcriptional and posttranscriptional regulation of CYP2E1 expression. Hepatology 35:263–273, 2002

Zand R, Nelson SD, Slattery JT, et al: Inhibition and induction of cytochrome P4502E1-catalyzed oxidation by isoniazid in humans. Clin Pharmacol Ther 54:142–149, 1993

Zimmerman H, Maddrey W: Acetaminophen (paracetamol) hepatotoxicity with regular intake of alcohol: analysis of instances of therapeutic misadventure. Hepatology 22:767–773, 1995

Study Case

A 26-year-old Caucasian man who had been binge drinking for several weeks decided to stop drinking. After 2 days of abstinence, because of a headache, he took five 650-mg tablets of acetaminophen. Forty-eight hours later, he was brought to the emergency room with symptoms of hepatitis.

Comment

2E1 induction: The patient's 2E1 enzymes were induced by the chronic, heavy ethanol consumption. When a large dose of acetaminophen is taken, the enhanced 2E1 enzymes metabolize acetaminophen into the hepatotoxic metabolite NAPQI. A theoretical therapeutic option is to give the patient a potent 2E1 inhibitor, such as disulfiram, to stop the metabolism of acetaminophen by 2E1.

Reference

Manyike PT, Kharasch ED, Kalhorn TF, et al: Contribution of CYP2E1 and CYP3A to acetaminophen reactive metabolite formation. Clin Pharmacol Ther 67:275–282, 2000

11

2A6, 2B6, and 2C8

Miia R. Turpeinen, M.D., Ph.D.

2A6

The gene for 2A6 (previously called 2A3) is located on chromosome 19 (see Entrez Gene review [National Center for Biotechnology Information 2008]). Although 2A6 is present in both brain and steroid-containing tissues (e.g., breast, ovary, testis, adrenal tissue), 2A6 occurs predominantly in the liver. Quantitatively, 2A6 constitutes only a minor portion of hepatic cytochromes and has few clinically relevant substrates (e.g., coumarin, nicotine). 2A6, however, is involved in oxidation of many precarcinogens, including nitrosamines and aflatoxin B_1 (Pelkonen et al. 2000). 2A6 polymorphisms currently identified include those present in poor, intermediate, and ultraextensive metabolizers. 2A6 polymorphisms may be associated with varying risk of cancers as well as variations in smoking habits (London et al. 1999; Nakajima et al. 2006; Oscarson 2001; Raunio et al. 2001; Tyndale and Sellers 2002). For example, poor metabolizers (PMs) typically smoke fewer cigarettes and have lower levels of plasma nicotine (Strasser et al. 2007) than more robust metabolizers.

2A6 specifically 7-hydroxylates coumarin and thus is sometimes referred to as the *coumarin 7-hydroxylase cytochrome*. This reaction may be used as a probe for 2A6 activity (Pelkonen et al. 2000). 2A6 is also the major enzyme involved in C-oxidation of nicotine (Hukkanen et al. 2005b), with secondary involvement of 2B6 in this reaction. In addition, 2A6 appears to be the major CYP involved in metabolism of several alkoxy ethers, such as methyl *tert*-butyl ether (MTBE), which have been used to decrease carbon monoxide production in motor vehicle exhaust (Hong et al. 1999; Le Gal et al. 2001).

Several potent inhibitors of 2A6 have been identified, including methoxsalen, tryptamine, and the monoamine oxidase inhibitor tranylcypromine (Zhang et al. 2001) (see Table 11–1). The use of 2A6 inhibitors to functionally create "2A6 PMs"—which would lead individuals to smoke less—has been proposed as one possibility to create drugs for smoking cessation (Sellers et al. 2000).

TABLE 11–1.   Some 2A6 substrates, inhibitors, and inducers

Substrates	Inhibitors	Inducers
Aflatoxin B_1	Methoxsalen	OCs
Coumarin[1]	Tranylcypromine	Phenobarbital
Cyclophosphamide[2]	Tryptamine	
ETBE		
Halothane[3]		
Ifosfamide[2]		
MTBE		
Nicotine[4]		
Nitrosamines[5]		
TAME		
Tegafur[6]		

Note. ETBE=ethyl *tert*-butyl ether; MTBE=*tert*-butyl ether; OC=oral contraceptive; TAME=*tert*-amyl methyl ether.
[1]7-hydroxylation is used widely as a 2A6 model reaction.
[2]2B6, 2C19, and 3A4 also contribute to the metabolism.
[3]Metabolized primarily by 2E1.
[4]Also metabolized by 2B6.
[5]Also metabolized by 2E1 and 1A2.
[6]Pro-drug 5FU (fluorouracil).

Induction of 2A6 has been reported for antiepileptics (phenobarbital) and oral contraceptives containing estrogen (Benowitz et al. 2006; Sotaniemi et al. 1995). This finding may account for the fact that women smokers have a higher risk of lung cancer than their male counterparts. 2A6 is induced via an estrogen receptor (Higashi et al. 2007), but it does not appear to vary during the menstrual cycle (Hukkanen et al. 2005a).

References (2A6)

Benowitz NL, Lessov-Schlaggar CN, Swan GE, et al: Female sex and oral contraceptive use accelerate nicotine metabolism. Clin Pharmacol Ther 79:480–488, 2006

Higashi E, Fukami T, Itoh M, et al: Human CYP2A6 is induced by estrogen via estrogen receptor. Drug Metab Dispos 35:1935–1941, 2007

Hong JY, Wang YY, Bondoc FY, et al: Metabolism of methyl tert-butyl ether and other gasoline ethers by human liver microsomes and heterologously expressed human cytochromes P450: identification of CYP2A6 as a major catalyst. Toxicol Appl Pharmacol 160:43–48, 1999

Hukkanen J, Gourlay SG, Kenkare S, et al: Influence of menstrual cycle on cytochrome P450 2A6 activity and cardiovascular effects of nicotine. Clin Pharmacol Ther 77:159–169, 2005a

Hukkanen J, Jacob P 3rd, Benowitz NL: Metabolism and disposition kinetics of nicotine. Pharmacol Rev 57:79–115, 2005b

Le Gal A, Dreano Y, Gervasi PG, et al: Human cytochrome P450 2A6 is the major enzyme involved in the metabolism of three alkoxyethers used in oxyfuels. Toxicol Lett 124:47–58, 2001

London SJ, Idle JR, Daly AK, et al: Genetic variation of CYP2A6, smoking, and risk of cancer. Lancet 353:898–899, 1999

Nakajima M, Fukami T, Yamanaka H, et al: Comprehensive evaluation of variability in nicotine metabolism and CYP2A6 polymorphic alleles in four ethnic populations. Clin Pharmacol Ther 80:282–297, 2006

National Center for Biotechnology Information: CYP2A6 cytochrome P450, family 2, subfamily A, polypeptide 6 [Homo sapiens]. PubMed Entrez Gene, 2008. Available at http://www.ncbi.nlm.nih.gov/sites/entrez?Db=gene&Cmd=Show-DetailView&TermToSearch=1548&ordinalpos=1&itool=EntrezSystem2.PEntrez.Gene.Gene_ResultsPanel.Gene_RVDocSum. Accessed July 17, 2008.

Oscarson M: Genetic polymorphisms in the cytochrome P450 2A6 (CYP2A6) gene: implications for interindividual differences in nicotine metabolism. Drug Metab Dispos 29:91–95, 2001

Pelkonen O, Rautio A, Raunio H, et al: CYP2A6: a human coumarin 7-hydroxylase. Toxicology 144:139–147, 2000

Raunio H, Rautio A, Gullsten H, et al: Polymorphisms of CYP2A6 and its practical consequences. Br J Clin Pharmacol 52:357–363, 2001

Sellers EM, Kaplan HL, Tyndale RF: Inhibition of cytochrome P450 2A6 increases nicotine's oral bioavailability and decreases smoking. Clin Pharmacol Ther 68:35–43, 2000

Sotaniemi EA, Rautio A, Backstrom M, et al: CYP3A4 and CYP2A6 activities marked by the metabolism of lignocaine and coumarin in patients with liver and kidney diseases and epileptic patients. Br J Clin Pharmacol 39:71–76, 1995

Strasser AA, Malaiyandi V, Hoffmann E, et al: An association of CYP2A6 genotype and smoking topography. Nicotine Tob Res 9:511–518, 2007

Tyndale RF, Sellers EM: Genetic variation in CYP2A6-mediated nicotine metabolism alters smoking behavior. Ther Drug Monit 24:163–171, 2002

Zhang W, Kilicarslan T, Tyndale RF, et al: Evaluation of methoxsalen, tranylcypromine, and tryptamine as specific and selective CYP2A6 inhibitors in vitro. Drug Metab Dispos 29:897–902, 2001

2B6

Appreciation of the importance of 2B6 has been changing over the past few years; it has moved from being an obscure P450 cytochrome to a more prominent one as more substrates, inhibitors, and inducers have been elucidated. CYP2B6 has been estimated to represent approximately 1%–10% of the total hepatic CYP content. The gene for 2B6 is located on chromosome 19 (see Entrez Gene review [National Center for Biotechnology Information 2008]). A notable interindividual variability in the expression of 2B6 has been reported and appears to be one of the most polymorphic P450 genes in humans. The possible significance of 2B6 pharmacogenetics for clinical drug treatment is just emerging. For example, the effects of 2B6 polymorphisms on plasma exposure and neurotoxicity with the anti-HIV drug efavirenz have been described (Owen et al. 2006; Zanger et al. 2007). It is conceivable that in the future, 2B6 genotyping will be recommended for individuals receiving efavirenz.

Several important drugs are substrates of 2B6. Bupropion, a widely used antidepressant and medical aid for smoking cessation, is hydroxylated mainly by 2B6 and can even be utilized as a specific probe for 2B6 (Faucette et al.

2000; Hesse et al. 2000). The bioactivation of the alkylating cancer chemo-therapy pro-drugs cyclophosphamide and ifosfamide is also mediated by 2B6 (2A6, 3A4 and others also participate in this metabolism). In addition, 2B6 is known to be important in the metabolism of the anesthetics propofol and ket-amine (Court et al. 2001; Turpeinen et al. 2006). Artemisinin, a drug partic-ularly useful in the treatment of chloroquine-resistant *Plasmodium falciparum* infection, is metabolized predominantly by 2B6 and 3A4 (Svensson and Ash-ton 1999). Other 2B6 substrates can be found in Table 11–2.

Potent inhibitors of 2B6 include the following:

- *The antiplatelet agents ticlopidine and clopidogrel.* These thienopyridine de-rivatives have been shown to be very potent inhibitors of CYP2B6 and to considerably decrease the plasma concentrations of 2B6 metabolite hy-droxybupropion as well as inhibit other P450 cytochromes (Turpeinen et al. 2005).
- *The antiretrovirals ritonavir, efavirenz, and nelfinavir.* Each of these drugs significantly decreased hydroxylation of bupropion, whereas other antiret-rovirals (indinavir, saquinavir, delavirdine, and nevirapine) had a negligible effect on 2B6 (Hesse et al. 2001; Park-Wyllie and Antoniou 2003).
- *Thiotepa, an alkylating agent.* This drug has been shown to be a specific and potent inhibitor of 2B6 (Turpeinen et al. 2004). Hence its use with the chemotherapy pro-drugs cyclophosphamide and ifosfamide, metabolized at 2B6, may lead to reduced efficacy of these cancer treatments.
- *The selective serotonin reuptake inhibitors paroxetine, fluoxetine, fluvoxam-ine, and sertraline.* In the study by Hesse et al. (2000), these drugs were shown to inhibit CYP2B6-mediated bupropion hydroxylation with IC50 values of 1.6 µM (paroxetine), 3.2 µM (sertraline), 6.1 µM (fluvoxamine), and 4.2 µM (fluoxetine through its major metabolite norfluoxetine), whereas venlafaxine and citalopram did not have any major effect on 2B6. In light of this inhibitory property, coadministration of paroxetine, sertra-line, fluvoxamine, or fluoxetine with a drug metabolized via 2B6 should be monitored for potential increased plasma levels causing toxicity.
- *Oral contraceptives and hormone replacement therapy.* These agents have been shown to effect bupropion (Palovaara et al. 2003) (see Chapter 12, "Gynecology: Oral Contraceptives").

2B6 is inducible by "classical" inducers such as rifampin, phenytoin, and phenobarbital (Gervot et al. 1999; Loboz et al. 2006).

TABLE 11–2. Some 2B6 substrates, inhibitors, and inducers

Substrates	Inhibitors	Inducers
Artemisinin[3]	**Clopidogrel**[6]	Lopinavir/ritonavir
Bupropion[3,9]	Efavirenz	Phenobarbital
Cyclophosphamide[1,3]	Fluoxetine[5,8]	Phenytoin
Efavirenz[3]	Fluvoxamine	Rifampin
Ifosfamide[1,3]	Memantine	
Ketamine[3]	Nelfinavir[6]	
Meperidine[3]	OCs and HRTs[6]	
Methadone[3]	Paroxetine[5]	
Nicotine[2]	Ritonavir[6]	
Propofol[3]	Sertraline	
Selegiline[3]	Thiotepa[7]	
Sertraline[3]	**Ticlopidine**[6]	
Tamoxifen[4]		
Testosterone[3]		

Note. **Bold** type indicates a potent inhibitor. HRT=hormone replacement therapy; OC=oral contraceptive.
[1]2B6 metabolizes these alkylating agents to active drugs.
[2]2B6 is secondary to 2A6 in C-oxidation.
[3]Other P450 cytochromes also participate.
[4]Metabolized by 2B6, 2C9, and 2D6 to a potent active antiestrogenic compound.
[5]Also a potent inhibitor of 2D6.
[6]Also inhibits other P450 cytochromes.
[7]Selective inhibitor of 2B6.
[8]Through norfluoxetine.
[9]Selective in vivo probe.

References (2B6)

Court MH, Duan SX, Hesse LM, et al: Cytochrome P-450 2B6 is responsible for interindividual variability of propofol hydroxylation by human liver microsomes. Anesthesiology 94:110–119, 2001

Faucette SR, Hawke RL, Lecluyse EL, et al: Validation of bupropion hydroxylation as a selective marker of human cytochrome P450 2B6 catalytic activity. Drug Metab Dispos 28:1222–1230, 2000

Gervot L, Rochat B, Gautier JC, et al: Human CYP2B6: expression, inducibility and catalytic activities. Pharmacogenetics 9:295–306, 1999

Hesse LM, Venkatakrishnan K, Court MH, et al: CYP2B6 mediates the in vitro hydroxylation of bupropion: potential drug interactions with other antidepressants. Drug Dispos Metab 28:1176–1183, 2000

Hesse LM, von Moltke LL, Shader RI, et al: Ritonavir, efavirenz, and nelfinavir inhibit CYP2B6 activity in vitro: potential drug interactions with bupropion. Drug Metab Dispos 29:100–102, 2001

Loboz KK, Gross AS, Williams KM, et al: Cytochrome P450 2B6 activity as measured by bupropion hydroxylation: effect of induction by rifampin and ethnicity. Clin Pharmacol Ther 80:75–84, 2006

National Center for Biotechnology Information: CYP2B6 cytochrome P450, family 2, subfamily B, polypeptide 6 [Homo sapiens]. PubMed Entrez Gene, 2008. Available at http://www.ncbi.nlm.nih.gov/sites/entrez?Db=gene&Cmd= ShowDetailView&TermToSearch=1555&ordinalpos=1&itool=Entrez System2.PEntrez.Gene.Gene_ResultsPanel.Gene_RVDocSum. Accessed July 17, 2008.

Owen A, Pirmohamed M, Khoo SH, et al: Pharmacogenetics of HIV therapy. Pharmacogenet Genomics 16: 693–703, 2006

Palovaara S, Pelkonen O, Uusitalo J, et al: Inhibition of cytochrome P450 2B6 activity by hormone replacement therapy and oral contraceptive as measured by bupropion hydroxylation. Clin Pharmacol Ther 74:326–333, 2003

Park-Wyllie LY, Antoniou T: Concurrent use of bupropion with CYP2B6 inhibitors, nelfinavir, ritonavir and efavirenz: a case series. AIDS 17:638–640, 2003

Svensson US, Ashton M: Identification of the human cytochrome P450 enzymes involved in the in vitro metabolism of artemisinin. Br J Clin Pharmacol 48:528–535, 1999

Turpeinen M, Nieminen R, Juntunen T, et al: Selective inhibition of CYP2B6-catalyzed bupropion hydroxylation in human liver microsomes in vitro. Drug Metab Dispos 32:626–631, 2004

Turpeinen M, Tolonen A, Uusitalo J, et al: Effect of clopidogrel and ticlopidine on cytochrome P450 2B6 activity as measured by bupropion hydroxylation. Clin Pharmacol Ther 77:553–559, 2005

Turpeinen M, Raunio H, Pelkonen O: The functional role of CYP2B6 in human drug metabolism: substrates and inhibitors in vitro, in vivo and in silico. Curr Drug Metab 7:705–714, 2006

Zanger UM, Klein K, Saussele T, et al: Polymorphic CYP2B6: molecular mechanisms and emerging clinical significance. Pharmacogenomics 8:743–759, 2007

2C8

2C8 is closely related to 2C9 and 2C19. The genes for all these three enzymes are located on chromosome 10 (see Entrez Gene review [National Center for Biotechnology Information 2008]). Like 2B6, the importance of 2C8 for drug metabolism has only recently been elucidated. A number of functional 2C8 polymorphisms, including some resulting in poor metabolizers, have been published during recent years, but more studies about their clinical relevance and the general role of CYP2C8 in drug metabolism are still needed. The endogenous compounds arachidonic and retinoic acid are known substrates of 2C8 (Totah and Rettie 2005). Exogenous compounds metabolized via 2C8 include the following:

- *The thiazolidinediones ("glitazones").* These oral hypoglycemic agents are metabolized to some extent by 2C8. Pioglitazone appears to utilize 2C8 much more than 3A4, whereas rosiglitazone is metabolized by 2C8 and 2C9 (Lebovitz 2000; Scheen 2007). Recommendations are provided to reduce their dosage by half when potent 2C8 inhibitor gemfibrozil is added (Scheen 2007).
- *The oral antidiabetic repaglinide.* The short-acting meglitinide analog repaglinide relies on 2C8 and 3A4 for metabolism (Bidstrup et al. 2003). Concomitant use of repaglinide with 2C8 inhibitor gemfibrozil has been shown to considerably prolong and enhance repaglinide's blood glucose-lowering effect (Niemi et al. 2003).
- *The antimalarial drug amodiaquine.* Metabolism of pro-drug amodiaquine into its primary metabolite, *N*-desethylamodiaquine, is mediated by 2C8 (Li et al. 2002). Very recently, the effects of 2C8 polymorphisms on the efficacy of amodiaquine have been reported. Since an estimated 1%–4% of Africans are 2C8 poor metabolizers, this population may experience increased side effects and toxicities from these drugs (Gil and Gil Berglund 2007; Parikh et al. 2007). Additionally, the antimalarials chloroquine and dapsone are also metabolized in part via 2C8 (Gil and Gil Berglund 2007).
- *The anticancer drug paclitaxel.* 2C8 and 3A4 are responsible for metabolism of paclitaxel (Cresteil et al. 2002). 2C8 pharmacogenetics is known to significantly alter paclitaxel metabolism, although the clinical relevance of this is uncertain (Dai et al. 2001; Spratlin and Sawyer 2007).

- *The antiarrhythmic drug amiodarone.* Metabolism of amiodarone to desethylamiodarone is known to be mediated via 2C8 and 3A4 (Ohyama et al. 2000) See Table 11–3 for other substrates of 2C8.

Inhibitors of 2C8 include the potent inhibitor gemfibrozil (Ogilvie et al. 2006) and the modest inhibitor trimethoprim (Niemi et al. 2004).

Induction of 2C8, as well as other 2C family members, seems to involve multiple factors (Ferguson et al. 2005; Gerbal-Chaloin et al. 2001). Rifampin is a known and clinically relevant inducer of 2C8 potentially causing subtherapeutic levels of 2C8 substrates when coadministered (Niemi et al. 2004). Clinical studies evaluating the effect of other inducers are not yet available.

TABLE 11–3.   Some 2C8 substrates, inhibitors, and inducers

Substrates	Inhibitors	Inducers
Amiodarone	**Gemfibrozil**	Rifampin
Amodiaquine	Trimethoprim	
Chloroquine		
Dapsone		
Loperamide		
Paclitaxel		
Pioglitazone		
Repaglinide		
Retinoic acid		
Rosiglitazone		

Note.   Many of these drugs are metabolized by other CYPs as well.
Bold type indicates a potent inhibitor.

References (2C8)

Bidstrup TB, Bjornsdottir I, Sidelmann UG, et al: CYP2C8 and CYP3A4 are the principal enzymes involved in the human in vitro biotransformation of the insulin secretagogue repaglinide. Br J Clin Pharmacol 56:305–314, 2003

Cresteil T, Monsarrat B, Dubois J, et al: Regioselective metabolism of taxoids by human CYP3A4 and 2C8: structure-activity relationship. Drug Metab Dispos 30:438–445, 2002

Dai D, Zeldin DC, Blaisdell JA, et al: Polymorphisms in human CYP2C8 decrease metabolism of the anticancer drug paclitaxel and arachidonic acid. Pharmacogenetics 11:597–607, 2001

Ferguson SS, Chen Y, LeCluyse EL, et al: Human CYP2C8 is transcriptionally regulated by the nuclear receptors constitutive androstane receptor, pregnane X receptor, glucocorticoid receptor, and hepatic nuclear factor 4alpha. Mol Pharmacol 68:747–757, 2005

Gerbal-Chaloin S, Pascussi JM, Pichard-Garcia L, et al: Induction of CYP2C genes in human hepatocytes in primary culture. Drug Metab Dispos 29:242–251, 2001

Gil J, Gil Berglund E: CYP2C8 and antimalaria drug efficacy. Pharmacogenomics 8:187–198, 2007

Lebovitz HE: Differentiating members of the thiazolidinedione class: a focus on safety. Diabetes Metab Res Rev 18 (Suppl 2):23–29, 2000

Li XQ, Bjorkman A, Andersson TB, et al: Amodiaquine clearance and its metabolism to N-desethylamodiaquine is mediated by CYP2C8: a new high affinity and turnover enzyme-specific probe substrate. J Pharmacol Exp Ther 300:399–407, 2002

National Center for Biotechnology Information: CYP2C8 cytochrome P450, family 2, subfamily C, polypeptide 8 [Homo sapiens]. PubMed Entrez Gene, 2008. Available at http://www.ncbi.nlm.nih.gov/sitesentrez?Db=gene&Cmd=ShowDetailView&TermToSearch=1558&ordinalpos=1&itool=EntrezSystem2.PEntrez.Gene.Gene_ResultsPanel.Gene_RVDocSum. Accessed July 17, 2008.

Niemi M, Backman JT, Neuvonen M, et al: Effects of gemfibrozil, itraconazole, and their combination on the pharmacokinetics and pharmacodynamics of repaglinide: potentially hazardous interaction between gemfibrozil and repaglinide. Diabetologia 46:347–551, 2003

Niemi M, Backman JT, Neuvonen PJ: Effects of trimethoprim and rifampin on the pharmacokinetics of the cytochrome P450 2C8 substrate rosiglitazone. Clin Pharmacol Ther 76:239–249, 2004

Ogilvie BW, Zhang D, Li W, et al: Glucuronidation converts gemfibrozil to a potent, metabolism-dependent inhibitor of CYP2C8: implications for drug-drug interactions. Drug Metab Dispos 34:191–197, 2006

Ohyama K, Nakajima M, Nakamura S, et al: A significant role of human cytochrome P450 2C8 in amiodarone N-deethylation: an approach to predict the contribution with relative activity factor. Drug Metab Dispos 28:1303–1310, 2000

Parikh S, Ouedraogo JB, Goldstein JA, et al: Amodiaquine metabolism is impaired by common polymorphisms in CYP2C8: implications for malaria treatment in Africa. Clin Pharmacol Ther 82:197–203, 2007

Scheen AJ: Pharmacokinetic interactions with thiazolidinediones. Clin Pharmacokinet 46:1–12, 2007

Spratlin J, Sawyer MB: Pharmacogenetics of paclitaxel metabolism. Crit Rev Oncol Hematol 8:222–229, 2007

Totah RA, Rettie AE: Cytochrome P450 2C8: substrates, inhibitors, pharmacogenetics, and clinical relevance. Clin Pharmacol Ther 77:341–352, 2005

PART III

Drug Interactions by Medical Specialty

Gary H. Wynn, M.D.
Editor

12

Gynecology: Oral Contraceptives

Jessica R. Oesterheld, M.D.

Reminder: This chapter is dedicated primarily to metabolic inter-actions. Interactions due to displaced protein-binding, alterations in absorption or excretion, and pharmacodynamics are not covered.

Nearly fifty years ago, the introduction of Enovid (norethynodrel and mestra-nol), which provided convenient and reliable contraception, revolutionized birth control. Over the years, reports of interactions between oral contracep-tives (OCs) and other drugs began to trickle into the literature. At first, these drug interactions appeared to be random and unrelated. Increased under-standing of P450 enzymes and phase II reactions of sulfation and glucu-ronidation has permitted preliminary categorization and assessment of the clinical relevance of these drug interactions.

Oral Contraceptives

Most OCs contain both estrogen and progestin. The estrogen suppresses ovulation while progestin suppresses luteinizing hormone, creating an environment unreceptive to sperm. In addition, progestins limit endometrial hyperplasia and decrease the likelihood of endometrial carcinoma. In recent years there has been a strong trend toward using lower-dose estrogen preparations to reduce the likelihood of estrogen-related complications (e.g., headache and thromboembolic disorders). Most OCs contain between 35 and 50 μg of estrogen, usually in the form of ethinyl estradiol (EE). Low-dose OCs contain only 20 μg of EE (e.g., Alesse, Levlite, Loestrin 1/20, Mircette). Only a few OCs contain the original estrogen, mestranol, the 3-methyl ether pro-drug of EE (e.g., Genora 1/50, Nelova 1/50M, Norinyl 1+50, Ortho-Novum 1/50).

Many formulations of OCs are now available. Monophasic preparations contain the same amount of EE and progestin and are taken for 21 days in each 28-day cycle. Biphasic and triphasic preparations take the form of two or three types of pills, with varying amounts of active ingredients. Biphasic and triphasic OCs have been formulated so that the amount of progestin is reduced and the effects correspond more closely to hormonal influences during natural menstrual cycles. Recently, several formulations of continuous daily regimens of EE (10 μg and 30 μg) and a progestin have entered the market. These formulations (e.g., Seasonale, Seasonique) allow withdrawal-bleeding periods only 4 times a year. A yearly no-cycling version of levonorgestrel and EE (Lybrel) received U.S. Food and Drug Administration approval in May 2007. There are a limited number of progestin-only contraceptives. These contraceptives are the minipill containing norethindrone, norgestrel, or levonorgestrel; a subdermal implant of norgestrel (Norplant II or etonogestrel [Implanon]); intramuscular and subcutaneous preparations of medroxyprogesterone acetate and norethindrone enanthate, administered every 3 months; and intrauterine devices that release progesterone and levonorgestrel.

Metabolism of Oral Contraceptives

The metabolism of OCs, which is very complicated, is incompletely understood. Mestranol is first metabolized by 2C9 to EE. After first-pass metabolism, about half of EE reaches the systemic circulation unchanged; the remainder is metabolized in the liver and gut wall. Although a variety of metabolic pathways exist, the major route of inactivation of EE is via 3A4, probably largely hepatic, and through 2C9 as a minor pathway (Guengerich 1990b; Zhang et al. 2007). An enterohepatic recirculation is also postulated for conjugated EE (but is not important for progestins). EE is hydrolyzed by gut bacteria (principally clostridia) back to free EE. Further metabolic steps include formation of estrone, estratriol, and catechol estrogens. Each estrogen is glucuronidated and sulfated via unique or overlapping conjugates (see Raftogianis et al. 2000 for a review of estrogen metabolism). Sulfated estrone, the major circulating form of estrogen, is desulfated, metabolized, and transported to free EE to act at estrogen receptors (Song 2001). SULT1E1 plays a major role in desulfation of sulfated estrone (Zhang et al. 2007). Transformation of EE and estrone to catechol estrogens and to quinones can result in genotoxic DNA adducts that may be very important in the development of breast and other cancers (Adjei and Weinshilboum 2002). EE is conjugated principally by UGT1A1 and perhaps by UGT1A8 and 1A9 (Zhang et al. 2007). The possible genetic polymorphisms at each "node" of this complex metabolic "web" may account for the enormous interindividual variations of each estrogen moeity.

Progestins are also metabolized via 3A4, including progestins containing desogestrel, a pro-drug that must be metabolized via 3A4 to the active form of progestin (Korhonen et al. 2005).

3A4 and UGT Induction of Oral Contraceptives

Induction of 3A4 or uridine 5'-diphosphate glucuronosyltransferase, UGT1A1 and others, may lead to increased clearance of EE and/or progestins and loss of clinical efficacy (see Table 12–1). Drug interactions resulting in spotting, breakthrough bleeding, or unwanted pregnancy have occurred in women taking OCs and griseofulvin (van Dijke and Weber 1984), rifampin (Joshi et al. 1980), rifabutin (Barditch-Crovo et al. 1999), troglitazone (Loi et al. 1999), or enzyme-inducing anticonvulsants: phenobarbital, primidone, phenytoin (Baciewicz 1985), or oxcarbazepine (Fattore et al. 1999). Carbamazepine can both render OCs ineffective and cause fetal neural tube defects. Topiramate (above 200 mg/d [Bialer et al. 2004]) and felbamate may also induce 3A4, and these two drugs increase clearance of EE and may also cause contraceptive failure (Rosenfeld et al. 1997; Saano et al. 1995).

Chronic use of ritonavir and ritonavir-boosted protease inhibitors may induce the metabolism of EE (Ouellet et al. 1998). The combination lopinavir or tipranavir with ritonavir can reduce EE area under curve (AUC) by as much as 50% via induction of EE metabolism. Nevirapine and nelfinavir have been shown to affect EE similarly (Mildvan et al. 2002; Viracept 2007). Although efavirenz is commonly listed in the literature as a 3A4 inducer, efavirenz may actually increase the plasma concentration of EE through an unknown mechanism (Sustiva 2007). St. John's wort has been shown to induce 3A4 (Moore et al. 2000; Murphy et al. 2005). Despite the fact that St. John's wort may cause contraceptive failure, studies have shown that health store and drugstore consumers are rarely told about this interaction (Sarino et al. 2007). Modafinil has also been shown to decrease the maximum observed plasma concentration of EE (Robertson et al. 2002). Some anticonvulsants/mood stabilizers, including valproate (Crawford et al. 1986), gabapentin (Eldon et al. 1998), lamotrigine (Holdich et al. 1991; Hussein and Posner 1997), zonisamide (Sills and Brodie 2007), vigabatrin (Bartoli et al. 1997), levetiracetam, tiagabine, and pregabalin (Harden and Leppik 2006), have been shown not to increase OC clearance in studies. (However, lamotrigine and valproic acid concentrations are reduced by OC induction. See section "CYP2A6 and Phase II Induction of and by Oral Contraceptives" in this chapter).

Aprepitant is a dose-related 3A inhibitor short term and a 3A inducer long term. Thus, despite an initial increase in EE, over time as much as an 18% AUC decrease in EE will occur (Emend 2008). Coadministration with bosentan and EE leads to decreases in EE AUC (van Giersbergen et al. 2006). Although a small decrease in AUC for EE (9%) occurs when isotretinoin is coadministered with OCs, coadministration of the two is unlikely to lead to contraceptive failure (Hendrix et al. 2004).

Because EE is believed to be metabolized largely by hepatic 3A4, agents that are known intestinal 3A inducers do not affect EE. Rifaximin, known to induce intestinal 3A, has been shown not to change the clearance of EE (Trapnell et al. 2007).

3A4 or UGT inducers also increase clearance of progestin-only contraceptives since they are substrates of 3A4. Although not as well documented as interactions involving EE, contraceptive failure of levonorgestrel

TABLE 12–1. Some drugs, herbs, and foods that affect oral contraceptives

Induce 3A4 and may cause contraceptive failure	May increase or prolong oral contraceptive activity
Aprepitant (Emend), long term	Acetaminophen (Tylenol)
Bosentan (Tracleer)	Amprenavir (Agenerase) and fosamprenavir (Lexiva)
Carbamazepine (Tegretol and others)	Atazanavir (Reyataz)
Felbamate (Felbatol)	Atorvastatin (Lipitor)
Griseofulvin (Fulvicin and others)	Dapsone
Isotretinoin (Accutane) (unlikely)	Delavirdine (Rescriptor)
Modafinil (Provigil)	Efavirenz (Sustiva and others)
Nelfinavir (Viracept)	Erythromycin, other macrolides
Nevirapine (Viramune)	Fluconazole (Diflucan)
Oxcarbazepine (Trileptal)	Fluoxetine (Prozac)
Phenobarbital (Luminal and others)	?Fluvoxamine (Luvox)
Phenytoin (Dilantin)	?Gestodene
Primidone (Mysoline)	Grapefruit juice
Rifabutin (Mycobutin)	Indinavir (Crixivan)
Rifampin (Rifadin and others)	Itraconazole (Sporanox)
Ritonavir (Norvir) and ritonavir-boosted protease inhibitors, long term	Ketoconazole (Nizoral)
	Nefazodone
St. John's wort, long term	Vitamin C
Topiramate (Topamax) (more than 200mg/d)	
Troglitazone (Rezulin)	

has been reported in women given phenobarbital (Shane-McWhorter et al. 1998) or phenytoin (Haukkamaa 1986; Odlind and Olsson 1986). Similarly, other 3A4 inducers known to affect EE will likely also decrease levels of progestin-only contraceptives.

There are reports of women becoming pregnant while taking both EE or progestin-only preparations and other drugs, but it is not known how frequently such pregnancies occur. Pharmacokinetic interactions between OCs and 3A4 or UGT inducers do occur, but how often do clinically significant pharmacodynamic outcomes result? Given the wide interindividual variation of 3A4 and UGTs, some women may be vulnerable to these drug interactions and others may experience no noticeable effects. A small clinical study of rifampin, rifabutin, and OCs substantiated the ability of rifampin and rifabutin to increase clearance of OCs, although none of the 12 women in the study ovulated (Barditch-Crovo et al. 1999). Currently, lower and lower dosing of EE is being used, and women who take OCs with only 20 µg of EE may be especially vulnerable. It is recommended that until clinicians can identify women at risk, patients taking enzyme-inducing anticonvulsants (except phenytoin; see the section "1A2, 2B6, 2C19, and 3A4 Inhibition by Oral Contraceptives" later in this chapter) with OCs take 50–100 µg of EE. Women taking enzyme-inducing drugs along with OCs should also be instructed to use barrier contraceptives midcycle to prevent pregnancy (Crawford et al. 1990). Another option is to substitute depot medroxyprogesterone acetate as a contraceptive, since it does not seem to be affected by inducing anticonvulsants (Teal and Ginosar 2007). Replacement of 3A4-inducing anticonvulsants with noninducing alternatives can also be considered (Guberman 1999). An enzyme inducer's effects can continue for a few weeks after administration of the inducer has ceased. As has been recommended with rifampin use ("Use of Rifampin and Contraceptive Steroids" 1999), after short-term treatments with 3A4 or UGT inducers are discontinued, patients taking OCs need to take extra contraceptive precautions for up to 4 weeks.

3A4 Inhibition of Oral Contraceptives

Because EE is a 3A4 substrate, potent inhibitors of 3A4 can increase or prolong estrogenic activity. Although not likely to impair contraceptive efficacy, potent 3A4 inhibitors can be expected to increase estrogen-related side effects (e.g., migraine headaches or thromboembolic events) in susceptible women (see Table 12–1). Drugs known to increase EE levels include the antifungals ketoconazole, itraconazole, voriconazole, and fluconazole (Hilbert et al. 2001; Sinofsky and Pasquale 1998) and potent 3A4-inhibiting macrolides (Meyer et al. 1990; Pessayre 1983; Wermeling et al. 1995). Dapsone has been shown to increase peak EE concentration (Joshi et al. 1984). Atorvastatin increases levels of EE by 20% (Lipitor 2007). Even grapefruit juice is known to increase EE levels (Weber et al. 1996). Although gestodene has been shown to increase EE *in vitro* (Guengerich 1990a), this interaction may not be clinically significant (Lawrenson and Farmer 2000). Nefazodone, known to potently inhibit 3A4, may also significantly affect EE clearance. If these 3A4 inhibitors are added to OCs that contain only 20 μg of EE, the low-dose OCs may be functionally converted to a higher-dose OC and cause adverse events (e.g., breast tenderness, bloating, weight gain). Adson and Kotlyar (2001) reported a case of increased side effects when nefazodone was added to Mircette.

OCs containing desogestrel (e.g., Apri, Cesia, Cyclessa, Desogen, Kariva, Mircette, Ortho-Cept, Solia, Velivet) and Cerazette (a progestin-only contraceptive) are also vulnerable to 3A inhibitors because these OCs are metabolized to their active forms by 3A (Korhonen et al. 2005).

Although the mechanism of action is not clear, several HIV drugs have been shown to increase the AUC of EE, including atazanavir, delavirdine, fosamprenavir, amprenavir, indinavir, and efavirenz.

2C9 Inhibition of Oral Contraceptives

Mestranol is demethylated to EE by 2C9. Thus, potent inhibitors of 2C9 may decrease EE levels in OCs such as Genora 1/50, Nelova 1/50M, Norinyl 1+50, and Ortho-Novum 1/50. Examples of 2C9 inhibitors are sulfaphenazole and other sulfonamide antibiotics, valproate, and fluconazole (see Chapter 8, "2C9").

1A2, 2B6, 2C19, and 3A4 Inhibition by Oral Contraceptives

Not only do other agents affect OC clearance, but OCs can also affect the clearance of other drugs (see Table 12–2). Current evidence suggests that OCs are moderate inhibitors of 1A2 and 2C19 and mild inhibitors of 2B6 and 3A4. OCs decrease the clearance of caffeine and theophylline, well-known substrates of 1A2 (Abernethy and Todd 1985; Gu et al. 1992; Roberts et al. 1983; Tornatore et al. 1982; Zhang and Kaminsky 1995). Because a 30% decrease in theophylline clearance occurs with OC use (Jonkman 1986), clinicians should consider reductions in theophylline dosing. Although OCs inhibit (Walle et al. 1996) the side-chain oxidation of propranolol via 1A2 (Yoshimoto et al. 1995), concomitant induction of propranolol glucuronidation by OCs prevents a clinically significant interaction (Walle et al. 1989, 1996). Two cases of interactions between OCs and psychotropic 1A2 substrates have been reported. Clozapine levels were significantly increased with the addition of an OC (Gabbay et al. 2002), while chlorpromazine levels increased 6-fold when coadministered with an OC (Chetty and Miller 2001). No reports of drug interactions between olanzapine and tacrine with OCs have been found in the literature, although these drug interactions are possible. Plasma concentrations of tacrine are increased by hormone replacement therapy (which delivers very low amounts of estrogen) (Laine et al. 1999b); therefore, OCs may well increase tacrine levels, but these interaction are not yet understood. A combination of EE and gestodene was shown to increase levels of tizanidine (a known 1A2 substrate) to clinically significant levels (Granfors et al. 2005).

An OC containing 30 µg of EE and 150 µg of desogestrel was shown to decrease the AUC of bupropion, a probe drug for 2B6, whereas hormone replacement therapy (with larger amounts of estrogen, 2 mg estradiol valerate) showed marked inhibition of 2B6 (Palovaara et al. 2003).

Support for OC inhibition of 2C19 has been provided by three population studies (Hagg et al. 2001; Laine et al. 2000; Tamminga et al. 1999). The activity of 2C19 in women who used OCs was decreased by 68% compared with that in women who did not use OCs (Tamminga et al. 1999). Although the focus of interest in the interaction between OCs and phenytoin has been on phenytoin's induction of OC metabolism (because of reduced efficacy of OCs), OCs have also been found to decrease the clearance of phenytoin (De Leacy et al. 1979). Thus, increasing OC dosing to overcome phenytoin induction of EE can result in phenytoin toxicity, and this drug combination is best avoided. The concentration of selegiline (a substrate of 2C19 and 2B6) is increased more than 10-fold when OCs are added (Laine et al. 1999a). The AUC of carisoprodol is increased 60% when this agent is coadministered with an OC (Bramness et al. 2005). Palovaara (2003) found a 38% increase in

TABLE 12–2. Some drugs whose clearance is changed by oral contraceptives

Clearance decreased by oral contraceptives	Clearance increased by oral contraceptives
Amitriptyline (Elavil, Endep)	Acetaminophen (Tylenol)
Bupropion	Clofibrate (Atromid-S)
Caffeine	Diflunisal (Dolobid)
Carisoprodol (Soma)	Lamotrigine (Lamictal)
?Chlordiazepoxide (Librium)	?Lorazepam (Ativan)
Chlorpromazine (Thorazine)	Nicotine
Clozapine (Clozaril)	?Oxazepam (Serax)
Cyclosporine (Neoral and others)	Phenprocoumon
Diazepam (Valium)	?Temazepam (Restoril)
Imipramine (Tofranil)	Valproic acid (Depakene, Depakote)
Midazolam (Versed)	
?Olanzapine (Zyprexa)	
Omeprazole (Prilosec)	
Phenytoin (Dilantin)	
Prednisolone	
Proguanil (pro-drug)	
Selegiline (Deprenyl and others)	
?Tacrine (Cognex)	
Theophylline (Slo-Bid and others)	
Tizanidine (Zanaflex)	
Voriconazole (Vfend)	

omeprazole AUC in combination with EE, though this interaction is unlikely to have clinical significance because of the wide margin of therapeutic index of omeprazole. Proguanil, a pro-drug, is increased by 34% with OCs, and it is recommended that its dosage be boosted by 50% to increase the concentration of the active metabolite (McGready et al. 2003).

Results of *in vitro* studies of EE effects are conflicting (Guengerich 1990a), but there is evidence to support the finding of very modest *in vivo* inhibition of 3A4 by EE (Belle et al. 2002; Palovaara et al. 2000). Although concentrations of EE are higher in women who take OCs for more than 6 months, clearance of EE in these women is reduced (Tornatore et al. 1982).

Other inferential evidence comes from several interactions between OCs and drugs known to be substrates of 3A4: cyclosporine (Deray et al. 1987), midazolam (Palovaara et al. 2000), prednisolone (Seidegard et al. 2000), and levonorgestrel (Fotherby 1991).

Inhibition of both 3A4 and 2C19 by OCs is likely responsible for several well-documented OC-drug interactions. Clearances of imipramine, amitriptyline, and diazepam are decreased by OCs (Abernethy et al. 1982, 1984; Edelbroek et al. 1987). 2C19 is a principal pathway for demethylation of these drugs; 3A4 is also involved (Jung et al. 1997; Koyama et al. 1997; Venkatakrishnan et al. 1998). Clearance of chlordiazepoxide may also be reduced by OCs (Patwardhan et al. 1983). Chlordiazepoxide's intermediate metabolite, nordiazepam, is known to be metabolized by 3A4 and 2C19 (Ono et al. 1996). Whether other benzodiazepines that are metabolized via this intermediate metabolite are also affected has not yet been investigated. The coadministration of OCs and voriconazole will result in mutual increases of both drugs given the similar pathways and inhibitory properties (Andrews et al. 2008).

Most of the studies above used OCs with EE and various progestins. Whether the progestin portion of the OC is also a CYP or UGT inhibitor or inducer is not clear. There is only fragmentary information on the contribution of individual progestins to CYP inhibition or induction. Some evidence suggests that *in vivo* parenteral progestins (medroxyprogesterone acetate) can induce CYP3A, in contrast to oral progestins, which do not induce CYP3A (Tsunoda et al. 1998). Several progestins are 2C9 and 2C19 inhibitors *in vitro* (Laine et al. 2003), but their *in vivo* effects have yet to be determined.

CYP2A6 and Phase 2 Induction of and by Oral Contraceptives

OCs have been shown to be associated with higher rates of nicotine clearance—an index of 2A6 activity (Benowitz et al. 2006). An increase in 2A6 activity may account for the finding that women smokers have higher rates of lung cancer than their male counterparts.

Although the specific UGTs remain to be identified, OCs increase glucuronidation clearance of the following drugs: clofibrate (Liu et al. 1991; Miners et al. 1984), propranolol (Walle et al. 1996) (see the preceding section of this chapter), phenprocoumon (Monig et al. 1990), diflunisal (Herman et al.

1994), and lamotrigine (Sabers et al. 2001). During the "washout week," the week off ethinyl estradiol, lamotrigine plasma levels were 84% higher (Christensen et al. 2007; Contin et al. 2006). A similar finding was made for valproic acid (Galimberti et al. 2006). In order to avoid fluctuating levels of lamotrigine and valproic acid when they are combined with these traditional OCs, it may be advantageous to keep women on a continuous daily regimen of OC therapy after monitoring and adjusting levels of the anticonvulsants/mood stabilizers. An alternative strategy (Sabers et al. 2008) for women taking lamotrigine is a dose reduction of 25% during "pill-free weeks." Additionally, a progestin-only OC can be considered, since it will not alter lamotrigine levels (Galimberti et al. 2006).

Data are conflicting with regard to changes in lorazepam, oxazepam, and temazepam (Abernethy et al. 1983; Patwardhan et al. 1983; Stoehr et al. 1984) when these agents are combined with OCs. Acetaminophen is conjugated by UGT1A1, UGT1A6, and UGT1A9 (Court et al. 2001), and OCs are known to increase glucuronidation of acetaminophen (Mitchell et al. 1983).

EE clearance decreases when vitamin C or acetaminophen is added. This interaction occurs because vitamin C and acetaminophen compete with EE in the gut wall (Rogers et al. 1987) for sulfation (via sulfotransferase SULT1A1 [Dooley 1998]).

Transporters and Oral Contraceptives

EE inhibits the P-glycoprotein (ABCB1) and may have effects on other transporters. However, there is no information on their clinical effects (Zhang et al. 2007).

Conclusion

When obtaining a woman's medication history, the clinician should inquire whether the patient is taking OCs. Clinicians often neglect to ask this question (Hocking and deMello 1997), despite the fact that OCs are so commonly used by women during their reproductive years. Clinicians need to be knowledgeable and vigilant about drug interactions involving OCs and should be particularly alert at times of addition or subtraction of other drugs.

References

Abernethy DR, Todd EL: Impairment of caffeine clearance by chronic use of low-dose oestrogen-containing oral contraceptives. Eur J Clin Pharmacol 28:425–428, 1985

Abernethy DR, Greenblatt DJ, Divoll M, et al: Impairment of diazepam metabolism by low-dose estrogen-containing oral-contraceptive steroids. N Engl J Med 306:791–792, 1982

Abernethy DR, Greenblatt DJ, Ochs HR, et al: Lorazepam and oxazepam kinetics in women on low-dose oral contraceptives. Clin Pharmacol Ther 33:628–632, 1983

Abernethy DR, Greenblatt DJ, Shader RI: Imipramine disposition in users of oral contraceptive steroids. Clin Pharmacol Ther 35:792–797, 1984

Adjei AA, Weinshilboum RM: Catecholestrogen sulfation: possible role in carcinogenesis. Biochem Biophys Res Commun 292:402–408, 2002

Adson DE, Kotlyar M: A probable interaction between a very low-dose oral contraceptive and the antidepressant nefazodone: a case report. J Clin Psychopharmacol 21:618–619, 2001

Andrews E, Damle BD, Fang A, et al: Pharmacokinetics and tolerability of voriconazole and a combination oral contraceptive co-administered in healthy female subjects. Br J Clin Pharmacol 65:531–539, 2008

Baciewicz AM: Oral contraceptive drug interactions. Ther Drug Monit 7: 26–35, 1985

Barditch-Crovo P, Trapnell CB, Ette E, et al: The effects of rifampin and rifabutin on the pharmacokinetics and pharmacodynamics of a combination oral contraceptive. Clin Pharmacol Ther 65:428–438, 1999

Bartoli A, Gatti G, Cipolla G, et al: A double-blind, placebo-controlled study on the effect of vigabatrin on in vivo parameters of hepatic microsomal enzyme induction and on the kinetics of steroid oral contraceptives in healthy female volunteers. Epilepsia 38:702–707, 1997

Belle DI, Callaghan JT, Gorski JC, et al: The effects of an oral contraceptive containing ethinylestradiol and norgestrel on CYP3A activity. Br J Clin Pharmacol 53:67–74, 2002

Benowitz NL, Lessov-Schlaggar CN, Swan GE, et al: Female sex and oral contraceptive use accelerate nicotine metabolism. Clin Pharmacol Ther 79:480–488, 2006

Bialer M, Doose DR, Murthy B, et al: Pharmacokinetic interactions of topiramate. Clin Pharmacokinet 43:763–780, 2004

Bramness JG, Skurtveit S, Gulliksen M, et al: The CYP2C19 genotype and the use of oral contraceptives influence the pharmacokinetics of carisoprodol in healthy human subjects. Eur J Clin Pharmacol 61:499–506, 2005

Chetty M, Miller R: Oral contraceptives increase the plasma concentrations of chlorpromazine. Ther Drug Monit 23:556–558, 2001

Christensen J, Petrenaite V, Atterman J, et al: Oral contraceptives induce lamotrigine metabolism: evidence from a double-blind, placebo-controlled trial. Epilepsia 48:484–489, 2007

Contin M, Albani F, Ambrosetto G, et al: Variation in lamotrigine plasma concentrations with hormonal contraceptive monthly cycles in patients with epilepsy. Epilepsia 47:1573–1575, 2006

Court MH, Duan SX, von Moltke LL, et al: Interindividual variability in acetaminophen glucuronidation by human liver microsomes: identification of relevant acetaminophen UDP-glucuronosyltransferase isoforms. J Pharmacol Exp Ther 299:998–1006, 2001

Crawford P, Chadwick D, Cleland P, et al: The lack of effect of sodium valproate on the pharmacokinetics of oral contraceptive steroids. Contraception 33:23–29, 1986

Crawford P, Chadwick DJ, Martin C, et al: The interaction of phenytoin and carbamazepine with combined oral contraceptive steroids. Br J Clin Pharmacol 30:892–896, 1990

De Leacy EA, McLeay CD, Eadie MJ, et al: Effects of subjects' sex, and intake of tobacco, alcohol and oral contraceptives on plasma phenytoin levels. Br J Clin Pharmacol 8:33–36, 1979

Deray G, le Hoang P, Cacoub P, et al: Oral contraceptive interaction with cyclosporin. Lancet 1:158–159, 1987

Dooley TP: Molecular biology of the human phenol sulfotransferase gene family. J Exp Zool 282:223–230, 1998

Edelbroek PM, Zitman FG, Knoppert-van der Klein EA, et al: Therapeutic drug monitoring of amitriptyline: impact of age, smoking and contraceptives on drug and metabolite levels in bulimic women. Clin Chim Acta 165:177–187, 1987

Eldon MA, Underwood BA, Randinitis EJ, et al: Gabapentin does not interact with a contraceptive regimen of norethindrone acetate and ethinyl estradiol. Neurology 50:1146–1148, 1998

Emend (package insert). Whitehouse Station, NJ, Merck & Co., Inc., 2008

Fattore C, Cipolla G, Gatti G, et al: Induction of ethinylestradiol and levonorgestrel metabolism by oxcarbazepine in healthy women. Epilepsia 40:783–787, 1999

Fotherby K: Intrasubject variability in the pharmacokinetics of ethynyloestradiol. J Steroid Biochem Mol Biol 38:733–736, 1991

Gabbay V, O'Dowd MA, Mamamtavrishvili M, et al: Clozapine and oral contraceptives: a possible drug interaction. J Clin Psychopharmacol 22:621–622, 2002

Galimberti CA, Mazzucchelli I, Arbasino C, et al: Increased apparent oral clearance of valproic acid during intake of combined contraceptive steroids in women with epilepsy. Epilepsia 47:1569–1572, 2006

Granfors MT, Backman JT, Laitila J, et al: Oral contraceptives containing ethinyl estradiol and gestodene markedly increase plasma concentrations and effects of tizanidine by inhibiting cytochrome P450 1A2. Clin Pharmacol Ther 78:400–411, 2005

Gu L, Gonzalez FJ, Kalow W, et al: Biotransformation of caffeine, paraxanthine, theobromine and theophylline by cDNA-expressed human CYP1A2 and CYP2E1. Pharmacogenetics 2:73–77, 1992

Guberman A: Hormonal contraception and epilepsy. Neurology 53 (suppl 1): 838–840, 1999

Guengerich FP: Inhibition of oral contraceptive steroid-metabolizing enzymes by steroids and drugs. Am J Obstet Gynecol 163:2159–2163, 1990a

Guengerich FP: Metabolism of 17 alpha-ethynylestradiol in humans. Life Sci 47:1981–1988, 1990b

Hagg S, Spigset O, Dahlqvist R: Influence of gender and oral contraceptives on CYP2D6 and CYP2C19 activity in healthy volunteers. Br J Clin Pharmacol 51:169–173, 2001

Harden CL, Leppik I: Optimizing therapy of seizures in women who use oral contraceptives. Neurology 67:S56–S58, 2006

Haukkamaa M: Contraception by Norplant subdermal capsules is not reliable in epileptic patients on anticonvulsant treatment. Contraception 33: 559–565, 1986

Hendrix CW, Jackson KA, Whitmore E, et al: The effect of isotretinoin on the pharmacokinetics and pharmacodynamics of ethinyl estradiol and norethindrone. Clin Pharmacol Ther 75:464–475, 2004

Herman RJ, Loewen GR, Antosh DM, et al: Analysis of polymorphic variation in drug metabolism, III: glucuronidation and sulfation of diflunisal in man. Clin Invest Med 17:297–307, 1994

Hilbert J, Messig M, Kuye O, et al: Evaluation of interaction between fluconazole and an oral contraceptive in healthy women. Obstet Gynecol 98:218–223, 2001

Hocking G, deMello WF: Taking a "drugs" history. Anaesthesia 52:904–905, 1997

Holdich T, Whiteman P, Orme M, et al: Effect of lamotrigine on the pharmacology of the combined oral contraceptive pill (abstract). Epilepsia 32 (suppl 1):96, 1991

Hussein Z, Posner J: Population pharmacokinetics of lamotrigine monotherapy in patients with epilepsy: retrospective analysis of routine monitoring data. Br J Clin Pharmacol 43:457–465, 1997

Jonkman JH: Therapeutic consequences of drug interactions with theophylline pharmacokinetics. J Allergy Clin Immunol 78:736–742, 1986

Joshi JV, Joshi UM, Sankolli GM, et al: A study of interaction of a low-dose combination oral contraceptive with anti-tubercular drugs. Contraception 21:617–629, 1980

Joshi JV, Maitra A, Sankolli G, et al: Norethisterone and ethinyl estradiol kinetics during dapsone therapy. J Assoc Physicians India 32:191–193, 1984

Jung F, Richardson TH, Raucy JL, et al: Diazepam metabolism by cDNA-expressed human 2C P450s: identification of P4502C18 and P4502C19 as low K(M) diazepam N-demethylases. Drug Metab Dispos 25:133–139, 1997

Korhonen T, Tolonen A, Uusitalo J, et al: The role of CYP2C and CYP3A in the disposition of 3-keto-desogestrel after administration of desogestrel. Br J Clin Pharmacol 60:69–75, 2005

Koyama E, Chiba K, Tani M, et al: Reappraisal of human CYP isoforms involved in imipramine N-demethylation and 2-hydroxylation: a study using microsomes obtained from putative extensive and poor metabolizers of S-mephenytoin and eleven recombinant human CYPs. J Pharmacol Exp Ther 281:1199–1210, 1997

Laine K, Anttila M, Helminen A, et al: Dose linearity study of selegiline pharmacokinetics after oral administration: evidence for strong drug interaction with female sex steroids. Br J Clin Pharmacol 47:249–254, 1999a

Laine K, Palovaara S, Tapanainen P, et al: Plasma tacrine concentrations are significantly increased by concomitant hormone replacement therapy. Clin Pharmacol Ther 66:602–608, 1999b

Laine K, Tybring G, Bertilsson L: No sex-related differences but significant inhibition by oral contraceptives of CYP2C19 activity as measured by the probe drugs mephenytoin and omeprazole in healthy Swedish white subjects. Clin Pharmacol Ther 68:151–159, 2000

Laine K, Yasar U, Widen J, et al: A screening study on the liability of eight different female sex steroids to inhibit CYP2C9, 2C19 and 3A4 activities in human liver microsomes. Pharmacol Toxicol 93:77–81, 2003

Lawrenson R, Farmer R: Venous thromboembolism and combined oral contraceptives: does the type of progestogen make a difference? Contraception 62 (suppl 2):S21–S28, 2000

Lipitor (package insert). New York, NY, Parke-Davis/Pfizer, 2007. Available at http://www.lipitor.com. Accessed May 15, 2007.

Liu HF, Magdalou J, Nicolas A, et al: Oral contraceptives stimulate the excretion of clofibric acid glucuronide in women and female rats. Gen Pharmacol 22:393–397, 1991

Loi CM, Stern R, Koup JR, et al: Effect of troglitazone on the pharmacokinetics of an oral contraceptive agent. J Clin Pharmacol 39:410–417, 1999

McGready R, Stepniewska K, Seaton E, et al: Pregnancy and use of oral contraceptives reduces the biotransformation of proguanil to cycloguanil. Eur J Clin Pharmacol 59:553–557, 2003

Meyer B, Muller F, Wessels P, et al: A model to detect interactions between roxithromycin and oral contraceptives. Clin Pharmacol Ther 47:671–674, 1990

Mildvan D, Yarish R, Marshak A, et al: Pharmacokinetic interaction between nevirapine and ethinylestradiol/norethindrone when administered concurrently to HIV-infected women. J Acquir Immune Defic Syndr 29:471–477, 2002

Miners JO, Robson RA, Birkett DJ: Gender and oral contraceptive steroids as determinants of drug glucuronidation: effects on clofibric acid elimination. Br J Clin Pharmacol 18:240–243, 1984

Mitchell MC, Hanew T, Meredith CG, et al: Effects of oral contraceptive steroids on acetaminophen metabolism and elimination. Clin Pharmacol 34:48–53, 1983

Monig H, Baese C, Heidemann HT, et al: Effect of oral contraceptive steroids on the pharmacokinetics of phenprocoumon. Br J Clin Pharmacol 30:115–118, 1990

Moore LB, Goodwin B, Jones SA, et al: St. John's wort induces hepatic drug metabolism through activation of the pregnane X receptor. Proc Natl Acad Sci U S A 97:7500–7502, 2000

Murphy PA, Kern SE, Stanczyk FZ, et al: Interaction of St. John's wort with oral contraceptives: effects on the pharmacokinetics of norethindrone and ethinyl estradiol, ovarian activity and breakthrough bleeding. Contraception 71:402–408, 2005

Odlind V, Olsson SE: Enhanced metabolism of levonorgestrel during phenytoin treatment in a woman with Norplant implants. Contraception 33:257–261, 1986

Ono S, Hatanaka T, Miyazawa S, et al: Human liver microsomal diazepam metabolism using cDNA-expressed cytochrome P450s: role of CYP2B6, 2C19 and the 3A subfamily. Xenobiotica 26:1155–1166, 1996

Ouellet D, Hsu A, Qian J, et al: Effect of ritonavir on the pharmacokinetics of ethinyl oestradiol in healthy female volunteers. Br J Clin Pharmacol 46:111–116, 1998

Palovaara S, Kivisto KT, Tapanainen P, et al: Effect of an oral contraceptive preparation containing ethinylestradiol and gestodene on CYP3A4 activity as measured by midazolam 1-hydroxylation. Br J Clin Pharmacol 50:333–337, 2000

Palovaara S, Pelkonen O, Uusitalo J, et al: Inhibition of cytochrome P450 2B6 activity by hormone replacement therapy and oral contraceptive as measured by bupropion hydroxylation. Clin Pharmacol Ther 74:326–333, 2003

Patwardhan RV, Mitchell MC, Johnson RF, et al: Differential effects of oral contraceptive steroids on the metabolism of benzodiazepines. Hepatology 3:248–253, 1983

Pessayre D: Effects of macrolide antibiotics on drug metabolism in rats and in humans. Int J Clin Pharmacol Res 3:449–458, 1983

Raftogianis R, Creveling C, Weinshilboum R, et al: Estrogen metabolism by conjugation. J Natl Cancer Inst Monogr 27:113–124, 2000

Roberts RK, Grice J, McGuffie C, et al: Oral contraceptive steroids impair the elimination of theophylline. J Lab Clin Med 101:821–825, 1983

Robertson P Jr, Hellriegel ET, Arora S, et al: Effect of modafinil on the pharmacokinetics of ethinyl estradiol and triazolam in healthy volunteers. Clin Pharmacol Ther 71:46–56, 2002

Rogers SM, Back DJ, Stevenson PJ, et al: Paracetamol interaction with oral contraceptive steroids: increased plasma concentrations of ethinyloestradiol. Br J Clin Pharmacol 23:721–725, 1987

Rosenfeld WE, Doose DR, Walker SA, et al: Effect of topiramate on the pharmacokinetics of an oral contraceptive containing norethindrone and ethinyl estradiol in patients with epilepsy. Epilepsia 38:317–323, 1997

Saano V, Glue P, Banfield CR, et al: Effects of felbamate on the pharmacokinetics of a low-dose combination oral contraceptive. Clin Pharmacol Ther 58:523–531, 1995

Sabers A: Pharmacokinetic interactions between contraceptives and antiepileptic drugs. Seizure 17:141–144, 2008

Sabers A, Buckholt JM, Uldall P, et al: Lamotrigine plasma levels reduced by oral contraceptives. Epilepsy Res 47:151–154, 2001

Sarino LV, Dang KH, Dianat N, et al: Drug interactions between oral contraceptives and St. John's wort: appropriateness of advice received from community pharmacists and health food store clerks. J Am Pharmacol Assoc (Wash DC) 47:42–47, 2007

Seidegard J, Simonsson M, Edsbacker S: Effect of an oral contraceptive on the plasma levels of budesonide and prednisolone and the influence on plasma cortisol. Clin Pharmacol Ther 67:373–381, 2000

Shane-McWhorter L, Cerveny JD, MacFarlane LL, et al: Enhanced metabolism of levonorgestrel during phenobarbital treatment and resultant pregnancy. Pharmacotherapy 18:1360–1364, 1998

Sills G, Brodie M: Pharmacokinetics and drug interactions with zonisamide. Epilepsia 48:435–441, 2007

Sinofsky FE, Pasquale SA: The effect of fluconazole on circulating ethinyl estradiol levels in women taking oral contraceptives. Am J Obstet Gynecol 178:300–304, 1998

Song WC: Biochemistry and reproductive endocrinology of estrogen sulfotransferase. Ann N Y Acad Sci 948:43–50, 2001

Stoehr GP, Kroboth PD, Juhl RP, et al: Effect of oral contraceptives on triazolam, temazepam, alprazolam, and lorazepam kinetics. Clin Pharmacol Ther 36:683–690, 1984

Sustiva (package insert). Princeton, NJ, Bristol-Myers Squibb Co., 2007

Tamminga WJ, Wemer J, Oosterhuis B, et al: CYP2D6 and CYP2C19 activity in a large population of Dutch healthy volunteers: indications for oral contraceptive-related gender differences. Eur J Clin Pharmacol 55:177–184, 1999

Teal SB, Ginosar DM: Contraception for women with chronic medical conditions. Obstet Gynecol Clin North Am 34:113–126, 2007

Tornatore KM, Kanarkowski R, McCarthy TL, et al: Effect of chronic oral contraceptive steroids on theophylline disposition. Eur J Clin Pharmacol 23:129–134, 1982

Trapnell CB, Connolly M, Pentikis H, et al: Absence of effect of oral rifaximin on the pharmacokinetics of ethinyl estradiol/norgestimate in healthy females. Ann Pharmacother 41:222–228, 2007

Tsunoda SM, Harris RZ, Mroczkowski PJ, al: Preliminary evaluation of progestins as inducers of cytochrome P450 3A4 activity in postmenopausal women. J Clin Pharmacol 38:1137–1143, 1998

Use of rifampin and contraceptive steroids. Br J Fam Plann 24:169–170, 1999

van Dijke CP, Weber JC: Interaction between oral contraceptives and griseofulvin. Br Med J (Clin Res Ed) 288:1125–1126, 1984

van Giersbergen PL, Halabi A, Dingemanse J: Pharmacokinetic interaction between bosentan and the oral contraceptives norethisterone and ethinyl estradiol. Int J Clin Pharmacol Ther 44:113–118, 2006

Venkatakrishnan K, Greenblatt DJ, von Moltke LL, et al: Five distinct human cytochromes mediate amitriptyline N-demethylation in vitro: dominance of CYP 2C19 and 3A4. J Clin Pharmacol 38:112–121, 1998

Viracept (package insert). La Jolla, CA, Agouron Pharmaceuticals, Inc., 2007

Walle T, Walle UK, Cowart TD, et al: Pathway-selective sex differences in the metabolic clearance of propranolol in human subjects. Clin Pharmacol Ther 46:257–263, 1989

Walle T, Fagan TC, Walle UK, et al: Stimulatory as well as inhibitory effects of ethinyloestradiol on the metabolic clearances of propranolol in young women. Br J Clin Pharmacol 41:305–309, 1996

Weber A, Jager R, Borner A, et al: Can grapefruit juice influence ethinylestradiol bioavailability? Contraception 53:41–47, 1996

Wermeling DP, Chandler MH, Sides GD, et al: Dirithromycin increases ethinyl estradiol clearance without allowing ovulation. Obstet Gynecol 86:78–84, 1995

Yoshimoto K, Echizen H, Chiba K, et al: Identification of human CYP isoforms involved in the metabolism of propranolol enantiomers: N-desisopropylation is mediated mainly by CYP1A2. Br J Clin Pharmacol 39:421–431, 1995

Zhang Z-Y, Kaminsky LS: Characterization of human cytochromes P450 involved in theophylline 8-hydroxylation. Biochem Pharmacol 50:205–211, 1995

Zhang H, Cui D, Wang B, et al: Pharmacokinetic drug interactions involving 17alpha-ethinylestradiol: a new look at an old drug. Clin Pharmacokinet 46:133–157, 2007

13

Internal Medicine

Scott G. Williams, M.D.

Gary H. Wynn, M.D.

Reminder: This chapter is dedicated primarily to metabolic and trans-porter interactions. Interactions due to displaced protein-binding, alterations in absorption or excretion, and pharmacodynamics are not covered.

A complete discussion of drugs used in general medicine would be incredibly lengthy and, in fact, many drugs have yet to be fully studied. This chapter examines drugs that alter the metabolism of other drugs via induction or inhibition of phase I and II metabolism as well as transport mechanisms. For drugs with narrow therapeutic windows, a more detailed discussion of P450 metabolic sites is presented.

Allergy, cardiovascular, gastrointestinal, and oral hypoglycemic drugs are covered. The chapter opens with a historical review of allergy medications such as H_1 receptor blockers, leukotriene antagonists, and xanthines (the drugs where the quest for more information about the P450 system really began). Cardiovascular agents presented here include antihypertensives, antiarrhythmics, anticoagulants, and antihyperlipidemics. The gastrointestinal drugs covered consist mainly of H_2 receptor blockers and proton pump inhibitors. The oral hypoglycemic agents include thiazolinediones, sulfonylureas, and some newer agents. Antimicrobials, which are used across medical and surgical specialties and produce the greatest number of P450-mediated drug interactions, have earned their own chapter (Chapter 14, "Infectious Diseases") and are discussed only if they have relevant interactions with the aforementioned agents. Analgesics are covered in Chapters 17 ("Pain Management I: Nonnarcotic Analgesics") and 18 ("Pain Management II: Narcotic Analgesics").

Allergy Drugs

The specialty of allergy and immunology involves management of some of the world's most frequent ailments: asthma and allergic rhinitis. Antihistamines are among the most widely prescribed medications in the world (Woosley 1996). The "first-generation," or "classic" antihistamines are often obtained over the counter and are also used as sleep aids. The first antihistamine was discovered more than 70 years ago, but the drug-drug interaction profiles have only recently been uncovered. Although most of the toxic nonsedating "second-generation" antihistamines have been removed from the market or reformulated, they are listed for historical purposes. The newer "third-generation" nonsedating antihistamines have much better safety profiles and combine the benefits of the first- and second-generation drugs (Armstrong and Cozza 2003). Theophylline, while no longer favored as a first-line agent, is still used as an alternative treatment of asthma and chronic obstructive pulmonary disease. Additionally, its sensitivity to altered metabolism via inhibition by other drugs makes it a useful tool in the study of drug interactions as a probe drug. Here, a review of allergy medications past and present serves as a way to discuss how an understanding of drug-drug interactions developed (see Table 13–1).

Histamine Subtype 1 (H$_1$) Receptor Blockers

First-Generation H$_1$ Receptor Blockers

Diphenhydramine (Benadryl), an over-the-counter antihistamine used for allergic rhinitis, urticaria, and insomnia and found in multiple combination preparations, is metabolized primarily by 2D6 with minor contributions from 1A2, 2C9, and 2C19 (Akutsu et al. 2007). Medications such as ritonavir (Norvir), quinidine (Quinidex) and some tricyclic antidepressants (TCAs) can increase plasma levels via 2D6 inhibition. Diphenhydramine itself is also a potent 2D6 inhibitor. Lessard et al. (2001) studied 15 healthy men—nine extensive metabolizers and six poor metabolizers at 2D6—and found that diphenhydramine alters the disposition of venlafaxine (Effexor) via 2D6 inhibition. Hamelin et al. (2000) studied the effect of diphenhydramine on metoprolol (Lopressor). They noted that in extensive metabolizers, the 2D6 inhibition produced significant pharmacokinetic and pharmacodynamic changes. The authors of both studies warned that some drugs with narrow therapeutic windows dependent on 2D6 for metabolism (TCAs, antiarrhythmics, beta-blockers, tramadol [Ultram], and some antipsychotics) can be significantly affected by diphenhydramine administration. Although diphenhydramine *in vitro* and *in vivo* seems to be a less potent inhibitor than quinidine at 2D6 (Hamelin et al. 1998), the fact that it is sold over the counter and is found in multiple cold remedies suggests that patients should be warned about the potential for increased side effects and toxicities.

Hydroxyzine (Atarax), another classic antihistamine, is metabolized to the nonsedating antihistamine cetirizine (Zyrtec) via 2D6. Like diphenhydramine, its plasma concentration may be increased via 2D6 inhibitors. Hydroxyzine itself is a 2D6 inhibitor. A study on the pharmacokinetics of etoposide revealed that hydroxyzine may also be an ABCB1 inhibitor (Kan 2001).

Chlorpheniramine (Chlor-trimeton) is metabolized primarily via 2D6. Yasuda et al. (2002) demonstrated quinidine, a 2D6 inhibitor, alters the pharmacokinetics of healthy volunteers who are extensive metabolizers at 2D6, resulting in significantly diminished enzymatic activity at 2D6. Quinidine coadministration with chlorpheniramine can result in potentially significantly increased levels of chlorpheniramine due to this inhibition. Similar to diphenhydramine and hydroxyzine, chlorpheniramine is also an inhibitor of

TABLE 13–1.   Allergy drugs

Drug	Metabolism site(s)	Enzyme(s) inhibited	ABCB1	
			Substrate	Inhibitor
H₁ receptor antagonists				
First generation				
Brompheniramine (Dimetapp)	Unknown	None known	No	Unknown
Chlorpheniramine (Chlor-trimeton)	2D6	2D6	No	Unknown
Diphenhydramine (Benadryl)[1]	2D6, 1A2, 2C9, 2C19	2D6	No	Unknown
Hydroxyzine (Atarax)	2D6	2D6	Possible	Yes
Second generation[2]				
Astemizole (Hismanal)	3A4[1]	None known	Yes	Yes
Cetirizine (Zyrtec)	Unknown	None known	Yes	Unknown
Ebastine (Ebastel)	3A4[1]	None known	Yes	Unknown
Loratadine (Claritin, Alavert)	3A4, 2D6	None known	Yes	Possible
Terfenadine (Seldane)	3A4[1]	None known	Yes	Yes
Third generation				
Desloratadine (Clarinex)	3A4, 2D6	None known	Yes	No
Fexofenadine (Allegra)[3]	Excreted unchanged	None known	Yes	Unknown
Levocetirizine (Xyzal)	Excreted unchanged	None known	Yes	Unknown

TABLE 13–1. Allergy drugs *(continued)*

Drug	Metabolism site(s)	Enzyme(s) inhibited	ABCB1 Substrate	ABCB1 Inhibitor
Leukotriene D_4/E_4 receptor antagonists				
Montelukast (Singulair)	3A4, 2C9	2C8	No	Unknown
Zafirlukast (Accolate)	2C9, 3A4	2C9, 2C8, 3A4, 1A2	No	Unknown
Xanthines				
Theophylline (Theo-Dur)	1A2, 2E1	3A4	No	Unknown

Note. H_1 = histamine, subtype 1.
[1] May be representative of all first-generation H_1 receptor antagonists (which also includes tripelennamine and promethazine).
[2] Arrhythmogenic parent compounds no longer marketed in the United States.
[3] Active metabolite of terfenadine.

2D6, although *in vitro* studies show a relatively weak level of inhibition (Hamelin et al. 1998).

Second-Generation H_1 Receptor Blockers

Drug interactions involving second-generation H_1 receptor antagonists (or nonsedating antihistamines) were some of the first-studied severe drug interactions relating to the P450 system. Interactions involving fluoxetine (Prozac) and TCAs were noted earlier but were not associated with significant morbidity or mortality. These newer antihistamines are ABCB1 substrates and subsequently are "pumped" out of the central nervous system (CNS) prior to reaching a significant level unless administered with an ABCB1 inhibitor (see Chapter 4, "Transporters"). Second-generation antihistamines are more H_1 selective than the older drugs, and this selectivity, in combination with "regulation" via ABCB1, results in less sedation, less weight gain, and overall fewer side effects outside of drug interactions. The first report of life-threatening cardiac arrhythmia associated with the use of terfenadine (Seldane) appeared in 1989, when a patient developed an arrhythmia after an intentional overdose (Davies et al. 1989). Monahan et al. (1990) wrote the first article on the cause of an arrhythmia in conjunction with terfenadine therapy. A woman taking terfenadine and cefaclor (Ceclor) developed *Candida* vaginitis and treated herself with ketoconazole (Nizoral) that she had remaining from treatment of a previous episode of vaginitis. She developed palpitations and later torsades de pointes, even though she was taking standard doses of all medications. These case reports led to *in vivo* studies involving healthy volunteers, which clearly demonstrated that the potent 3A4 inhibitor ketoconazole increases levels of unmetabolized terfenadine and is associated with QT interval prolongation (Woosley 1996; Yap and Camm 1999).

Terfenadine is a pro-drug that usually undergoes a rapid and nearly complete first-pass hepatic biotransformation, producing the active metabolite at 3A4. Terfenadine is cardiotoxic in overdose or when its first-pass metabolism is significantly impaired by another compound. Terfenadine seems to block potassium channels and be as potent as quinidine in inhibiting the delayed rectifier potassium channel in cardiac tissue (Yap and Camm 1999), causing prolongation of the QT interval. Carboxyterfenadine (fexofenadine), terfenadine's active metabolite, is not cardiotoxic and has no known hepatic metabolism (it is eliminated unchanged in urine). The pro-drug terfenadine has been voluntarily removed by the manufacturer from the U.S. market and has been replaced by fexofenadine (Allegra).

Astemizole (Hismanal), which has also been taken off the U.S. market, and ebastine (Ebastel), which is available only in Europe, also are arrhythmogenic at high doses. This effect may be heightened when either of these drugs is administered with potent inhibitors of 3A4, including nefazodone (Serzone), cyclosporine (Neoral and others), cimetidine (Tagamet), some macrolide antibiotics, azole antifungals, antiretrovirals, some selective serotonin reuptake inhibitors (SSRIs), and grapefruit juice (Renwick 1999; Slater et al. 1999; Woosley 1996; Yap and Camm 1999) (see Table 13–1; also see Chapter 6, "3A4").

Cetirizine is the carboxylic acid metabolite of hydroxyzine and has minimal hepatic metabolism. In a study evaluating the effect of the "pan-inhibitor" ritonavir on cetirizine's pharmacokinetics, area under the plasma concentration-time curve (AUC) increased by 42%, but there were no serious drug-related adverse events (Peytavin et al. 2005). Ritonavir pharmacokinetics were not altered, and there is no known inhibition or induction of hepatic isozymes that might affect other medications. Cetirizine also appears to be safe when taken in high concentrations. During a study in which healthy volunteers were administered cetirizine at three times the recommended dose, no effect on QT interval was found.

Loratadine (Claritin) is hepatically metabolized via 3A4 and 2D6. According to loratadine's package insert, 3A4 inhibitors can increase levels of loratadine, but 2D6 takes over when this occurs, and no mention is made of QTc prolongation (Claritin 2002). Multiple *in vivo* drug interaction studies involving potent 3A4 inhibitors such as erythromycin and ketoconazole did not demonstrate statistically significant QTc prolongation (Chaikin et al. 2005; Renwick 1999; Woosley 1996; Yap and Camm 1999). Kosoglou et al. (2000) reported that in healthy volunteers, coadministration of loratadine and ketoconazole significantly increased plasma concentrations of loratadine and its major metabolite, desloratadine (Clarinex), without significantly affecting the QTc. Coadministration of cimetidine (a potent 3A4 and 2D6 inhibitor) and loratadine significantly increased plasma loratadine concentrations, but not those of desloratadine, and did not significantly alter the QTc. The investigators concluded that although a statistically significant drug interaction occurred, there were no clinically significant QTc changes in healthy adult volunteers. In contrast, Abernethy et al. (2001) studied healthy volunteers who received nefazodone (a potent 3A4 inhibitor) 300 mg every 12 hours (the manufacturer's recommended maximum dose) plus either terfenadine 60 mg twice a day, loratadine 20 mg once a day, or placebo. Terfenadine's levels increased, and the average QTc increased 42.4 milliseconds when the drug was used with nefazodone. Concomitant use of nefazodone and loratadine yielded an average increase in QTc of 21.6 milliseconds. Loratadine alone did not increase the QTc, and they pointed out that loratadine is often used at higher doses than 20 mg/day. Surprisingly, one of the authors later wrote a letter to the journal's editor questioning whether the methods of obtaining and processing the data might be inaccurate, which furthers the debate on loratadine's safety profile (Barbey 2002). One case of torsades de pointes (Atar et al. 2003) has been reported in an elderly female on chronic amiodarone therapy for atrial fibrillation recently started on loratadine. These conflicting data suggest that when loratadine and potent 3A4 and 2D6 inhibitors are used, careful monitoring is required, especially given that loratadine is available over the counter. *In vitro* studies by Wang et al. (2001) have shown that loratadine may also be an inhibitor of ABCB1, causing an increase in adenosine triphosphatase activity to above baseline levels (by inhibiting the adenosine triphosphate–binding transporters), but the clinical importance of this interaction is not fully understood.

Third-Generation H_1 Receptor Blockers

Third-generation H_1 receptor antagonists are the active metabolites or the active enantiomers of second-generation drugs. They have a much more favorable safety profile and are not subject to as many drug interactions.

Fexofenadine has been found to be without the problems that occur with its pro-drug, terfenadine. Even at 10 times the recommended dose, its safety is well established. Fexofenadine is not metabolized, meaning that more than 95% of a dose can be found in urine and feces. Long-term postmarketing surveillance further indicates its safety, as did an observational cohort study (Craig-McFeely et al. 2001). Uptake transporters such as the intestinal OATP1A2 (Glaeser et al. 2007) and efflux transporters such as ABCB1 are likely involved in fexofenadine's "regulation," and study is ongoing to determine if there are any clinically relevant interactions. Fexofenadine has been used as a probe drug for ABCB1 study and is affected by both ABCB1 inhibitors and inducers such as St. John's wort (Wang et al. 2002) and rifampin (Hamman et al. 2001) (see Chapter 4, "Transporters").

Levocetirizine (Xyzal) is the eutomer of cetirizine. Eutomers are the active or more active enantiomers of a racemic compound. Levocetirizine seems to have a smaller volume of distribution than cetirizine, providing for better safety and efficacy. To date, there are no reports of cardiotoxicity or drug interactions associated with levocetirizine therapy (Baltes et al. 2001; Walsh 2006).

Desloratadine, the orally active major metabolite of loratadine, is metabolized via 3A4 and 2D6. Desloratadine is 15 times more potent than loratadine at the H_1 receptor and seems to have a more rapid onset of action (McClellan and Jarvis 2001). There are reportedly no adverse cardiac effects in healthy volunteers, even at 10 times the recommended dose. Like loratadine, desloratadine neither inhibits nor induces metabolism of other medications via the P450 system, and no clinically significant P450-mediated drug interactions have been reported to date (Banfield et al. 2002; Gupta et al. 2004). Desloratadine is a substrate of ABCB1 but does not appear to induce or inhibit the ABCB1 system (Geha and Meltzer 2001; Wang et al. 2001).

Leukotriene D_4/E_4 Receptor Antagonists

Zafirlukast (Accolate) is a leukotriene D_4/E_4 receptor antagonist used in prophylactic and chronic treatment of asthma. Zafirlukast antagonizes the contractile activity of leukotrienes, suppressing airway responses to antigens such as pollen and cat dander by inhibiting bronchoconstriction. Zafirlukast is metabolized by 2C9 with contributions from 3A4 and is a moderate inhibitor of 2C9, 2C8, and 3A4. The inhibition of 3A4 has been postulated to be responsible for idiosyncratic hepatotoxicity (Kassahun et al. 2005). In one case, zafirlukast's activity at 1A2 was implicated in increasing serum theophylline (Uniphyl) levels. Zafirlukast has also been reported to increase the half-life of warfarin (Coumadin) (Katial et al. 1998).

Montelukast (Singulair), another leukotriene D_4/E_4 receptor antagonist, is metabolized at 3A4 and 2C9. *In vitro* data have shown it is a potent inhibitor of 2C8 and suggest it has the potential for clinically relevant drug interactions with 2C8 substrates such as rosiglitazone, repaglinide, paclitaxel, and cerivastatin (Walsky et al. 2005). These *in vitro* data have not been confirmed with *in vivo* data, however, and multiple studies have shown no significant pharmacokinetic or pharmacodynamic interactions with repaglinide (Kajosaari et al. 2006), warfarin (Van Hecken et al. 1999), or digoxin (Depre et al. 1999).

Xanthines

Aminophylline and its active metabolite theophylline are thought to cause bronchodilation, or smooth muscle relaxation, and suppression of airway response to antigens or irritants in asthmatic patients. Clinical efficacy is achieved at serum levels of 5–20 µg/mL. Serum concentrations greater than 20 µg/mL produce adverse reactions including nausea, vomiting, headache, tremor, seizures, cardiac arrhythmias, and death. With such a narrow therapeutic window and such dangerous toxic adverse reactions, this hepatically metabolized compound is very susceptible to drug interactions.

Theophylline is known to be metabolized at 1A2, with some activity at 2E1. Inhibitors of 1A2 such as fluvoxamine (Luvox) and fluoroquinolone antibiotics are well known to increase serum theophylline levels (Batty et al. 1995; Rasmussen et al. 1995; Upton 1991). Despite many *in vitro* and *in vivo* studies, and warnings in package inserts, cases of theophylline toxicity are still reported. Andrews (1998) reported a case of theophylline toxicity in which the addition of ciprofloxacin (Cipro) to an asthmatic patient's regimen to treat productive cough resulted in hospitalization and hemodialysis. DeVane et al. (1997) reported that a patient at a residential care facility who was being treated with fluvoxamine for depression with psychotic features was prescribed theophylline for chronic obstructive pulmonary disease by her primary care physician. The woman experienced confusion, lack of energy, reduced sleep, and nausea and vomiting, all leading to an acute hospital admission. The half-life of theophylline in this case was increased 5-fold. Serum levels of theophylline are decreased by compounds that induce 1A2, such as omeprazole (Prilosec), rifampin (Rifadin), carbamazepine (Tegretol), phenytoin (Dilantin), barbiturates, caffeine, cigarette smoke (Zevin and Benowitz 1999), and even some foods (see Chapter 7, "1A2"). Some metabolism of theophylline occurs at 2E1 (Rasmussen et al. 1997), so the potential for drug interactions exists here as well. Because theophylline metabolism is so sensitive to inhibition and induction and theophylline levels are easily measured, the drug is used as a probe in determining whether other compounds are metabolized at 1A2. Theophylline is also a moderate inhibitor of 3A4. Boubenider et al. (2000) reported an interaction between theophylline and the 3A4 substrate tacrolimus (Prograf) in a renal transplant patient.

Summary: Allergy Drugs

First-generation antihistamines such as diphenhydramine are inhibitors of 2D6 and may alter the metabolism of drugs dependent on 2D6 for metabolism, such as venlafaxine, TCAs, some antipsychotics, beta-blockers, antiarrhythmics, and tramadol. The second-generation antihistamines have a known potential for cardiotoxicity and include terfenadine, astemizole, and ebastine. If used in overdose or when administered with other compounds that inhibit 3A4, these drugs could lead to palpitations, syncope, or fatal arrhythmias. Loratadine in combination with potent 3A4 inhibitors may increase the QTc interval, but more studies are needed before conclusions can be drawn. Third-generation antihistamines appear to be safe and effective, with minimal drug interactions. Zafirlukast inhibits several enzymes *in vitro*, and there are case reports of interactions involving warfarin and theophylline via 1A2 inhibition. Theophylline is metabolized almost exclusively at 1A2 and has a narrow therapeutic index. Therefore, whenever possible, use of potent inhibitors of 1A2 (which include the fluoroquinolones and the SSRI fluvoxamine) should be avoided. If it is necessary to administer a fluoroquinolone, the least-potent drug of that class must be chosen, and serum theophylline levels must be monitored. As a rule to thumb, some clinicians suggest decreasing the theophylline daily dose by two-thirds.

Cardiovascular Agents

The body of evidence supporting aggressive control of cardiovascular risk factors such as blood pressure and cholesterol has resulted in polypharmacy becoming the norm. Beyond cardiovascular risk factors, other comorbidities such as depression (with incidences as high as 31% in the first year post myocardial infarction) add to the tendency toward polypharmacy. Many cardiovascular medications are substrates, inducers, or inhibitors of the cytochrome P450 (CYP) system, leaving significant potential for drug interactions. The following is a review of the known and potential interactions among many commonly prescribed cardiovascular agents, with specific emphasis on antihypertensive agents, antiarrhythmics, anticoagulants, and cholesterol-lowering agents.

Antiarrhythmics

The antiarrhythmics are a potent and diverse group of medications used for a variety of cardiovascular disorders. The interactions between psychotropic medications and antiarrhythmics can be some of the most clinically important, owing to narrow therapeutic indices and potentially life-threatening side effects. Caution should be used when coadministering agents from the classes detailed below, especially if there is concern for additive QTc prolongation (see Table 13–2).

Quinidine (Quinidex)

Quinidine is a class IA antiarrhythmic drug used in atrial fibrillation and ventricular arrhythmias and is associated with multiple drug interactions of significance. Quinidine is metabolized primarily via 3A4. 3A4 inhibitors such as paroxetine, nefazodone, diltiazem, ketoconazole, clarithromycin, erythromycin, ritonavir, and grapefruit juice may increase quinidine levels, resulting in toxicity. Quinidine is a potent inhibitor of 2D6, with full inhibition of the enzyme occurring when just one-sixth of the usual antiarrhythmic dose is given. This can lead to elevated plasma concentrations of 2D6 substrates such as digoxin, codeine, fluvoxamine, paroxetine, venlafaxine, TCAs, and haloperidol (Brown et al. 2006; Damkier et al. 1999; Young et al. 1993). In particular, secondary amine TCAs are strongly contraindicated with quinidine. Quinidine also exerts potent ABCB1 effects, and studies are being conducted *in vivo* to assess the importance of this finding (Weiss et al. 2003). For an excellent review of this drug and a full table with references to pertinent literature, see Grace and Camm (1998).

Lidocaine (Xylocaine)

Lidocaine is a class 1B antiarrhythmic, used for ventricular arrhythmias, metabolized primarily by 1A2, with minor contributions from 2B6, 3A4, and 3A5. Medications that inhibit 1A2, such as fluvoxamine (Orlando et al. 2004) and propofol (Inomata et al. 2003), have been shown to increase lidocaine plasma levels. Mexiletine has also been shown to increase plasma lidocaine

TABLE 13–2. Antiarrhythmics

Drug	Metabolism site(s)	Enzyme(s) inhibited[a]	Enzyme(s) induced
Amiodarone (Cordarone)	2C8, 3A4	1A2, 2C9, 2D6, 3A4	None known
Digoxin (Lanoxin)	Unknown	None known	None known
Dofetilide (Tikosyn)[1]	3A4	None known	None known
Encainide (Enkaid)	2D6	None known	None known
Flecainide (Tambocor)[1]	2D6, 1A2	2D6	None known
Ibutilide (Corvert)	Unknown	None known	None known
Lidocaine (Xylocaine)	1A2, 2B6, 3A4	1A2[c]	None known
Mexiletine (Mexitil)[1]	2D6, 1A2	1A2	None known
Moricizine (Ethmozine)[2]	3A4	None known	1A2, 3A4[3]
Propafenone (Rythmol)	2D6, 1A2, 3A4	1A2, 2D6	None known
Quinidine (Quinidex)	3A4	**2D6**	None known
Tocainide (Tonocard)	None known	1A2[c]	None known

[1]Also significant renal metabolism and elimination.
[2]Induces antipyrine metabolism and its own metabolism.
[3]Possible induction.

[a]**Bold** type indicates potent inhibition.
[b]Moderate inhibition or induction.
[c]Mild inhibition.

concentration, but the mechanism is still being investigated (Maeda et al. 2002). Fluoroquinolones such as ciprofloxacin may also inhibit lidocaine metabolism via 1A2, but there are no clinical data to support this at present. Amiodarone inhibits lidocaine metabolism via actions at 3A4 (Ha et al. 1996). Lidocaine itself inhibits 1A2, but it appears to be a less potent inhibitor than mexiletine (Wei et al. 1999). Lidocaine is also an inhibitor of the ABCB1 transporter.

Mexiletine (Mexitil)

Mexiletine is a class IB antiarrhythmic agent with a narrow therapeutic index used for controlling ventricular arrhythmias. Mexiletine undergoes metabolism via 2D6 with some metabolism through 1A2. Kusumoto et al. (2001) found that concomitant administration of mexiletine with fluvoxamine, a 1A2 and 2D6 inhibitor, resulted in significantly increased mexiletine peak concentrations. Additionally, *in vitro* data suggest possible interactions when mexiletine is coadministered with paroxetine, fluoxetine, desipramine, and thioridazine, but clinical data are lacking (Hara et al. 2005). Mexiletine is known to inhibit 1A2, potentially increasing serum levels of 1A2 substrates such as theophylline, caffeine, and warfarin (Konishi et al. 1999; Wei et al. 1999). Mexiletine is not known to induce any hepatic enzymes.

Tocainide (Tonocard)

Tocainide is a class 1B antiarrhythmic used primarily for control of ventricular arrhythmias. Although tocainide is metabolized via an incompletely understood mechanism, hepatic oxidation is not thought to play a significant role. Glucuronidation is thought to play a role, but again, the magnitude is yet to be determined. There are no significant drug interactions that increase or decrease serum tocainide levels. Tocainide has a modest inhibitory effect on 1A2, but the magnitude of the effect was shown to be much smaller than that of mexilitine (Loi et al. 1993).

Encainide (Enkaid)

Encainide is a class 1C antiarrhythmic, used for supraventricular and ventricular tachyarrhythmias, metabolized primarily by 2D6. There are putative interactions with 2D6 inhibitors such as quinidine, ritonavir, metoclopramide, and amiodarone, but case reports and *in vivo* literature are sparse. Encainide is not known to inhibit or induce hepatic metabolism. As with all antiarrhythmics, caution is advised when co-prescribing any medication that can prolong the QT interval.

Flecainide (Tambocor)

Flecainide, a class IC antiarrhythmic used for a variety of supraventricular and ventricular arrhythmias, undergoes oxidative metabolism via 2D6 with some minor activity from 1A2. Inhibition of 2D6 by medications such as paroxetine, fluoxetine, sertraline, and citalopram may result in elevated flecainide levels (Ereshefsky et al. 1995). Flecainide may inhibit 2D6 (Triola and Kowey 2006), although the significance of this is unknown. Flecainide does not induce any P450 enzymes.

Propafenone (Rythmol)

Propafenone is a class IC antiarrhythmic drug metabolized by 2D6 with minor contributions from 1A2 and 3A4. As with flecainide, SSRIs and other medications that inhibit 2D6 may lead to elevated propafenone levels (Trujillo and Nolan 2000). Propafenone is also an inhibitor of 1A2 and 2D6 and may significantly alter the pharmacokinetics of mexiletine, leading to reduced clearance (Labbe et al. 2000). Katz (1991) reported a case of desipramine toxicity after the addition of propafenone, likely resulting from 2D6 inhibition. Propafenone is also an ABCB1 inhibitor.

Moricizine (Ethmozine)

Moricizine is a class I antiarrthymic metabolized by 3A4. In a study by Shum et al. (1996), diltiazem (Tiazac) increased moricizine plasma concentration and AUC by inhibition of 3A4. Oral clearance was reduced by 54%. In the same study, moricizine was shown to be a 3A4 inducer and decreased the plasma concentration of diltiazem. Via 3A4 and 1A2 induction, moricizine also induces the metabolism of theophylline and the probe antipyrine (Benedek et al. 1994; Pieniaszek et al. 1993).

Amiodarone (Cordarone, Pacerone)

Amiodarone is a class III antiarrhythmic, used for a variety of arrhythmias, metabolized extensively by 2C8 and 3A4. Grapefruit juice and other 3A4 inhibitors can raise amiodarone serum levels. Amiodarone inhibits 1A2, 2C9,

2D6, and 3A4, and there are many reported drug interactions in the literature. Schmidt et al. (2007) reported severe rhabdomyolysis when simvastatin (Zocor) was administered with amiodarone and atazanavir (Reyataz), both 3A4 inhibitors. Other groups have reported similar interactions with clarithromycin. Some important pharmacokinetic drug interactions include metronidazole, clonazepam, and lidocaine (via 3A4 inhibition), digoxin and sirolimus (via 3A4 and ABCB1), metoprolol and flecainide (via 2D6 inhibition), warfarin (via 2C9 inhibition), phenytoin (via 2C9 and 2C19 inhibition), and theophylline (via 1A2 inhibition). Trazodone is contraindicated with amiodarone because there are case reports of polymorphic ventricular tachycardia and torsades de pointes secondary to QT prolongation from the coadministration of these two medications, although the mechanism is not yet fully understood (Antonelli et al. 2005; Mazur 1995). The same may be true for tricyclic antidepressants, so caution should be exercised if these agents are used together. For a more complete review of amiodarone drug interactions, please see the review article by Trujillo and Nolan (2000).

Dofetilide (Tikosyn)

Dofetilide is a class III antiarrhythmic agent that is approved for conversion and maintenance of sinus rhythm in patients with atrial fibrillation and flutter. Although dofetilide is partly metabolized via 3A4, no significant interactions have been reported. Dofetilide is excreted mainly unchanged in the urine via the cationic transport system.

Ibutilide (Corvert)

Ibutilide is a class III antiarrhythmic commonly used to convert atrial fibrillation to normal sinus rhythm. The metabolic profile is not completely understood, but there is no evidence of significant CYP450 or other interaction.

Digoxin (Lanoxin)

Digoxin is a class IV antiarrhythmic agent with a narrow therapeutic window, used for increasing ionotropy in congestive heart failure as well as for control of supraventricular arrhythmias. The metabolism of digoxin is not well understood, but only a small fraction occurs via hepatic oxidation. Known interactions with benzodiazepines occur through mechanisms that are also not well understood. Although an interaction has been reported with alprazolam, other benzodiazepines may also produce changes in the serum concentration of digoxin (Castillo-Ferrando et al. 1983). Insofar as digoxin is a substrate of the ABCB1 transporter, plasma concentrations may be increased when fluoxetine, paroxetine, or nefazodone are added. These increases are likely to be attributable to inhibition of ABCB1 by these drugs, thus increasing intestinal absorption of digoxin (Dockens et al. 1996). Increases in digoxin concentration of up to 30% have been seen with nefazodone and paroxetine, whereas fluoxetine has shown as much as a 3-fold increase in digoxin levels (Leibovitz et al. 1998).

Summary: Antiarrhythmics

The effectiveness of antiarrhythmics and the morbidity and mortality associated with their administration vary across patient groups. Very narrow therapeutic windows also make drug interactions all the more difficult to predict. Buchert and Woosley (1992) and Trujillo and Nolan (2000) carefully reviewed patient variability with regard to response to this group of drugs, as well as the genetically determined differences in metabolism. Many antiarrhythmics are metabolized via 2D6, and its genetic polymorphisms play a role in many interactions involving this class of drugs. Amiodarone and quinidine have a multitude of interactions via P-450 inhibition. Mexiletine and propafenone are potent inhibitors of 1A2, and caution should be used when these drugs are coadministered with 1A2-dependent drugs such as theophylline, tacrine, caffeine, and the newer atypical antipsychotics. Caution is advised when prescribing any medication that can prolong the QT interval, such as TCAs, erythromycin, and some antipsychotics.

Anticoagulants

With an aging population, the increasing risk of myocardial infarction and stroke has led to an increasing number of patients receiving anticoagulant therapy. The most commonly prescribed anticoagulant is warfarin, but many newer agents are being used with increasing frequency for a variety of disease processes, including claudication, prevention of in-stent restenosis following percutaneous interventions, and the treatment of heparin-induced thrombocytopenia (see Table 13–3).

TABLE 13–3. Anticoagulants

Drug	Metabolism site(s)	Enzyme(s) inhibited[a]	Enzyme(s) induced
Argatroban	3A4	None known	None known
Cilostazol (Pletal)	3A4, 2C19	None known	None known
Clopidogrel (Plavix)	3A4, 1A2, 2C9, 2C19	2B6, 2C19[b], 2C9[c]	None known
Ticlopidine (Ticlid)	3A4?	2B6, **2C19**, 2D6[b] 1A2[c], 2C9[c]	None known
Warfarin (Coumadin)	2C9, 1A2, 2C8, 2C18, 2C19, 3A4	None known	None known

[a]**Bold** type indicates potent inhibition.
[b]Moderate inhibition.
[c]Mild inhibition.

Warfarin (Coumadin)

Warfarin has a narrow therapeutic index and a greater than 20-fold variation in maintenance dose, requiring diligent monitoring of international normalized ratios (INRs). In addition to age, gender, vitamin K intake, and co-administration of other medicines, genetic factors are a major contributor to dose variation and patient hazards (Higashi et al. 2002). DNA sequence polymorphisms in 2C9, which metabolizes the active S form, are found in 35% of patients and cause a doubling or tripling of the warfarin serum half-life (Linder et al. 2002). The considerably less active R form is metabolized by multiple enzymes, including 1A2, 2C8, 2C18, 2C19, and 3A4. A combination of genetic testing for CYP2C9 polymorphism and the vitamin K receptor (VKORC1), warfarin's site of action (Rieder et al. 2005), may allow a clinical provider a more accurate means of estimating dosing requirements (Sconce et al. 2005).

The *Physicians' Desk Reference* (2002) lists three pages of possible interactions with warfarin. Of these, a variety of psychotropics, including fluvoxamine, fluoxetine, and sertraline, may lead to increased warfarin levels via inhibition of 2C9. However, clinically important interactions have been shown only with fluoxetine and fluvoxamine (Sayal et al. 2000). Fluvoxamine has the potential to reduce activity of 1A2, 2C9, and 3A4. This interaction has led to sustained, elevated inhibition of the clotting cascade (Limke et al. 2002). Coadministration of warfarin with 3A4 inducers such as carbamazepine should be avoided because the interaction could lead to decreased efficacy with increased coagulation (Kendall and Boivin 1981). Paroxetine has been shown to increase bleeding with warfarin via an undetermined mechanism. In one study, 5 of 27 healthy volunteers developed clinically significant bleeding without associated changes in their prothrombin times or their INRs (Bannister et al. 1989).

Antiplatelet Drugs

The thienopyridines ticlopidine (Ticlid) and clopidogrel (Plavix) are inhibitors of platelet function. These agents reduce the incidence of atherosclerotic events in patients with histories of myocardial infarction, peripheral vascular disease, or stroke.

The metabolism of ticlopidine is incompletely understood but is thought to be related to 3A4. Ticlopidine is an inhibitor of many isozymes, including (in descending order of potency) 2B6, 2C19, 2D6, 1A2, and 2C9 (Ha-Duong et al. 2001; Ko et al. 2000; Richter et al. 2004). Donahue et al. (1999) reported on ticlopidine's effects on phenytoin clearance via 2C19. Brown and Cooper (1997) reported carbamazepine toxicity in an epileptic patient, but a P450 interaction was not established. The relative clinical importance of the inhibition of these isozymes is still being studied, and *in vivo* data are lacking.

Clopidogrel is metabolized via 3A4, 2C9, 2C19, and 1A2 with some ABCB1 activity. Clopidogrel is a potent inhibitor of 2B6, also inhibiting of 2C19 and 2C9. No reports were found in the literature of *in vivo* randomized, double-blind pharmacokinetic studies of clopidogrel, but *in vitro* studies did show an increase in AUC of bupropion (Wellbutrin) with both clopidogrel and ticlopidine (Turpienen et al. 2005). In theory, clopidogrel could also increase serum concentrations, and hence side-effect profiles, of phenytoin, tamoxifen, tolbutamide, warfarin, fluvastatin, zafirlukast, oral hypoglycemics, and many nonsteroidal anti-inflammatory drugs (Plavix 2007).

Cilostazol, an antiplatelet agent metabolized by 3A4 and 2C19, does not appear to have any significant interactions via hepatic metabolism.

Other Anticoagulant Agents

Currently, no available published data suggest a potential for pharmacokinetic interactions between heparin and other medications. Newer drugs such as argatroban, a direct thrombin inhibitor, also have not shown any potential for P450-mediated interactions, despite metabolism via 3A4.

Beta-Blockers

Beta-blockers are some of the most widely prescribed antihypertensives and are the standard of care in patients with congestive heart failure. The metabolism of these medications varies throughout the class, but only those that have significant CYP450 system interactions will be discussed in detail here (see Table 13–4).

Carvedilol (Coreg)

Carvedilol, a nonselective beta-blocker indicated for the treatment of mild to moderate congestive heart failure, has also been used extensively to reduce mortality following myocardial infarction (Kopecky 2006). Carvedilol is primarily metabolized via 2D6, with minor contributions from 1A2, 2C9,

TABLE 13–4. Beta-blockers

Drug	Metabolism site(s)	Enzyme(s) inhibited[a]	Enzyme(s) induced[a]
Carvedilol (Coreg)	2D6, 1A2, 2C9, 2C19, 2E1, 3A4	None known	None known
Metoprolol (Lopressor)	2D6	None known	None known
Propranolol (Inderal)	2D6, 1A2, 2C19	2D6[b]	None known

[a]**Bold** type indicates potent inhibition or induction.
[b]Moderate inhibition.
[c]Mild inhibition or induction.

2C19, 2E1, and 3A4, but it does not appear to create any inhibition or induction of metabolic enzymes. Fluoxetine, a potent 2D6 inhibitor, has been shown to increase plasma concentrations of carvedilol *in vivo* by 77%, although this increase appears to be a stereospecific isomer effect, with inactive *R*-carvedilol being the affected enantiomer. Clinically, this effect did not result in significant blood pressure or heart rate changes (Strain et al. 2002). More worrisome, given carvedilol's multienzyme metabolism, would be an interaction with a "pan-inhibitor" such as ritonavir, likely resulting in significantly elevated levels of carvedilol.

Propranolol (Inderal)

Propranolol is a nonselective beta-blocker used for the treatment of a variety of disorders, including hypertension, essential tremor, migraines, performance anxiety, and neuroleptic-induced akithesia. Propranolol is metabolized via 2D6, with minor contribution of 1A2 and 2C19. Fluvoxamine, a potent inhibitor of 1A2 and 2C19 and a mild inhibitor of 2D6, can increase plasma levels of propranolol, since all metabolic routes are inhibited (van Harten 1995). This interaction could potentiate the effect of propranolol, resulting in hypotension or other deleterious side effects. Quinidine has also been shown to potentiate the effects of propranolol, likely through a similar mechanism. There are also potentially serious interactions between propranolol and phenothiazines such as thioridazine (Mellaril), chlorpromazine (Thorazine), and the antiemetic promethazine (Phenergan) (Markowitz et al. 1995). The phenothiazine class is primarily metabolized at 2D6, with potent autoinhibition of 2D6. Concomitant use of a phenothiazine and propranolol can result in elevated levels of both medications due to enzymatic inhibition. Elevated levels of a phenothiazine and propranolol may cause hypotension, prolong the QT interval, or cause other cardiac arrhythmias. While such a regimen is not contraindicated, caution is advised when coadministering these medications (Nakamura et al. 1996; Shand and Oates 1971). Propranolol and its 4-hydroxy propranolol metabolite are also moderate inhibitors of 2D6. This inhibition may lead to increased levels of psychotropic medications dependent on 2D6, such as the aforementioned chlorpromazine and promethazine, though clinical data to assess such an interaction are scarce.

Metoprolol (Lopressor, Toprol XL)

Metoprolol is a cardioselective beta-blocker shown to reduce mortality in patients with heart disease (Hjalmarson et al. 2000). Metoprolol is metabolized primarily via 2D6. Potent inhibitors of 2D6 such as quinidine and ritonavir may greatly reduce metoprolol's metabolism and enhance its effects. Hemeryck et al. (2000) studied concomitant metoprolol and paroxetine administration in healthy volunteers and found that with repeated dosing of paroxetine, a potent 2D6 inhibitor, plasma levels of metoprolol were significantly elevated and time to elimination prolonged. Metoprolol's active (*S*)-enantiomer was found to accumulate in these subjects, accumulation can lead to prolonged bradycardia, exercise intolerance, and an increase in side effects such as fatigue, headache, and gastrointestinal symptoms. Konig et al. (1996) reported an interaction between metoprolol and paroxetine leading to increased beta blockade and bradycardia. There is also a case report of severe bradycardia with concomitant bupropion and metoprolol administration thought to be secondary to bupropion's 2D6 inhibition (McCollum et al. 2004). Although there are some *in vitro* data suggestive of an interaction between metoprolol and fluoxetine or fluvoxamine, there are few data to support any significant *in vivo* effect. Metoprolol is not known to inhibit or induce any P450 enzymes.

Summary: Beta-Blockers

Beta-blockers are very common antihypertensives metabolized via 2D6. 2D6 inhibitors such as the antidepressants fluoxetine, paroxetine, fluvoxamine, and bupropion can increase plasma concentrations, leading to significant pharmacodynamic effects. Other 2D6 inhibitors such as quinidine and ritonavir may also interact with beta-blockers. Caution is advised when prescribing phenothiazines and beta-blockers concurrently.

Calcium Channel Blockers

Calcium channel blockers (CCBs) act through voltage-gated L calcium channels via the alpha-1 subunit. Dihydropyridine CCBs, including nifedipine, antagonize voltage-dependent interactions, whereas diltiazem and verapamil antagonize frequency-dependent interactions. Calcium channel blockade is not uniformly seen across tissue types, and certain CCBs have differential affinities for various tissues. In addition to its antihypertensive effects, calcium channel modulation is important in seizure prevention, mood stabilization, and pain control (Levy and Janicak 2000). A majority of CCBs are metabolized primarily through 3A4, making them sensitive to potent 3A4 inhibitors like clarithromycin as well as 3A4 inducers like carbamazepine and rifampin, although data to demonstrate such interactions are not available for all CCBs (DeVane and Markowitz 2000). Only CCBs with known clinical interactions will be discussed (see Table 13–5).

Nifedipine (Adalat, Procardia)

Nifedipine is a first-generation dihydropyridine used for hypertension and the prevention of anginal symptoms. Primarily metabolized by 3A4, nifedipine is vulnerable to drug interactions with potent 3A4 inhibitors such as erythromycin, clarithromycin, ketoconazole, ritonavir, nefazodone, and grapefruit juice. Several case reports detailing these effects can be found in the literature. For instance, Azaz Livshits and Danenberg (1997) discuss an elderly woman taking nifedipine for hypertension who was started on fluoxetine for symptoms of fatigue, depression, and weakness. This combination led to significant worsening of weakness, orthostatic hypotension, and tachycardia, resulting in hospitalization. The authors concluded that fluoxetine's inhibition of 3A4 led to her condition. Nifedipine is also susceptible to interactions with 3A4 inducers such as rifampin (Holtbecker et al. 1996) (see "Antimycobacterials and Antitubercular Agents" in Chapter 14, "Infectious Diseases").

Nisoldipine (Sular)

Nisoldipine is also metabolized via 3A4, with some contribution from 2D6. Ketoconazole, a potent 3A4 inhibitor, increased the AUC and maximal drug concentration of nisoldipine and its active metabolite in healthy volunteers in a randomized crossover trial (Heinig et al. 1999).

Diltiazem (Cardizem, Tiazac)

Diltiazem, a benzothiazepine derivative, is a non-dihydropyridine CCB used for hypertension and atrial fibrillation. Diltiazem is primarily metabolized via oxidation at 3A4 with minor contributions from 2D6 and 2E1. Diltiazem is also a potent inhibitor of 3A4, leading to interactions with cisapride (Propulsid) (Thomas et al. 1998), lovastatin, pravastatin (Azie et al. 1998), and

TABLE 13–5. Calcium channel blockers

Drug	Metabolism site(s)	Enzyme(s) inhibited[a]	Enzyme(s) induced
Amlodipine (Norvasc)	3A4	None known	None known
Diltiazem (Cardizem, Tiazac)	3A4, 2D6, 2E1	3A4	None known
Felodipine (Plendil)	3A4	None known	None known
Isradipine (DynaCirc)	None known	None known	None known
Nicardipine (Cardene)	3A4	None known	None known
Nifedipine (Adalat, Procardia)	3A4	None known	None known
Nimodipine (Nimotop)	3A4	None known	None known
Nisoldipine (Sular)	3A4, 2D6	None known	None known
Verapamil (Calan, Covera, Verelan)	3A4, ABCB1	3A4	None known

[a]**Bold** type indicates potent inhibition.
[b]Moderate inhibition or induction.
[c]Mild inhibition or induction.

simvastatin (Mousa et al. 2000). The combination of diltiazem with 3A4-dependent triazolobenzodiazepines such as midazolam, alprazolam, or triazolam may prolong the half-life of these sedatives. Ahonen et al. (1996) studied the combination of midazolam and diltiazem in coronary artery bypass patients and found the measured half-life of midazolam was 43% longer when combined with diltiazem, leading to a significantly longer time to extubation. Diltiazem, and to a lesser extent verapamil, have been shown to increase the AUC of buspirone and potentiate its side effects in a randomized, placebo-controlled, three-phase crossover study of nine healthy volunteers (Lamberg et al. 1998). This interaction was thought to be due to the modulation of first-pass metabolism via CYP 3A4 inhibition.

Nimodipine (Nimotop)

Nimodipine is a dihydropyridine CCB used in the treatment of subarachnoid hemorrhage. Nimodipine is primarily metabolized by 3A4, and thus affected by medications that induce or inhibit this enzyme. Tartara et al. (1991) investigated the effects of several anticonvulsant medications with coadministration of nimodipine. 3A4 inducers carbamazepine and phenytoin decreased the AUC of nimodipine by 7-fold. Valproate, a 3A4 inhibitor, was found to increase the AUC of nimodipine by about 50%. Nimodipine must be prescribed with caution with a 3A4 inhibitor or inducer. Nimodipine is not known to inhibit or induce any enzymes.

Verapamil (Calan, Covera, Verelan)

Verapamil is a phenylalkylamine CCB metabolized exclusively by 3A4. There are no significant 3A4 inhibitors that have been shown to increase plasma verapamil concentration, but carbamazepine can act as a 3A4 inducer resulting in decreased plasma concentrations of verapamil. In addition, verapamil inhibits 3A4, which can increase carbamazepine levels, resulting in potential toxicity (MacPhee et al. 1986). 3A4 inhibition can also enhance the effect of 3A4-dependent triazolobenzodiazepines such as alprazolam and triazolam, leading to increased plasma levels and oversedation (Kosuge et al. 1997). Grapefruit juice causes an increase in the AUC and peak plasma concentration of verapamil; however, intravenous administration of CCBs in combination with grapefruit juice has no such effect. Compared with 3A4 inhibition by a potent inhibitor such as erythromycin, grapefruit juice does not prolong the half-lives of CCBs, indicating there is no hepatic interaction and the interaction is likely mediated in the gut wall. Johnson et al. (2001) showed that ABCB1 efflux significantly affected the overall gut metabolism of verapamil and greatly affected levels of medication available.

Summary: Calcium Channel Blockers

CCBs have potentially noxious side effects that may be dose-limiting. Inhibition of CCB metabolism by 3A4 inhibitors, including grapefruit juice, can lead to orthostasis, hypotension, and tachycardia. Diltiazem and verapamil are also inhibitors of 3A4 metabolism and cause increases in side effects of triazolobenzodiazepines, buspirone, alfentanil, and potentially any other 3A4 substrate.

Angiotensin-Converting Enzyme (ACE) Inhibitors

Angiotensin-converting enzyme (ACE) inhibitors are named for their interference of the conversion of angiotensin I to angiotensin II. Their metabolism is not fully understood, but the CYP450 system is thought to play an insignificant role. Therefore, CYP450 drug interactions are not common (see Table 13–6).

Captopril (Capoten)

White (1986) published a case report of postural hypotension and syncope in a patient receiving chlorpromazine and captopril. The mechanism of action remains unclear, and this may have been an idiosyncratic reaction, but the proposed synergistic effects led to greater than 80 mm Hg drop in supine systolic arterial blood pressure when compared with chlorpromazine alone, with exaggerated postural hypotension. There is a theoretical risk of this interaction with other ACE inhibitors, but few supporting published data.

Enalapril (Vasotec)

In another case report, by Aronowitz et al. (1994), detailing the additive hypotensive effects of ACE inhibitors and antipsychotic medications, patients receiving clozapine with enalapril experienced orthostatic symptoms and syncope. ACE inhibitors have been shown to have interactions with tricyclic antidepressants as well. Despite an incomplete knowledge of the metabolism of enalapril, it has been shown that the addition of clomipramine to enalapril at steady state can result in clomipramine toxicity (Strain et al. 1999).

TABLE 13–6.   ACE inhibitors and angiotensin receptor blockers

Drug	Metabolism site(s)	Enzyme(s) inhibited	Enzyme(s) induced
ACE inhibitors			
Captopril (Capoten)	Unknown	None known	None known
Enalapril (Vasotec)	Unknown	None known	None known
Lisinopril (Zestril)	Unknown	None known	None known
Angiotensin receptor blockers			
Irbesartan (Avapro)	2C9	None known	None known
Losartan (Cozaar)	2C9	None known	None known
Valsartan (Diovan)	2C9	None known	None known

Note.   ACE = angiotensin converting enzyme.

Angiotensin Receptor Blockers

Angiotensin receptor blockers (ARBs) are some of the newest antihypertensives on the market today. They include candesartan (Atacand), irbesartan (Avapro), olmesartan (Benicar), losartan (Cozaar), and valsartan (Diovan). ARBs are used frequently in patients with chronic kidney disease and heart failure, especially when ACE inhibitors are not tolerated because of side effects such as chronic cough. Compared with other classes of agents, ARBs have a low potential for drug interactions. Within this class, losartan and irbesartan have the greatest affinity for the P450 system; along with valsartan, they undergo metabolism via 2C9. Although there is the potential for psychotropic drug interactions (fluoxetine's inhibition of 2C9; carbamazepine as an inducer of 2C9), to date there have been no studies or reports to indicate a clinically significant metabolic interaction would limit the use of ARBs in conjunction with psychotropics (Unger and Kaschina 2003). ARBs are not known to inhibit or induce any CYP isoenzymes.

Hydroxymethylglutaryl–Coenzyme A Reductase Inhibitors (Statins)

Most statins, except pravastatin and rosuvastatin, are metabolized primarily by 3A4 (see Table 13–7). Statins primarily metabolized via 3A4, including simvastatin, lovastatin, and atorvastatin, may present an increased risk of toxicity, including muscle aches, myopathy, and rhabdomyolysis, when combined with potent 3A4 inhibitors such as itraconazole (Sporanox), ketoconazole, ritonavir, ciprofloxacin, clarithromycin (Biaxin), erythromycin, cyclosporine, efavirenz (Sustiva), lopinavir/ritonavir (Kaletra), nefazodone, and grapefruit juice (Armstrong et al. 2002; Corsini et al. 1999; Jacobson et al. 1997). Many moderate or mild inhibitors of 3A4, including beta-blockers and calcium channel blockers, are simultaneously prescribed with statins more frequently than are the potent inhibitors. Rosuvastatin and fluvastatin are either wholly or partially dependent on 2C9 for metabolism. Potent inhibitors of 2C9 include fluconazole (Diflucan), fluvoxamine (Luvox) and ritonavir; these may also lead to statin toxicity. Although atorvastatin, fluvastatin, and simvastatin show moderate to potent inhibition of 2C9, the remaining statins do not inhibit or induce any hepatic enzyme.

TABLE 13–7. Hydroxymethylglutaryl–coenzyme A reductase inhibitors ("statins")

Drug	Metabolism site(s)	Enzyme(s) inhibited[a]	Enzyme(s) induced
Atorvastatin (Lipitor)	3A4	2C9, 3A4	None known
Fluvastatin (Lescol)	2C9, 2C8, 3A4	2C9	None known
Lovastatin (Mevacor)	3A4	None known	None known
Pravastatin (Pravachol)	3A4[1]	None known	None known
Rosuvastatin (Crestor)	2C9[1]	None known	None known
Simvastatin (Zocor)	3A4, 2D6	2C9	None known

[1]Minimal metabolism at this site.
[a]**Bold** type indicates potent inhibition.
[b]Moderate inhibition or induction.
[c]Mild inhibition or induction.

Gastrointestinal Agents

General Drug Interactions

Metoclopramide (Reglan)

Metoclopramide is frequently prescribed for conditions previously treated with cisapride, which is now off the market. In a study using human liver microsomes and recombinant human cytochromes, Desta et al. (2002) showed that metoclopramide is both metabolized by and a potent inhibitor of 2D6, with minor metabolic contribution by 1A2.

Histamine Subtype 2 (H$_2$) Receptor Blockers

Cimetidine (Tagamet), an over-the-counter antacid, is involved in numerous drug interactions at multiple enzyme sites. Although cimetidine's metabolism is not mediated by the P450 system, it is a "pan-inhibitor" of this system. Because most of cimetidine's premarketing research was conducted in the early 1980s, microsomal and drug-interaction studies are incomplete and results are sometimes conflicting. Cimetidine has been documented as causing decreased clearance, unwanted side effects, and toxic levels of nonsteroidal anti-inflammatory drugs and warfarin (via 2C9 inhibition), theophylline and olanzapine (via 1A2 inhibition) (Szymanski et al. 1991), and beta-blockers and TCAs (via 2D6 inhibition) (see Table 13–8).

Ranitidine (Zantac) is primarily oxidatively metabolized by flavin-containing monooxygenases (FMO$_3$ and FMO$_5$). Minor metabolic contributions occur via 2C19, 1A2, and 2D6. There is also some evidence that ranitidine is a substrate of the ABCB1 transporter. Ranitidine weakly inhibits 1A2, 2C9, 2D6, and 2C19, and there are case reports of interactions between ranitidine and phenytoin (Tse et al. 1993). Clinically insignificant increases in plasma concentrations of metoclopramide (via 2D6 inhibition) and omeprazole (via 2C19 inhibition) have been demonstrated when coadministered with ranitidine.

Proton Pump Inhibitors

Proton pump inhibitors act by irreversibly blocking the hydrogen/potassium adenosine triphosphatase enzyme system on the apical surface of parietal cells in the stomach, thereby preventing secretion of gastric acid. All proton pump inhibitors are extensively metabolized via 2C19 and 3A4 (Stedman and

TABLE 13–8.　Gastrointestinal agents

Drug	Metabolism site(s)	Enzyme(s) inhibited[a]	Enzyme(s) induced
Prokinetics			
Cisapride (Propulsid)[1]	3A4, 2A6	2D6, 3A4	None known
H2 receptor blockers			
Cimetidine (Tagamet)	Renal, P-glycoprotein (ABCB1)	3A4, 2D6, 1A2, 2C9, 2C19	None known
Famotidine (Pepcid)	Renal, ABCB1	None known	None known
Metoclopramide (Reglan)	2D6, 1A2	2D6	None known
Nizatidine (Axid)	Renal	None known	None known
Ranitidine (Zantac)	FMO3, 5, 2C19, 1A2, 2D6, ABCB1	1A2, 2C9, 2C19[b], 2D6[c]	None known
Proton pump inhibitors			
Esomeprazole (Nexium)	2C19, 3A4	2C19	None known
Lansoprazole (Prevacid)	2C19, 3A4	**2C19**, 2D6, 2C9, 3A4	1A2[c]
Omeprazole (Prilosec)	2C19, 3A4	3A4, 2C19, 2C9	1A2
Pantoprazole (Protonix)	2C19, 3A4	3A4	None known
Rabeprazole (Aciphex)	2C19, 3A4	None known	None known

[1]Cardiotoxic pro-drug; removed from the U.S. market in July 2000.
[a]**Bold** type indicates potent inhibition.
[b]Moderate inhibition.
[c]Mild inhibition or induction.

Barclay 2000); the variation in potential for drug interactions is based on differences in enzyme inhibition and induction. Genetic polymorphisms of 2C19 also affect the metabolism of proton pump inhibitors and thus their metabolic profiles (Ishizaki and Horai 1999; McColl and Kennerley 2002). 2C19 inhibition occurs with administration of omeprazole (Prilosec) and, more markedly, its S-enantiomer, esomeprazole (Nexium) (McColl and Kennerley 2002). St. John's wort, a 2C19 and 3A4 inducer, significantly decreases plasma concentrations of omeprazole. Omeprazole and lansoprazole (Prevacid) induce 1A2 and may affect theophylline and caffeine levels. However, a study by Dilger and colleagues (1999) demonstrated no interactions with coadministration of omeprazole, pantoprazole (Protonix), or lansoprazole with theophylline. Rabeprazole and pantoprazole seem to have none of these effects, a conclusion drawn from a review of the literature and findings of randomized studies involving healthy volunteers (Andersson et al. 2001; Hartmann et al. 1999; Humphries and Merritt 1999). Rabeprazole does not appear to have any metabolic interactions (Robinson 2001).

Psychotropic Drug Interactions

H$_2$ Receptor Blockers

In a comparative study, Greenblatt et al. (1984) found no difference in cognitive and psychomotor function between subjects taking both diazepam and cimetidine and control subjects taking cimetidine only. Plasma concentrations were, however, significantly higher in the test group (up to 57% higher than those in the control group). Klotz et al. (1985) also showed an increase in plasma concentrations of midazolam when it was coadministered with cimetidine, but similarly reported no cognitive or psychomotor effects. Sanders et al. (1993) conducted a head-to-head study of ranitidine and cimetidine. Midazolam was added to steady-state H2 receptor blockers in a randomized, double-blind crossover study involving healthy volunteers. The investigators found a significant difference in impairment (pharmacodynamic effect; 2.5 hours) in the cimetidine treatment group compared with the ranitidine treatment group, and the decrement was evident in cognitive and psychomotor functions but was not subjectively reported. Given the findings of these studies, one can conclude that cimetidine affects plasma concentrations of benzodiazepines in healthy volunteers but the effect may be without clinical relevance.

Proton Pump Inhibitors

Omeprazole, and more markedly its S-enantiomer, esomeprazole, are inhibitors of 2C19. Omeprazole's effect on diazepam levels was studied in poor and extensive metabolizers of omeprazole. In a double-blind crossover study, diazepam was administered intravenously after steady state had been achieved with omeprazole therapy (Andersson et al. 1990). Diazepam metabolism was slowed significantly in the extensive metabolizers (i.e., the extensive metabolizers became phenotypic poor metabolizers) by omeprazole. The poor metabolizers showed no apparent interaction. The clinical relevance of this finding is unclear.

In a study of low-dose fluvoxamine administered with omeprazole, Christensen et al. (2002) found marked inhibition of omeprazole metabolism at 2C19 by fluvoxamine.

Oral Hypoglycemics

Thiazolidinedione Agents

The thiazolidinedione antidiabetic agents have been designed to decrease insulin resistance in patients with type II diabetes (diabetes mellitus). Troglitazone (Rezulin), a known 3A4 inducer, has been shown to decrease cyclosporine concentrations in renal transplant recipients (Kaplan et al. 1998) and may decrease serum concentrations of other 3A4 substrates such as protease inhibitors and oral contraceptives. Troglitazone hepatotoxicity was the primary reason for withdrawal from the U.S. market ("Rezulin to Be Withdrawn From the Market" 2000).

Rosiglitazone (Avandia) is primarily metabolized via 2C8, with minor contribution at 2C9 (Baldwin et al. 1999). The manufacturer's randomized, open-label crossover study of 28 healthy volunteers found no clinically significant alterations in nifedipine (3A4 probe drug) pharmacokinetics. Pioglitazone (Actos) is at least partially metabolized by 3A4 and 2C8, although understanding of pioglitazone metabolism and interactions remains unclear (see Table 13–9).

Sulfonylureas

The oral hypoglycemics of the sulfonylurea class include tolbutamide (Orinase), glipizide (Glucotrol), glyburide (Glynase), and glimepiride (Amaryl).

TABLE 13–9. Oral hypoglycemics

Drug	Metabolism site(s)	Enzyme(s) inhibited	Enzyme(s) induced
Thiazolidinediones			
Pioglitazone (Actos)	2C8	None known	None known
Rosiglitazone (Avandia)	2C8	None known	None known
Troglitazone (Rezulin)[1]	2C8, 2C9	None known	3A4
Sulfonylureas			
Glimepiride (Amaryl)	2C9	None known	None known
Glipizide (Glucotrol)	2C9	None known	None known
Glyburide (Micronase)	2C9, 2C19, 3A4	None known	None known
Tolbutamide (Orinase)	2C9	None known	None known
Metglitinides			
Repaglinide (Prandin)	2C8, 3A4	None known	None known

[1]No longer available.

All of these drugs are primarily metabolized at 2C9. Coadministration with 2C9 inhibitors—fluvoxamine (Luvox), fluconazole (Diflucan), or ritonavir (Norvir), for example, may cause significant hypoglycemia (Kidd et al. 1999; Kirchheiner et al. 2002a, 2002b; Niemi et al. 2002). Some members of this class of drugs may induce 3A4, but the clinical importance of such an interaction remains unclear (van Giersbergen et al. 2002), and further research is required to clarify this possible effect.

Meglitinides

Repaglinide (Prandin) belongs to a new class of oral hypoglycemic agents, called meglitinides, which stimulate pancreatic islet beta cell insulin release. Repaglinide is metabolized by 2C8 and 3A4. Inhibitors of 2C8 such as the antibiotic telithromycin have been shown to increase the peak plasma concentration and AUC, and to prolong the hypoglycemic effects of repaglinide (Kajosaari et al. 2006). Repaglinide does not appear to inhibit or induce enzymatic activity.

References

Abernethy DR, Barbey JT, Franc J, et al: Loratadine and terfenadine interaction with nefazodone: both antihistamines are associated with QTc prolongation. Clin Pharmacol Ther 69:96–103, 2001

Ahonen J, Olkkola KT, Salmenpera M, et al: Effect of diltiazem on midazolam and alfentanil disposition in patients undergoing coronary artery bypass grafting. Anesthesiology 85:1246–1252, 1996

Akutsu T, Kobayashi K, Sakurada K, et al: Identification of human cytochrome p450 isozymes involved in diphenhydramine N-demethylation. Drug Metab Dispos 35:72–78, 2007

Andersson T, Cederberg C, Edvardsson G, et al: Effect of omeprazole treatment on diazepam plasma levels in slow versus normal rapid metabolizers of omeprazole. Clin Pharmacol Ther 47:79–85, 1990

Andersson T, Hassan-Alin M, Hasselgren G, et al: Drug interaction studies with esomeprazole, the (S)-isomer of omeprazole. Clin Pharmacokinet 40:523–537, 2001

Antonelli D, Atar S, Freedberg N, et al: Torsade de pointes in patients on chronic amiodarone treatment: contributing factors and drug interactions. Isr Med Assoc J 7:163–165, 2005

Andrews PA: Interactions with ciprofloxacin and erythromycin leading to aminophylline toxicity. Nephrol Dial Transplant 13:1006–1008, 1998

Armstrong SC, Cozza KL, Oesterheld JR: Med-psych drug-drug interactions update. Psychosomatics 43:79–81, 2002

Armstrong SC, Cozza KL: Med-psych drug-drug interactions update. Psychosomatics 44:430–434, 2003

Aronowitz J, Chakos M, Safferman A, et al: Syncope associated with the combination of clozapine and elalapril. J Clin Psychopharmacol 14:429–430, 1994

Atar S, Freedberg NA, Antonelli D, et al: Torsades de pointes and QT prolongation due to a combination of loratadine and amiodarone. Pacing Clin Electrophysiol 26:785–786, 2003

Azaz-Livshits T, Danenberg H: Tachycardia, orthostatic hypotension and profound weakness due to concomitant use of fluoxetine and nifedipine. Pharmacopsychiatry 30:274–275, 1997

Azie NE, Brater DC, Becker PA, et al: The interaction of diltiazem with lovastatin and pravastatin. Clin Pharmacol Ther 64:369–377, 1998

Baldwin SJ, Clarke SE, Chenery RJ: Characterization of the cytochrome P450 enzymes involved in the in vitro metabolism of rosiglitazone. Br J Clin Pharmacol 48:424–432, 1999

Baltes E, Coupez R, Giezek H, et al: Absorption and disposition of levocetirizine, the eutomer of cetirizine, administered alone or as cetirizine to healthy volunteers. Fundam Clin Pharmacol 15:269–277, 2001

Banfield C, Hunt T, Reyderman L, et al: Lack of clinically relevant interaction between desloratadine and erythromycin. Clin Pharmacokinet 41 (suppl 1):29–35, 2002

Bannister S, Houser V, Hulse J, et al: Evaluation of the potential interactions of paroxetine with diazepam, cimetidine, warfarin and digoxin. Acta Psychiatr Scand 80 (suppl 350):102–106, 1989

Barbey J: Loratadine/nefazodone interaction. Clin Pharmacol Ther 71:403, 2002

Batty KT, Davis TM, Ilett KF, et al: The effect of ciprofloxacin on theophylline pharmacokinetics in healthy subjects. Br J Clin Pharmacol 39:305–311, 1995

Benedek IH, Davidson AF, Pieniaszek HJ Jr: Enzyme induction by moricizine: time course and extent in healthy subjects. J Clin Pharmacol 34:167–175, 1994

Brown HS, Gatelin A, Hallifax D, et al: Prediction of in vivo drug-drug interactions from in vitro data: factors affecting prototypic drug-drug interactions involving CYP2C9, CYP2D6 and CYP3A4. Clin Pharmacokinet 45:1035–1050, 2006

Brown R, Cooper T: Ticlopidine-carbamazepine interaction in a coronary stent patient. Can J Cardiol 13:853–854, 1997

Boubenider S, Vincent I, Lambotte O, et al: Interaction between theophylline and tacrolimus in a renal transplant patient. Nephrol Dial Transplant 15:1066–1068, 2000

Buchert E, Woosley RL: Clinical implications of variable antiarrhythmic drug metabolism. Pharmacogenetics 2:2–11, 1992

Castillo-Ferrando J, Carrasco Prieto A, De La Torre Brasas F: Effects of benzodiazepines on digoxin tissue concentrations and plasma protein binding. J Pharm Pharmacol 35:462–463, 1983

Chaikin P, Gillen MS, Malik M, et al: Co-administration of ketoconazole with H_1-antagonists ebastine and loratadine in healthy subjects: pharmacokinetic and pharmacodynamic effects. Br J Clin Pharmacol 59:346–354, 2005

Christensen M, Tybring G, Mihara K, et al: Low daily 10-mg and 20-mg doses of fluvoxamine inhibit the metabolism of both caffeine (cytochrome P4501A2) and omeprazole (cytochrome P4502C19). Clin Pharmacol Ther 71:141–152, 2002

Claritin (package insert). Memphis, TN, Schering-Plough HealthCare Products, Inc., 2002

Corsini A, Bellosta S, Baetta R, et al: New insights into the pharmacodynamic and pharmacokinetic properties of statins. Pharmacol Ther 84:413–428, 1999

Craig-McFeely PM, Acharya NV, Shakir SA: Evaluation of the safety of fexofenadine from experience gained in general practice use in England in 1997. Eur J Clin Pharmacol 57:313–320, 2001

Damkier P, Hansen L, Brosen K: Effect of fluvoxamine on the pharmacokinetics of quinidine. Eur J Clin Pharmacol 55:451–456, 1999

Davies AJ, Harindra V, McEwan A, et al: Cardiotoxic effect with convulsions in terfenadine overdose. BMJ 298:325, 1989

Depre M, Van Hecken A, Verbesselt R, et al: Effect of multiple doses of montelukast, a CysLT1 receptor antagonist, on digoxin pharmacokinetics in healthy volunteers. J Clin Pharmacol 39:941–944, 1999

Desta Z, Wu GM, Morocho AM, et al: The gastroprokinetic and antiemetic drug metoclopramide is a substrate and inhibitor of cytochrome P450 2D6. Drug Metab Dispos 30:336–343, 2002

DeVane CL, Markowitz JS, Hardesty SJ, et al: Fluvoxamine-induced theophylline toxicity. Am J Psychiatry 154:1317–1318, 1997

DeVane CL, Markowitz JS: Avoiding psychotropic drug interactions in the cardiovascular patient. Bull Menninger Clin 64:49–59, 2000

Dilger K, Zheng Z, Klotz U: Lack of drug interaction between omeprazole, lansoprazole, pantoprazole and theophylline. Br J Clin Pharmacol 48:438–444, 1999

Dockens R, Greene D, Barbhaiya R: Assessment of pharmacokinetic and pharmaco-dynamic drug interactions between nefazodone and digoxin in healthy male volunteers. J Clin Pharmacol 36:160–167, 1996

Donahue S, Flockhart DA, Abernethy DR: Ticlopidine inhibits phenytoin clearance. Clin Pharmacol Ther 66:563–568, 1999

Ereshefsky L, Reisenman C, Lam Y: Antidepressant drug interaction and the cytochrome P450 system: the role of cytochrome P450 2D6. Clin Pharmacokinet 29 (suppl 1):10–18, 1995

Flockhart DA, Desta Z, Mahal SK: Selection of drugs to treat gastro-esophageal reflux disease: the role of drug interactions. Clin Pharmacokinet 39:295–309, 2000

Geha RS, Meltzer EO: Desloratadine: a new, nonsedating, oral antihistamine. J Allergy Clin Immunol 107:751–762, 2001

Glaeser H, Bailey DG, Dresser GK, et al: Intestinal drug transporter expression and the impact of grapefruit juice in humans. Clin Pharmacol Ther 81:362–370, 2007

Grace AA, Camm AJ: Quinidine. N Engl J Med 338:35–44, 1998

Greenblatt DJ, Abernethy DR, Morse DS, et al: Clinical importance of the interaction of diazepam and cimetidine. N Engl J Med 310:1639–1643, 1984

Gupta S, Banfield C, Kantesaria B, et al: Pharmcokinetics/ pharmacodynamics of desloratadine and fluoxetine in healthy volunteers. J Clin Pharmacol 44:1252–1259, 2004

Ha HR, Candinas R, Stieger B, et al: Interaction between amiodarone and lidocaine. J Cardiovasc Pharmacol 28:533–539, 1996

Ha-Duong NT, Dijols S, Macherey AC, et al: Ticlopidine as a selective mechanism-based inhibitor of human cytochrome P450 2C19. Biochemistry 40:12112–12122, 2001

Hara Y, Nakajima M, Miyamoto K, et al: Inhibitory effects of psychotropic drugs on mexiletine metabolism in human liver microsomes: prediction of in vivo drug interactions. Xenobiotica 35:549–560, 2005

Hamelin BA, Bouayad A, Drolet B, et al: In vitro characterization of cytochrome P450 2D6 inhibition by classic histamine H1 receptor antagonists. Drug Metab Dispos 26:536–539, 1998

Hamelin BA, Bouayad A, Methot J, et al: Significant interaction between the nonprescription antihistamine diphenhydramine and the CYP2D6 substrate metoprolol in healthy men with high or low CYP2D6 activity. Clin Pharmacol Ther 67:466–477, 2000

Hamman MA, Bruce MA, Haehner-Daniels BD, et al: The effect of rifampin administration on the disposition of fexofenadine. Clin Pharmacol Ther 69:114–121, 2001

Hartmann M, Zech K, Bliesath H, et al: Pantoprazole lacks induction of CYP1A2 activity in man. Int J Clin Pharmacol Ther 37:159–164, 1999

Heinig R, Adelmann HG, Ahr G: The effect of ketoconazole on the pharmacokinetics, pharmacodynamics and safety of nisoldipine. Eur J Clin Pharmacol 55:57–60, 1999

Hemeryck A, Lefebvre R, De Vriendt C, et al: Paroxetine affects metoprolol pharmacokinetics and pharmacodynamics in healthy volunteers. Clin Pharmacol Ther 67:283–291, 2000

Higashi M, Veenstra D, Kondo L, et al: Association between CYP2C9 genetic variants and anticoagulation-related outcomes during warfarin therapy. JAMA 287:1690–1698, 2002

Hjalmarson A, Goldstein S, Fagerberg B, et al: Effects of controlled-release metoprolol on total mortality, hospitalizations, and well-being in patients with heart failure: the Metoprolol CR/XL Randomized Intervention Trial in congestive heart failure (MERIT-HF). MERIT-HF Study Group. JAMA 283:1295–1302, 2000

Holtbecker N, Fromm MF, Kroemer HK, et al: The nifedipine-rifampin interaction: evidence for induction of gut wall metabolism. Drug Metab Dispos 24:1121–1123, 1996

Hsyu PH, Schultz-Smith MD, Lillibridge JH, et al: Pharmacokinetic interactions between nelfinavir and 3-hydroxy-3-methylglutaryl coenzyme A reductase inhibitors atorvastatin and simvastatin. Antimicrob Agents Chemother 45:3445–3450, 2001

Humphries TJ: Famotidine: a notable lack of drug interactions. Scand J Gastroenterol Suppl 134:55–60, 1987

Humphries TJ, Merritt GJ: Review article: drug interactions with agents used to treat acid-related diseases. Aliment Pharmacol Ther 13 (suppl 3):18–26, 1999

Inomata S, Nagashima A, Osaka Y, et al: Propofol inhibits lidocaine metabolism in human and rat liver microsomes. J Anesth 17:246–250, 2003

Ishizaki T, Horai Y: Review article: cytochrome P450 and the metabolism of proton pump inhibitors—emphasis on rabeprazole. Aliment Pharmacol Ther 13 (suppl 3):27–36, 1999

Jacobson R, Wang P, Gluek C: Myositis and rhabdomyolysis associated with concurrent use of simvastatin and nefazodone. JAMA 277:296–297, 1997

Johnson BM, Charman WN, Porter CJ: The impact of p-glycoprotein efflux on enterocyte residence time and enterocyte-based metabolism of verapamil. J Pharm Pharmacol 53:1611–1619, 2001

Kajosaari LI, Niemi M, Backman JT, et al: Telithromycin, but not montelukast, increases plasma concentrations and effects of the cytochrome P450 3A4 and 2C8 substrate repaglinide. Clin Pharmacol Ther 79:231–242, 2006

Kan WM, Liu YT, Hsiao CL, et al: Effect of hydroxyzine on the transport of etoposide in rat small intestine. Anticancer Drugs 12:267–273, 2001

Kaplan B, Friedman G, Jacobs M, et al: Potential interaction of troglitazone and cyclosporine. Transplantation 27:1399–1400, 1998

Kassahun K, Skordos K, McIntosh I, et al: Zafirlukast metabolism by cytochrome P450 3A4 produces an electrophilic alpha,beta-unstaturated iminium species that results in the selective mechanism-based inactivation of the enzyme. Chem Res Toxicol 18:1427–1437, 2005

Katial RK, Stelzle RC, Bonner MW, et al: A drug interaction between zafirlukast and theophylline. Arch Intern Med 158:1713–1715, 1998

Katz MR: Raised serum levels of desipramine with the antiarrhythmic propafene. J Clin Psychiatry 52:432–433, 1991

Kendall A, Boivin M: Warfarin-carbamazepine interaction. Ann Intern Med 94:280, 1981

Kidd RS, Straughn AB, Meyer MC, et al: Pharmacokinetics of chlorpheniramine, phenytoin, glipizide and nifedipine in an individual homozygous for the CYP2C9*3 allele. Pharmacogenetics 9:71–80, 1999

Kirchheiner J, Bauer S, Meineke I, et al: Impact of CYP2C9 and CYP2C19 polymorphisms on tolbutamide kinetics and the insulin and glucose response in healthy volunteers. Pharmacogenetics 12:101–109, 2002a

Kirchheiner J, Brockmoller J, Meineke I, et al: Impact of CYP2C9 amino acid polymorphisms on glyburide kinetics and on the insulin and glucose response in healthy volunteers. Clin Pharmacol Ther 71:286–296, 2002b

Klotz U, Arvela P, Rosenkranz B: Effect of single doses of cimetidine and ranitidine on the steady state plasma levels of midazolam. Clin Pharmacol Ther 38:652–655, 1985

Ko JW, Desta Z, Soukhova NV, et al: In vitro inhibition of the cytochrome P450 (CYP450) system by the antiplatelet drug ticlopidine: potent effect on CYP2C19 and CYP2D6. Br J Clin Pharmacol 49:343–351, 2000

Konig F, Hafele M, Hauger B, et al: Bradycardia after beginning therapy with metoprolol and paroxetine. Psychiatr Prax 23:244–245, 1996

Konishi H, Morita K, Minouchi T, et al: Preferential inhibition of CYP1A enzymes in hepatic microsomes by mexiletine. Eur J Drug Metab Pharmacokinet 24:149–153, 1999

Kopecky SL: Effect of beta blockers, particularly carvedilol, on reducing the risk of events after acute myocardial infarction. Am J Cardiol 98:1115–1119, 2006

Kosoglou T, Salfi M, Lim JM, et al: Evaluation of the pharmacokinetics and electrocardiographic pharmacodynamics of loratadine with concomitant administration of ketoconazole or cimetidine. Br J Clin Pharmacol 50:581–589, 2000

Kosuge K, Nishimoto M, Kimura M, et al: Enhanced effect of triazolam with diltiazem. Br J Clin Pharmacol 43:367–372, 1997

Kusumoto M, Ueno K, Oda A, et al: Effect of fluvoxamine on the pharmacokinetics of mexiletine in healthy Japanese men. Clin Pharmacol Ther 69:104–107, 2001

Labbe L, O'Hara G, Lefebvre M, et al: Pharmacokinetic and pharmacodynamic interaction between mexiletine and propafenone in human beings. Clin Pharmacol Ther 68:44–57, 2000

Lamberg TS, Kivisto KT, Neuvonen PJ: Effects of verapamil and diltiazem on the pharmacokinetics and pharmacodynamics of buspirone. Clin Pharmacol Ther 63:640–645, 1998

Lee CR, Goldstein JA, Pieper JA: Cytochrome P450 2C9 polymorphisms: a comprehensive review of the in-vitro and human data. Pharmacogenetics 12:251–263, 2002

Leibovitz A, Bilchinsky T, Gil I, et al: Elevated serum digoxin level associated with coadministered fluoxetine. Arch Intern Med 158:1152–1153, 1998

Lessard E, Yessine MA, Hamelin BA, et al: Diphenhydramine alters the disposition of venlafaxine through inhibition of CYP2D6 activity in humans. J Clin Psychopharmacol 21:175–184, 2001

Levy N, Janicak P: Calcium channel antagonists for the treatment of bipolar disorder. Bipolar Disord 2:108–119, 2000

Limke K, Shelton A, Elliot E: Fluvoxamine interaction with warfarin. Ann Pharmacother 36:1890–1892, 2002

Linder M, Looney S, Adams J, et al: Warfarin Dose Adjustments Based on CYP2C9 Genetic Polymorphisms. J Thromb Thrombolysis 14:227–232, 2002

Loi CM, Wei X, Parker BM, et al: The effect of tocainide on theophylline metabolism. Br J Clin Pharmacol 35:437–440, 1993

Loi CM, Young M, Randinitis E, et al: Clinical pharmacokinetics of troglitazone. Clin Pharmacokinet 37:91–104, 1999

MacPhee G, McInnes G, Thompson G, et al: Verapamil potentiates carbamazepine neurotoxicity: a clinically important inhibitory interaction. Lancet 1:700–703, 1986

Maeda Y, Funakoshi S, Nakamura M, et al: Possible mechanism for pharmacokinetic interaction between lidocaine and mexiletine. Clin Pharmacol Ther 71:389–397, 2002

Markowitz JS, Wells BG, Carson WH: Interactions between antipsychotic and antihypertensive drugs. Ann Pharmacother 29:603–609, 1995

Mazur A, Strasburg B, Kusneic J, et al: QT prolongation and polymorphous ventricular tachycardia associated with trasodone-amiodarone combination. Int J Cardiol 10:27–29, 1995

McClellan K, Jarvis B: Desloratadine. Drugs 61:789–796, 2001

McColl KE, Kennerley P: Proton pump inhibitors-differences emerge in hepatic metabolism. Dig Liver Dis 34:461–467, 2002

McCollum D, Greene J, McGuire D: Severe sinus bradycardia after initiation of bupropion therapy: a probable drug interaction with metoprolol. Cardiovasc Drugs Ther 18:329–330, 2004

Molden E, Asberg A, Christensen H: CYP2D6 is involved in O-demethylation of diltiazem: an in vitro study with transfected human liver cells. Eur J Clin Pharmacol 56:575–579, 2000

Molden E, Johansen PW, Boe GH, et al: Pharmacokinetics of diltiazem and its metabolites in relation to CYP2D6 genotype. Clin Pharmacol Ther 72:333–342, 2002

Monahan BP, Ferguson CL, Killeavy ES, et al: Torsades de pointes occurring in association with terfenadine use. JAMA 264:2788–2790, 1990

Mousa O, Brater DC, Sunblad KJ, et al: The interaction of diltiazem with simvastatin. Clin Pharmacol Ther 67:267–274, 2000

Nakamura K, Yokoi T, Inoue K, et al: CYP2D6 is the principal cytochrome P450 responsible for metabolism of the histamine H_1 antagonist promethazine in human liver microsomes. Pharmacogenetics 6:449–457, 1996

Negro RD: Pharmacokinetic drug interactions with anti-ulcer drugs. Clin Pharmacokinet 35:135–150, 1998

Niemi M, Cascorbi I, Timm R, et al: Glyburide and glimepiride pharmacokinetics in subjects with different CYP2C9 genotypes. Clin Pharmacol Ther 72:326–332, 2002

Orlando R, Piccoli P, DeMartin S, et al: Cytochrone P450 1A2 is a major determinant of lidocaine metabolism in vivo: effects of liver function. Clin Pharmacol Ther 75:80–88, 2004

Peytavin G, Gautran C, Otoul C, et al: Evaluation of pharmacokinetic interaction between cetirizine and ritonavir, an HIV-1 protease inhibitor, in healthy male volunteers. Eur J Clin Pharmacol 61:267–273, 2005

Physicians' Desk Reference, 56th Edition. Montvale, NJ, Medical Economics, 2002

Pieniaszek HJ Jr, Davidson AF, Benedek IH: Effect of moricizine on the pharmacokinetics of single-dose theophylline in healthy subjects. Ther Drug Monit 15:199–203, 1993

Plavix (package insert). New York, Sanofi/Bristol-Myers Squibb Co., 2007

Rasmussen BB, Maenpaa J, Pelkonen O, et al: Selective serotonin reuptake inhibitors and theophylline metabolism in human liver microsomes: potent inhibition by fluvoxamine. Br J Clin Pharmacol 39:151–159, 1995

Rasmussen BB, Jeppesen U, Gaist D, et al: Griseofulvin and fluvoxamine interactions with the metabolism of theophylline. Ther Drug Monit 19:56–62, 1997

Renwick AG: The metabolism of antihistamines and drug interactions: the role of cytochrome P450 enzymes. Clin Exp Allergy 29 (suppl 3):116–124, 1999 [This entire supplement of Clinical Experimental Allergy is devoted to antihistamines, with many articles relevant to this topic for psychiatrists.]

Rezulin to be withdrawn from the market. HHS News, US Department of Health and Human Services, March 21, 2000. Available at http://www.fda.gov/bbs/topics/NEWS/NEW00721.html.

Richter T, Murdter TE, Heinkele G, et al: Potent mechanism-based inhibition of human CYP2B6 by clopidogrel and ticlopidine. J Pharmacol Exp Ther 308:189–197, 2004

Rieder M, Reiner A, Gage B, et al: Effect of VKORC1 haplotypes on transcriptional regulation and warfarin dose. N Engl J Med 352:2285–2293, 2005

Robinson M: New-generation proton pump inhibitors: overcoming the limitations of early generation agents. Eur J Gastroenterol Hepatol 13 (suppl 1):S43–S47, 2001

Sanders LD, Whitehead C, Gildersleve CD: Interaction of H2-receptor antagonists and benzodiazepine sedation: a double-blind placebo-controlled investigation of the effects of cimetidine and ranitidine on recovery after intravenous midazolam. Anaesthesia 48:286–292, 1993

Sayal K, Duncan-McConnell D, McConnell H, et al: Psychotropic interactions with warfarin. Acta Psychiatr Scand 102:250–255, 2000

Schmidt GA, Hoehns JD, Purcell JL, et al: Severe rhabdomyolysis and acute renal failure secondary to concomitant use of simvastatin, amiodarone, and atazanavir. J Am Board Fam Med 20:411–416, 2007

Sconce E, Khan T, Wynne H, et al: The impact of CYP2C9 and VKORC1 genetic polymorphism and patient characteristics upon warfarin dose requirements: proposal for a new dosing regimen. Blood 106:2329–2333, 2005

Shand DG, Oates JA: Metabolism of propranolol by rat liver microsomes and its inhibition by phenothiazine and tricyclic antidepressant drugs. Biochem Pharmacol 20:1720–1723, 1971

Sharma A, Hamelin BA: Classic histamine H1 receptor antagonists: a critical review of their metabolic and pharmacokinetic fate from a bird's eye view. Curr Drug Metab 4:105–129, 2003

Shum L, Pieniaszek HJ Jr, Robinson CA: Pharmacokinetic interactions of moricizine and diltiazem in healthy volunteers. J Clin Pharmacol 36:1161–1168, 1996

Slater JW, Zechnich AD, Haxby DG: Second generation antihistamines: a comparative review. Drugs 57:31–47, 1999

Stedman CA, Barclay ML: Review article: comparison of the pharmacokinetics, acid suppression and efficacy of proton pump inhibitors. Aliment Pharmacol Ther 14:963–978, 2000

Strain J, Caliendo G, Alexis J, et al: Cardiac drug and psychotropic drug interactions: significance and recommendations. Gen Hosp Psychiatry 21:408–429, 1999

Strain J, Karim A, Caliendo G, et al: Cardiac drug-psychotropic drug update. Gen Hosp Psychiatry 24:283–289, 2002

Strik JJ, Lousberg R, Cheriex EC, et al: One year cumulative incidence of depression following myocardial infarction and impact on cardiac outcome. J Psychosom Res 56:59–66, 2004

Szymanski S, Lieberman JA, Picou D, et al: A case report of cimetidine-induced clozapine toxicity. J Clin Psychiatry 52:21–22, 1991

Tanaka M, Ohkubo T, Otani K, et al: Stereoselective pharmacokinetics of pantoprazole, a proton pump inhibitor, in extensive and poor metabolizers of S-mephenytoin. Clin Pharmacol Ther 69:108–113, 2001

Tartara A, Galimberti C, Manni R, et al: Differential effects of valproic acid and enzyme-inducing anticonvulsants on nimodipine pharmacokinetics in epileptic patients. Br J Clin Pharmacol 32:335–340, 1991

Thomas AR, Chan LN, Bauman JL, et al: Prolongation of the QT interval related to cisapride-diltiazem interaction. Pharmacotherapy 18:381–385, 1998

Triola BR, Kowey PR: Antiarrhythmic drug therapy. Curr Treat Options Cardiovasc Med 8:362–370, 2006

Trujillo TC, Nolan PE: Antiarrhythmic agents: drug interactions of clinical significance. Drug Saf 23:509–532, 2000

Tse CS, Akinwande KI, Biallowons K: Phenytoin concentration elevation subsequent to ranitidine administration. Ann Pharmacother 27:1448–1451, 1993

Turpeinen M, Tolonen A, Uusitalo J, et al: Effect of clopidogrel and ticlopidine on cytochrome P450 2B6 activity as measured by bupropion hydroxylation. Clin Pharmacol Ther 77:553–559, 2005

Unger T, Kaschina E: Drug interactions with angiotensin receptor blockers: a comparison with other antihypertensives. Drug Saf 26:707–720, 2003

Upton RA: Pharmacokinetic interactions between theophylline and other medication (part I). Clin Pharmacokinet 20:66–80, 1991

van Giersbergen PL, Treiber A, Clozel M, et al: In vivo and in vitro studies exploring the pharmacokinetic interaction between bosentan, a dual endothelin receptor antagonist, and glyburide. Clin Pharmacol Ther 71:253–262, 2002

van Harten J: Overview of the pharmacokinetics of fluvoxamine. Clin Pharmacokinet 29 (Suppl 1):1–9, 1995

Van Hecken A, Depre M, Verbesselt R, et al: Effect of montelukast on the pharmacokinetics and pharmacodynamics of warfarin in healthy volunteers. J Clin Pharmacol 39:495–500, 1999

Walsky RL, Gaman EA, Obach RS: Examination of 209 drugs for inhibition of cytochrome P450 2C8. J Clin Pharmacol 45:68–78, 2005

Walsh GM: Levocetirizine: an update. Curr Med Chem 13:2711–2715, 2006

Wang E, Casciano CN, Clement RP, et al: Evaluation of the interaction of loratadine and desloratadine with p-glycoprotein. Drug Metab Dispos 29:1080–1083, 2001

Wang Z, Hamman MA, Huang SM, et al: Effect of St. John's wort on the pharmacokinetics of fexofenadine. Clin Pharmacol Ther 71:414–420, 2002

Wei X, Dai R, Zhai S, et al: Inhibition of human liver cytochrome P450 1A2 by the class 1B antiarrhythmics mexiletine, lidocaine, and tocainide. J Pharmacol Exp Ther 289:853–858, 1999

Weiss J, Dormann S, Martin-Facklam M, et al: Inhibition of P-glycoprotein by newer antidepressants. J Pharmacol Exp Ther 305:197–204, 2003

White W: Hypotension with postural syncope secondary to the combination of chlorpromazine and captopril. Arch Intern Med 146:1833–1834, 1986

Williams D, Feely J: Pharmacokinetic-pharmacodynamic drug interactions with HMG-CoA reductase inhibitors. Clin Pharmacokinet 41:343–370, 2002

Woosley RL: Cardiac actions of antihistamines. Annu Rev Pharmacol Toxicol 36:233–252, 1996

Yap TG, Camm AJ: The current cardiac safety situation with antihistamines. Clin Exp Allergy 29 (suppl 1):15–24, 1999

Yasuda SU, Zannikos P, Young AE, et al: The roles of CYP2D6 and stereoselectivity in the clinical pharmacokinetics of chlorpheniramine. Br J Clin Pharmacol 53:519–525, 2002

Young D, Midha K, Fossler MJ, et al: Effect of quinidine on the interconversion kinetics between haloperidol and reduced haloperidol in humans: implications for the involvement of cytochrome P450 2D6. Eur J Clin Pharmacol 44:433–438, 1993

Zevin S, Benowitz NL: Drug interactions with tobacco smoking: an update. Clin Pharmacokinet 36:425–438, 1999

14

Infectious Diseases

Joseph E. Wise, M.D.

Kelly L. Cozza, M.D., F.A.P.M., F.A.P.A.

Reminder: This chapter is dedicated primarily to metabolic and transporter interactions. Interactions due to displaced protein-binding, alterations in absorption or excretion, and pharmacodynamics are not fully covered.

Antimicrobials are the second most commonly prescribed class of medications in the United States. These drugs are associated with many interactions. Antimalarials and antiparasitic drugs are not covered in this book because they are not regularly encountered in general medical practice in the United States. The antimicrobials with significant drug-interaction activity, as both inducers and inhibitors of enzymes, include macrolide, ketolide, fluoroquinolone, streptogramin, and linezolid antibiotics; imidazole antifungals; antimycobacterials; and antiretrovirals. Each of these antimicrobial classes is dealt with in a separate section below. The section on antiretrovirals contains the latest updates from *Psychosomatics* (Wynn et al. 2004; Zapor et al. 2004).

257

Antibiotics

Macrolides and Ketolides

General Drug Interactions

Macrolide antibiotics are known for inhibiting the metabolism of drugs dependent on 3A4 for metabolism—erythromycin and clarithromycin (Biaxin) being the most potent (see Table 14–1). Azithromycin (Zithromax) appears to be a very mild inhibitor of 3A4, and dirithromycin (Dynabac) does not seem to affect P450 metabolism at all (Watkins et al. 1997).

Telithromycin (Ketek) is the first ketolide antibiotic. It is a substrate of 3A4. ABCB1 (P-gp) and ABCC2 (MRP2) play a role in telithromycin transport. In one *in vitro* study, telithromycin, clarithromycin, erythromycin, roxithromycin (Rulide), but not azithromycin, inhibited drug transporters (Seithel et al. 2007). (See Chapter 4, "Transporters," for a complete review of P-glycoproteins.)

Telithromycin is a potent inhibitor of 3A4. Telithromycin was not affected by grapefruit juice, which inhibits intestinal but not hepatic 3A4 (Shi et al. 2005). Combining telithromycin with the beta-blocker metoprolol (Lopressor) increased the concentration of metoprolol by 1.4-fold. A clinical study cited in the package insert shows increases of metoprolol C_{max} and AUC of about 35%. Careful monitoring of side effects, heart rate, and blood pressure and potentially dose reduction of metoprolol may be necessary (Ketek 2007). Metoprolol is metabolized primarily by 2D6, so this interaction may be due to telithromycin's P-glycoprotein (P-gp/ABCB1) inhibition (Seithel et al. 2007).

Calcium channel blockers.   Calcium channel blocker metabolism is inhibited by macrolides (see "Cardiovascular Agents" in Chapter 13, "Internal Medicine"), and this inhibition increases the risk of orthostasis, falls, and arrhythmia. One 79-year-old woman who tolerated 480 mg of verapamil (Calan, Verelan, Isoptin) per day (the maximum dose approved by the U.S. Food and Drug Administration [FDA]) without complications developed fatigue and dizziness after administration of erythromycin (2,000 mg/day) for a new-onset productive cough (Goldschmidt et al. 2001). She developed complete atrioventricular heart block and QTc prolongation. Verapamil and erythromycin are both inhibitors of 3A4 and transporters.

TABLE 14–1. Macrolides and ketolides

Drug	Metabolism site(s)	Enzyme(s) inhibited[a]	Enzyme(s) induced
Azithromycin (Zithromax)	3A4	3A4[c]	None known
Clarithromycin (Biaxin)	3A4	3A4	None known
Dirithromycin (Dynabac)	3A4	None known	None known
Erythromycin	3A4	3A4	None known
Rokitamycin (Ricamycin)	3A4	3A4	None known
Telithromycin (Ketek)[1]	3A4, others	3A4	None known
Troleandomycin (Tao)	3A4	3A4	None known

[1]Ketolide antibiotic.
[a]**Bold** type indicates potent inhibition.
[b]Moderate inhibition or induction.
[c]Mild to no inhibition.

Warfarin.   In one trial in which 24 healthy males were coadministered warfarin (Coumadin) and telithromycin, there were no significant effects on coagulation times (Scholtz et al. 2000).

Digoxin.   Digoxin is a substrate for P-gp. Macrolides may inhibit intestinal and renal P-gp, increasing digoxin exposure. In one study, digoxin plus telithromycin increased digoxin AUC by 37% (Montay et al. 2002).

Theophylline.   In a study of 10 normal volunteers coadministered theophylline and clarithromycin, theophylline levels remained therapeutic, but there was a 20% increase in theophylline AUC. This indicates closer monitoring of theophylline levels may be necessary (Abbott Laboratories, data on file). Telithromycin may affect theophylline levels less than erythromycin and clarithromycin. When telithromycin was coadministered with theophylline to 20 male and female healthy volunteers, there were no serious adverse events and the theophylline concentrations remained within the therapeutic window (Bhargava et al. 2002),

Fluconazole.   Fluconazole (Diflucan) is an azole antibiotic dependent upon 3A4 for metabolism. In 20 healthy subjects, coadministration of fluconazole and clarithromycin resulted in an AUC increase of fluconazole by 18%, but with limited clinical significance (Gustavson et al. 1996).

Colchicine.   Colchicine is a substrate for 3A4 and P-gp, which are inhibited by clarithromycin. The manufacturer suggests monitoring for colchicine toxicity (Biaxin 2007). Several cases are reported of colchicine toxicity in patients with renal failure when colchicine was coadministered with macrolides (Akdag et al. 2006). Hung and others (2005), in a retrospective study of patients who had been coadministered colchicine and clarithromycin, recommend avoiding coprescribing these two agents, especially for patients with renal failure.

Cyclosporine.   Severe cyclosporine toxicity occurred in a transplant recipient who was concomitantly given clarithromycin (Spicer et al. 1997). (See Chapter 20, "Transplant Surgery and Rheumatology: Immunosupprressants," for more on cyclosporine toxicity.)

Ergots.   Ausband and Goodman (2001) reported a case of ergotism associated with clarithromycin use. A 41-year-old woman who each day took omeprazole for gastrointestinal upset, a beta-blocker for migraine prophylaxis, and an oral caffeine-ergotamine combination for treatment of migraines was prescribed clarithromycin for a nonproductive cough that had lasted 10 days. Shortly after starting clarithromycin therapy, she developed burning leg pain, paresthesias, and edema in both legs, symptoms that progressed to pallor and coolness over 7 days. Ergots are metabolized at 3A4, and their metabolism is inhibited by clarithromycin. Lingual ischemia has also been reported with 2 mg of ergotamine tartrate and clarithromycin (Horowitz et al. 1996).

Hydroxymethylglutaryl–Coenzyme A reductase inhibitors ("statins"). The lipid-lowering agents of the hydroxymethylglutaryl–coenzyme A (HMG-CoA) reductase inhibitor class, often called *statins*, are greatly affected by P450 inhibition. Erythromycin increases the AUC and C_{max} of simvastatin (Zocor), as evidenced by a randomized, double-blind study involving healthy volunteers. Most statins are metabolized at 3A4, and toxicity leads to muscle pain, fatigue, and, potentially, rhabdomyolysis (Armstrong et al. 2002). In a study of 30 healthy men coadministered telithromycin and simvastatin, there was a 761% increase in simvastatin AUC (Harris and Leroy 2001). (See Chapter 13, "Internal Medicine," for more details on lipid-lowering agents.)

Statins are substrates of 3A4. Lovastatin (Mevacor) and simvastatin are very susceptible to 3A4 inhibition, atorvastatin mildly so, and pravastatin (Pravachol) and rosuvastatin (Crestor) are not significantly metabolized at 3A4 (Armstrong and Cozza 2002).

Oral hypoglycemics.   In a randomized, double-blind crossover study, nine healthy volunteers tolerated a single dose of tolbutamide (an oral hypoglycemic of the sulfonylurea class) when given on its own, but when a single dose of clarithromycin was coadministered, seven of the nine volunteers in part one of the study and nine of nine volunteers in part two complained of side effects consistent with hypoglycemia. There was a mean increase in the C_{max} of tolbutamide when the drug was administered with clarithromycin, and the mean bioavailability increased, as did the mean absorption rate constant. Given that the elimination half-life of tolbutamide increased only slightly, it seems that clarithromycin increased the rate of absorption of tolbutamide. Macrolides are known to increase gastric emptying, but clarithromycin may also inhibit P-gp/ABCB1 in the gut (Jayasagar et al. 2000), allowing more tolbutamide into the cells and therefore into the body. (See Chapter 4, "Transporters," for a complete review of P-glycoproteins, and the section "Oral Hypoglycemics" in Chapter 13, "Internal Medicine," for further discussion.)

Rifamycins. Clarithromycin can increase serum levels of rifamycins (rifampin), causing neutropenia and uveitis (Griffith et al. 1995).

Sildenafil. Sildenafil (Viagra) is metabolized at 3A4. Sildenafil's package insert (Viagra 2000) states that erythromycin increases sildenafil's AUC by 182%. The clinical significance of this increase is unknown. Muirhead et al. (2002) evaluated the effects of multiple doses of erythromycin and azithromycin on sildenafil in a placebo-controlled study. As expected, azithromycin had no effect. Erythromycin caused statistically significant changes in the AUC and C_{max} of sildenafil (increasing both more than 2-fold in healthy volunteers) and a slight increase in the AUC of sildenafil's primary metabolite.

Psychotropic Interactions

Clozapine. A single dose of clozapine (Clozaril) given with a single dose of erythromycin did not result in any significant interaction in healthy volunteers (Hagg et al. 1999). Clozapine is known to be metabolized mostly at 1A2 and 2D6, with some 3A4 involvement. An earlier case report implied that clozapine toxicity (somnolence, slurred speech, incoordination, incontinence, and increased serum levels) occurred when erythromycin was added for several days to a stable clozapine regimen (Cohen et al. 1996).

Pimozide. There have been three reports of sudden death with coadministration of pimozide (Orap), an antipsychotic approved for use in patients with Tourette's syndrome (motor and vocal tics), and clarithromycin. Pimozide prolongs QT intervals; therefore, electrocardiographic monitoring is required with pimozide use. Pimozide is a potent inhibitor of 2D6 and a moderate inhibitor of 3A4. The drug is metabolized mainly at 3A4, with some 1A2 contribution (Desta et al. 1998, 1999; Flockhart et al. 2000). We recommend that pimozide not be used with any macrolide antibiotics, and patients must be warned about other moderate to potent 3A4 inhibitors, including grapefruit juice.

Triazolobenzodiazepines. Prospective, double-blind clinical studies involving small numbers of healthy volunteers have revealed that erythromycin increases the AUCs, decreases oral clearance, and prolongs the elimination half-lives of midazolam (Versed) and alprazolam (Xanax) (Yasui et al. 1996; Yeates et al. 1997). Most of these formal studies did not reveal significant adverse or prolonged pharmacodynamic effects (oversedation or decreased motor performance), although the duration of saccadic eye movements changed from 15 minutes to 6 hours in one early study (Olkkola et al. 1993). In one controlled, double-blind study, women showed more extensive interaction with clarithromycin and midazolam than did men (Gorski et al. 1998).

Case reports suggest that some patients may be at risk for prolonged or exaggerated effects when triazolobenzodiazepines and macrolides are combined. Delirium was reported when a patient taking triazolam was given erythromycin (Tokinaga et al. 1996). A healthy male who had been taking flunitrazepam and fluoxetine regularly for depression and sleep apnea developed delirium when clarithromycin was added to his regimen for a respiratory infection (Pollak et al. 1995). The authors of the report thought that the delirium was due to fluoxetine intoxication (which rarely causes delirium), but we believe that the clarithromycin probably increased the effects of the benzodiazepine, which is partially metabolized at 3A4. A child experienced prolonged unconsciousness after minor surgery when administered midazolam while taking prophylactic erythromycin (Hiller et al. 1990). Benzodiazepines that are glucuronidated (and therefore not dependent on phase I metabolism), such as lorazepam (Ativan), oxazepam (Serax), and temazepam (Restoril), do not produce problems of this type.

Azithromycin and dirithromycin mildly inhibit 3A4 and are also without these interactions, as demonstrated in clinical trials and as evidenced by the lack of clinical reports of difficulties (Rapp 1998; Watkins et al. 1997). Rokitamycin (Ricamycin) needs further clinical study; *in vitro* human liver enzyme studies indicate that this agent may inhibit 3A4-catalyzed triazolam α-hydroxylation (Zhao et al. 1999). Greenblatt et al. (1998) studied troleandomycin (Tao), azithromycin, erythromycin, and clarithromycin in the presence of triazolam (Halcion) *in vitro*, using human liver microsomes, and then conducted an *in vivo* study of pharmacodynamic effects of interactions. As expected, troleandomycin, erythromycin, and clarithromycin were potent inhibitors of triazolam metabolism, and erythromycin and clarithromycin enhanced the effects of triazolam. Azithromycin had effects similar to those of placebo.

Midazolam is exclusively metabolized by 3A4. Combination of telithromycin and midazolam resulted in increased midazolam serum levels comparable to clarithromycin's inhibition but less than ketoconazole's inhibition (Shi et al. 2005).

Buspirone.   In a randomized, double-blind study involving healthy volunteers, Kivisto et al. (1997) found that erythromycin increases the AUC of buspirone (BuSpar) 6-fold, causing psychomotor impairment and an increase in reported buspirone side effects.

Carbamazepine.   Carbamazepine (Tegretol) toxicity with concomitant use of erythromycin was first reported by Carranco et al. in 1985. Carbamazepine is metabolized at 3A4, which may be inhibited by macrolides and other 3A4 inhibitors (Spina et al. 1996). Yasui et al. (1997) reported on a series of seven psychiatric inpatients receiving 600 mg of carbamazepine a day who then received 400 mg of clarithromycin a day for pneumonia. Four of the seven patients developed carbamazepine toxicity (drowsiness, ataxia) associated with plasma carbamazepine levels that were about twice as high as clarithromycin levels.

Summary: Macrolides and Ketolides

The macrolide antibiotics clarithromycin and erythromycin, and potentially the newer ketolide telithromycin, are potent inhibitors of 3A4 and have been shown to increase unwanted side effects and toxicity of medications that are dependent on this enzyme for metabolism. Clarithromycin, erythromycin, and telithromycin should not be administered with any medication that both requires metabolism at 3A4 and has known cardiac or other life-threatening effects (i.e., a narrow therapeutic window). Azithromycin and dirithromycin do not seem prone to these interactions at 3A4.

Fluoroquinolones

The fluoroquinolones are a large group of antibiotics with a broad spectrum of activity. Between 5% and 8% of patients have difficulty with side effects, gastrointestinal and central nervous system effects being the most common. The latter effects of this class of drugs include headache, sleep disturbance, irritability, seizures, and psychosis. De Sarro and De Sarro (2001) and Fish (2001) wrote particularly good reviews of the central nervous system–related pharmacodynamic and pharmacokinetic interactions of this class of drugs. Fluoroquinolone antibiotics are also known for being inhibitors of 1A2, with ciprofloxacin (Cipro) and enoxacin (Penetrex) being the most potent (see Table 14–2). Drugs dependent on 1A2 for more than 50% of their metabolism include theophylline, caffeine, clozapine, amitriptyline (Elavil), imipramine (Tofranil), cyclobenzaprine (Flexeril), mirtazapine (Remeron), frovatriptan (Frova), zolmitriptan (Zomig), and naproxen (Aleve, Naprosyn) (see Chapter 7, "1A2," for full listings). Ciprofloxacin and norfloxacin (Noroxin) potently inhibit both 1A2 and 3A4 and thus may affect many drugs. Gemifloxacin (Factive) has very limited hepatic metabolism (Davy et al. 1999a, 1999b). The CYP enzymes do not play an important role and are not inhibited by gemifloxacin (Factive 2005). Moxifloxacin (Avelox) is also free of 1A2 interaction and likely other significant P450 interactions (Stass and Kubitza 2001a, 2001b). The absorption of flouroquinolones is reduced by coadministration with antacid or other compounds containing di/trivalent cations. Ciprofloxacin bioavailability may be reduced by as much as 90% (Cipro 2007; Factive 2005). Clinafloxacin also significantly inhibits 1A2 (Randinitis et al. 2001).

General Drug Interactions

Xanthines. Fluoroquinolones affect the metabolism of the xanthine derivatives theophylline, caffeine, and, to some extent, theobromine. These compounds are among the most widely consumed compounds in beverages and pharmaceutical preparations (Robson 1992). Decreased theophylline and caffeine clearance with administration of quinolone antibiotics has been reported since the 1980s (Mizuki et al. 1996). Fish (2001) ranked fluoroquinolones in terms of the potential for interacting with theophylline, as follows (from strongest to weakest potential): enoxacin, ciprofloxacin, norfloxacin, and ofloxacin (Floxin), levofloxacin (Levaquin), trovafloxacin (Trovan), and moxifloxacin. Caffeine metabolism has been found to be decreased by ciprofloxacin and enoxacin, leading to increased side effects (Cipro 2007; Davis et al. 1996; Fish 2001). Gemifloxacin did not affect theophylline pharmacokinetics in healthy male volunteers (Factive 2005). In one study of 12 healthy volunteers age 21–30, there was no significant interaction when theophylline was coadministered with moxifloxacin (Staas and Kubitza 2001a).

When theophylline and clinafloxacin were coadministered to 12 healthy mixed-gender control subjects, theophylline clearance was significantly

TABLE 14–2. Fluoroquinolones

Drug	Metabolism	Enzyme(s) inhibited[a]	Enzyme(s) induced
Ciprofloxacin (Cipro)	3A4	1A2, 3A4	None known
Enoxacin (Penetrex)	?	1A2	None known
Levofloxacin (Levaquin)	Renal excretion	None known	None known
Moxifloxacin (Avelox)	Not P450	None known	None known
Norfloxacin (Noroxin)	Renal excretion/?	1A2, 3A4	None known
Ofloxacin (Floxin)	Renal excretion	1A2[c]	None known
Sparfloxacin (Zagam)[1]	3A4, phase II	1A2, 3A4[c]	None known
Trovafloxacin (Trovan)	Phase II	None known	None known

Note. ?=unknown.
[1]Prolongs the QTc interval and was removed from the U.S. market.
[a]**Bold** type indicates potent inhibition.
[b]Moderate inhibition or induction.
[c]Mild inhibition.

reduced. On the basis of this study, Randinitis et al. (2001) suggested reducing the theophylline dose by half in those receiving theophylline and clinafloxacin. Coadministration of clinafloxacin and caffeine in 12 healthy volunteers reduced caffeine clearance by 84%; this indicates that caffeine should be avoided in those receiving clinafloxacin (Randinitis et al. 2001). In a study of 16 healthy volunteers who were coadministered theophylline and grepafloxacin, the subjects showed increased serum theophylline concentration after theophylline reached steady state (Efthymiopoulos 1997). The study authors recommended halving the theophylline dose during grepafloxacin administration.

In a study of 12 healthy nonsmokers, there was no interaction when moxifloxacin was coadministered with theophylline. This result confirms moxifloxacin's absence of 1A2 metabolism. (Staas and Kubitza 2001a). In a randomized double-blind, crossover study in 15 healthy volunteers, coadministration of gemifloxacin did not cause clinically significant change in theophylline levels (Allen et al. 2001).

Warfarin. The FDA reported enhanced warfarin anticoagulation with norfloxacin coadministration, but controlled studies involving healthy volunteers (subjects with no acute illnesses requiring antibiotics) found no such effect. The *R*-enantiomer of warfarin, which is the less potent enantiomer, is metabolized at 1A2. Ravnan and Locke (2001) reported that two patients taking both levofloxacin and warfarin had increased International normalized ratios (INR). Levofloxacin is the *S*-enantiomer of ofloxacin. Ofloxacin has also been found to increase INR when administered with warfarin. The enhanced anticoagulation might not be merely a P450-mediated effect; warfarin-drug interactions may be due to displaced protein-binding, altered vitamin K–producing gut flora, or other metabolic drug interactions (Ravnan and Locke 2001). The prothrombin time and INR of warfarin should be monitored whenever these antibiotics are coadministered.

In a study of 16 healthy volunteers on warfarin with INR 1.5–2.5 who were coadministered grepafloxacin, INR did not change, suggesting that grepafloxacin does not affect the pharmacokinetics of warfarin (Efthymiopoulos 1997). In a double-blind, placebo-controlled study of 35 healthy subjects, gemifloxacin and warfarin administration had no clinical effect on INR (Davy et al. 1999b).

Glyburide. The coadministration of ciprofloxacin (Cipro) and glyburide (Glynase and others) has in some cases resulted in severe hypoglycemia (Cipro

2007). There have been several case reports of gatifloxacin affecting glucose metabolism (Ball et al. 2004). The mechanism of fluorquinolones and hypo/hyperglycemia is not understood and probably involves action on the pancreatic beta cell (Park-Wyllie et al. 2006).

Digoxin. In a double-blind, randomized, placebo-controlled, two-way crossover study of 20 elderly volunteers (mean age 65) who were coadministered digoxin and gemifloxacin, there was no clinically significant change in digoxin levels. The authors pointed out that this result confirmed lack of interaction between flouroquinolones and digoxin (Vousden et al. 1999).

Phenytoin. Clinafloxacin has mild inhibition of 2C19, which is partly responsible for the metabolism of phenytoin (Dilantin). In a small study of healthy volunteers, clinafloxacin and phenytoin coadministration resulted in a lower clearance of phenytoin, but this was not felt to be of clinical significance (Randinitis et al. 2001).

Psychotropic Interactions

Tricyclic antidepressants. N-Demethylation of doxepin (Sinequan) and imipramine occurs at 1A2, as does partial metabolism of other tricyclics (clomipramine [Anafranil] and cyclobenzaprine [Flexeril]). Even so, there are no reports of adverse interactions between tricyclic compounds and fluoroquinolones.

Clozapine. In a case report, Markowitz et al. (1997) suggested that adding ciprofloxacin to the regimen of a patient taking clozapine increased the clozapine levels by 80%. The same authors later reported that ciprofloxacin led to increased plasma concentrations of olanzapine (Markowitz and DeVane 1999). A randomized, double-blind, placebo-controlled study involving schizophrenic patients taking clozapine revealed that ciprofloxacin increased mean serum concentrations of clozapine and its active metabolite, N-desmethylclozapine, by about 30% (Raaska and Neuvonen 2000). This study found no clinical signs or symptoms of toxicity, but the patients were not originally taking high-dose clozapine (mean dose, 304 mg/day), only seven patients were studied, and no objective testing of cognitive functioning was performed. We suspect that in the general population, ciprofloxacin would cause significant toxicity, particularly in elderly individuals and chronically medically ill patients (see also Case 3 in the study cases section of Chapter 7, "1A2").

Diazepam. In a controlled, double-blind study in which 12 healthy volunteers were administered ciprofloxacin for 7 days, serum concentrations of diazepam (Valium) were increased after a single dose of ciprofloxacin (Kamali et al. 1993). This effect was most likely due to inhibition of diazepam metabolism by ciprofloxacin, because diazepam has not been shown to be metabolized by 1A2.

Methadone. Ciprofloxacin has been found to inhibit metabolism of methadone, leading to profound sedation (reversed by naloxone therapy) (Herrlin et al. 2000). Methadone is metabolized at 3A4 and 2D6. Ciprofloxacin's inhibition of 3A4 leaves the responsibility of metabolism to low-capacity/high-affinity 2D6, which may become overwhelmed, leading to toxicity.

Summary: Fluoroquinolones

Most fluoroquinolones are inhibitors of 1A2, and ciprofloxacin and norfloxacin are potent inhibitors of both 1A2 and 3A4. Drugs with significant adverse reactions or toxicity that are dependent on 1A2 for metabolism include clozapine, warfarin, theophylline, caffeine, and the tertiary tricyclic antidepressant (TCA) doxepin. Ciprofloxacin can inhibit metabolism of methadone and the aforementioned drugs. For full lists of drugs dependent on 3A4 and 1A2 for metabolism, please see Chapters 6 ("3A4") and 7 ("1A2").

Streptogramins

The combination antibiotic quinupristin/dalfopristin (Synercid) is administered intravenously and used to treat infections caused by resistant gram-positive organisms. Vancomycin was the mainstay for treatment of these infections but is now ineffective in a growing number of patients. Quinupristin/dalfopristin is a potent inhibitor of 3A4 (Allington and Rivey 2001). In *in vivo* studies involving healthy volunteers, quinupristin/dalfopristin increased the serum levels of cyclosporine, midazolam, and nifedipine (Quinupristin/Dalfopristin 1999). This antibiotic does not significantly inhibit 1A2, 2A6, 2C9, 2C19, 2D6, 2E1 (Synercid 2003).

Linezolid

Linezolid (Zyvox) is a member of a new class of antibiotics, the oxazolidinones, which are used against gram-positive bacteria, especially some of the resistant strains. Linezolid is a mild, reversible monoamine oxidase A inhibitor. It may interact with drugs that are adrenergic or serotonergic, leading to serotonin syndrome or hypertensive crisis (Fung et al. 2001). Tyramine-containing foods, meperidine (Demerol), selective serotonin reuptake inhibitors (SSRIs), TCAs, over-the-counter sympathomimetics such as pseudoephedrine, and other medications may interact pharmacodynamically with linezolid (Fung et al. 2001; Hendershot et al. 2001). *In vitro* and *in vivo* studies have found that linezolid is not metabolized by and has no effect on the P450 system (Zyvox 2005).

Antifungals: Azoles and Terbinafine

Azoles

Azole antifungals were developed in the 1980s and 1990s as oral systemic antifungal agents and are associated with far fewer side effects and toxicities than other antifungals such as nystatin and amphotericin B. Unfortunately, azoles have their own problems, which include significant P450 interactions and P-glycoprotein activity (see Table 14–3). Most azole antifungals are inhibitors of 3A4; ketoconazole (Nizoral) is often used as a probe drug to determine indirectly whether a new drug is metabolized at 3A4. Fluconazole (Diflucan) inhibits 3A4 moderately but is a potent inhibitor of 2C9 *in vivo* and has been identified as an inhibitor of 2C19 *in vitro* (Venkatakrishnan et al. 2000). Fluconazole therapy has led to increased phenytoin levels, and healthy volunteers receiving fluconazole and warfarin have had increased prothrombin times. Fluconazole may also inhibit the metabolism of nonsteroidal anti-inflammatory drugs, potentially worsening gastrointestinal side effects. Itraconazole (Sporanox) and ketoconazole inhibit P-glycoprotein function, whereas fluconazole seems to have no activity on P-glycoprotein (Wang et al. 2002). Azole metabolism can be induced at 3A4 (i.e., their plasma concentrations can be reduced) by carbamazepine, phenobarbital, phenytoin, and rifampin, resulting in a reduction in antifungal efficacy.

Voriconazole (Vfend) is a substrate and inhibitor of 2C19, 2C9, and 3A4, causing multiple drug interactions. It is metabolized primarily by 2C19.

TABLE 14–3. Antifungals

Drug	Metabolism site(s)	Enzyme(s) inhibited[a]	Enzyme(s) induced
Azoles			
Fluconazole (Diflucan)	3A4	2C9, 3A4[b]	None known
Itraconazole (Sporanox)	3A4	3A4	None known
Ketoconazole (Nizoral)	3A4	3A4	None known
Miconazole	3A4	3A4	None known
Other antifungals			
Terbinafine (Lamisil)	?	2D6	None known

Note. ?=unknown.
[a]**Bold** type indicates potent inhibition.
[b]Moderate inhibition.
[c]Mild inhibition or induction.

Because 2C19 exhibits extensive genetic polymorphism, poor metabolizers at this cytochrome may develop 4-fold higher voriconazole concentrations (Theuretzbacher et al. 2006). Cytochrome 2C19 is deficient in 5% of Caucasians and in up to 20% of Japanese. In those with poor metabolism at 2C19, there is an increased role of 3A4 (Hyland et al. 2003).

General Drug Interactions

Cyclosporine.　Ketoconazole has been found to inhibit metabolism of cyclosporine via inhibition of 3A4. Ketoconazole inhibition has been used in an attempt to reduce the amount spent on cyclosporine in transplant recipients, with mixed results (see Chapter 20, "Transplant Surgery and Rheumatology: Immunosuppressants").

Dexamethasone.　Itraconazole and intravenous and oral dexamethasone were the focus of a randomized, double-blind, placebo-controlled crossover study involving healthy volunteers. Dexamethasone is metabolized at 3A4. The adrenal suppressant effect of dexamethasone was increased in all itraconazole phases (Varis et al. 2000).

HMG-CoA reductase inhibitors ("statins").　Azoles affect the pharmacokinetics of simvastatin and lovastatin. Small doses of itraconazole greatly increase plasma concentrations of lovastatin and simvastatin but not fluvastatin. These increased serum levels of HMG-CoA reductase inhibitors can lead to skeletal muscle toxicity (Kivisto et al. 1998; Maxa et al. 2002; Neuvonen et al. 1998). In the itraconazole manufacturer's package insert, it was reported that rhabdomyolysis occurred in renal transplant recipients given itraconazole and HMG-CoA reductase inhibitors (Sporanox 2000). In two randomized, double-blind crossover studies, in which healthy volunteers were administered fluconazole and fluvastatin (one study) or fluconazole and pravastatin (the other study), fluconazole inhibited the metabolism of fluvastatin, via interaction at 2C9, and did not interact with pravastatin (Kantola et al. 2000). See "Hydroxymethylglutaryl–Coenzyme A Reductase Inhibitors ('Statins')" in Chapter 13 ("Internal Medicine") for further discussion.

Psychotropic Interactions

Buspirone.　In a randomized, double-blind crossover study, healthy volunteers received buspirone and either itraconazole or erythromycin (Kivisto et al.

1997). Psychomotor tests were administered, and plasma concentrations, half-lives, and AUCs were determined. All subjects had increased serum buspirone concentrations and significant psychomotor impairments and buspirone-related side effects. In a later study, itraconazole increased serum buspirone levels 14.5-fold and decreased the AUC of buspirone's primary active metabolite by 50% (Kivisto et al. 1999).

Micafungin.　Micafungin (Mycamine) inhibits glucan synthesis, which disrupts fungal cell walls. Micafungin is more than 99% protein-bound to albumin. Micafungin is a substrate for and weak inhibitor of CYP3A *in vitro*, but this is not a major pathway *in vivo*. Micafungin is not a substrate or inhibitor of P-gp (Mycamine 2008).

Triazolobenzodiazepines.　Triazolobenzodiazepines require metabolism at 3A4, and therefore their metabolism may be inhibited by the azoles, resulting in increased drowsiness or sedation and increases in other effects of these medications (Backman et al. 1998).

Summary: Azole Antifungals

The azole antifungal's mechanism of effect is to inhibit P450 fungal enzyme systems. In humans, the antifungals ketoconazole and itraconazole are potent inhibitors of 3A4 and may increase plasma concentrations of cyclosporine, triazolobenzodiazepines, warfarin, vinca alkaloids (antineoplastic agents), felodipine (Plendil), methylprednisolone, glyburide, phenytoin, rifabutin, ritonavir (Norvir), saquinavir (Invirase), nevirapine (Viramune), tacrolimus (Prograf and others), some HMG-CoA reductase inhibitors, and any other drugs dependent on 3A4 for metabolism (Albengres et al. 1998). Fluconazole is the only systemic azole antifungal with mild to moderate inhibition at 3A4 and is thus clinically less problematic at this enzyme. However, fluconazole is a potent inhibitor at 2C9 and may increase the risk of toxicity of phenytoin, warfarin, nonsteroidal anti-inflammatory drugs, and oral hypoglycemics.

Terbinafine

Terbinafine (Lamisil) is a topical and systemic antifungal used to treat dermatophytoses. Abdel-Rahman et al. (1999) reported that healthy extensive metabolizers were "converted" to poor metabolizers of dextromethorphan by terbinafine—*in vivo* evidence that this drug may inhibit 2D6. Concomitant use with other drugs dependent on 2D6 for metabolism (beta-blockers, TCAs, codeine) may increase the risk of toxicity of these drugs. The manufacturer notes that terbinafine is an inhibitor of 2D6 and recommends careful monitoring in drugs metabolized by 2D6 (Lamisil 2005).

Antimycobacterials and Antitubercular Agents

Rifamycins

Rifampin (also called rifampicin) and rifabutin were discovered in 1957, and by 1965 they were in wide use for the treatment of tuberculosis. It was quickly learned that these agents *decreased* serum levels of many medications. These drugs are now used in combination therapy for *Mycobacterium avium* complex infection, human immunodeficiency virus (HIV)–related tuberculosis, and community-acquired tuberculosis. Where the rifamycins are metabolized has not yet been fully determined, but hepatic and gut wall 3A4 is involved (see Table 14–4). The rifamycins induce most P450 enzymes. They also induce the uridine 5′-diphosphate glucuronosyltransferase (UGT) enzyme system (Gallicano et al. 1999) and P-glycoprotein/ABCB1 transport systems (Finch et al. 2002).

Rifamycins have been reported to increase (induce) the P450-mediated metabolism of the antiemetics ondansetron and dolasetron (Finch et al. 2002; Villikka et al. 1999), codeine and methadone (Caraco et al. 1997; Holmes 1990; Kreek et al. 1976), tacrolimus, and cyclosporine (Finch et al. 2002). Rifampin caused SSRI withdrawal syndrome when given with sertraline (Markowitz and DeVane 2000). Coadministration of rifampin and nortriptyline resulted in decreased antidepressant effect (Bebchuk and Stewart 1991; Self et al. 1996). Once rifampin therapy was discontinued, nortriptyline levels increased over a 2-week period to toxic range. The efficacy of buspirone (Kivisto et al. 1999; Lamberg et al. 1998), triazolam (Villikka et al. 1997a), and zolpidem (Villikka et al. 1997b) is considerably reduced via induction by rifampin. Coadministration of rifampin and haloperidol resulted in subtherapeutic haloperidol concentrations and marked clinical ineffectiveness (Kim et al. 1996; Takeda et al. 1986).

TABLE 14–4. Antimycobacterials

Drug	Metabolism site(s)	Enzyme(s) inhibited	Enzyme(s) induced[a]
Isoniazid (INH; Nydrazid and others)	?	2C9, 2E1	2E1
Rifabutin (Mycobutin)	3A4	None known	3A4, 1A2, 2C9, 2C19[1]
Rifampin (Rifadin)	3A4	None known	3A4, 1A2, 2C9, 2C19[1]
Rifapentine (Priftin)	3A4/?	None known	3A4, 2C9

Note. ?=unknown.

[1]**Bold** type indicates potent induction. Rifampin and rifabutin induce all enzymes in humans except 2D6 and 2E1.

[a]**Bold** type indicates potent induction.

[b]Moderate inhibition or induction.

[c]Mild inhibition or induction.

Reminder: Induction usually takes a few days and may not wear off for several weeks after discontinuation of the inducing drug. Induction of drugs may lead to ineffectiveness and increase morbidity and mortality. Induction may also result in the production of more toxic metabolites.

Rifamycins are metabolized at 3A4, and their plasma concentrations can be increased by drugs that inhibit 3A4, especially clarithromycin, ketoconazole, and the potent protease inhibitors ritonavir (Norvir) (acutely) and indinavir (Crixivan). Uveitis is one of the toxicities of rifamycins. It is recommended that rifamycin doses be decreased (which may also reduce induction by these medications) when they are administered with potent inhibitors of 3A4 (Strayhorn et al. 1997).

Rifampin induces P-gp/ABCB1 transporters in the duodenum and can decrease levels of digoxin (a P-glycoprotein substrate). Greiner et al. (1999) found that rifampin decreased the AUC and C_{max} of digoxin and reduced digoxin's bioavailability. They noted that intestinal P-glycoprotein levels were 3.5 times greater after rifampin administration. Digoxin may lose its antiarrhythmic effect when taken with rifampin.

Isoniazid

Isoniazid (INH) is an old drug (it was developed in the 1950s), and its mechanism of action is still poorly understood. Slow and fast acetylators were first described with reference to this drug. The half-life of isoniazid is influenced by genetic differences in acetyltransferase activity; isoniazid's elimination half-life in rapid acetylators is 50% that in slow acetylators (Douglas and McLeod 1999). Isoniazid has also been found to inhibit the metabolism of phenytoin, implicating isoniazid as an inhibitor of 2C9 or 2C19 (Kay et al. 1985). Isoniazid may have a biphasic effect at 2E1, greatly affecting acetaminophen metabolism. When first administered with acetaminophen, isoniazid acts as an inhibitor and decreases metabolism to the toxic metabolite *N*-acetyl-*p*-benzoquinone. After 2 weeks, isoniazid becomes an inducer and may increase production of acetaminophen's toxic metabolite and cause hepatotoxicity (Chien et al. 1997; Self et al. 1999). Isoniazid has been found to increase serum haloperidol levels in schizophrenic patients (Takeda et al. 1986), but the mechanism has not been fully elucidated.

In human liver microsomes, pyrazinamide and ethionamide were observed to have no remarkable effects on CYP activities. In contrast, isoniazid inhibited 1A2, 2A6, 2C19, 3A, and possibly 2C9 and 2E1. The inhibition of 2C19 was irreversible (Nishimura et al. 2004).

Summary: Antimycobacterials and Antitubercular Agents

In general, the antitubercular drugs rifampin, rifabutin, and rifapentine induce metabolism at most P450 enzymes and can significantly decrease the efficacy of drugs dependent on the active parent drug for effect. In the case of drugs that are metabolized to toxic metabolites, induction may lead to more rapid or severe toxicity. Rifampin is known also to induce phase II UGT metabolism and P-glycoproteins. Isoniazid is a peculiar drug, with many food and drug interactions, and careful monitoring is needed when isoniazid is administered with drugs metabolized at 2C9 and 2E1.

Antiretrovirals: Combination Antiretroviral Therapy (CART)[1]

This synopsis is intended to provide information regarding the potential side effects, toxicities, and drug interactions of medications commonly used in the management of the HIV-infected patient. Up-to-the-minute information on these medications, including dosing strategies, may be found at Web sites such as http://www.medscape.com/hiv-home?src=pdown, http://hivinsite.ucsf.edu, and http://www.hopkins-hivguide.org.

HIV-positive patients are frequently administered multiple drugs (typically, the lower the CD4 count, the more medications a patient takes). Polypharmacy places this group at great risk for drug-drug interactions. *Combination antiretroviral therapy* (CART) became available in 1996 and generally includes at least three different antiretrovirals administered in combination

---

[1]Adapted and updated from Wynn GH, Zapor MJ, Smith BL, et al.: "Antiretrovirals, Part 1: Overview, History, and Focus on Protease Inhibitors." *Psychosomatics* 45:262–270, 2004; Zapor MJ, Cozza KL, Wynn GH, et al.: "Antiretrovirals, Part II: Focus on Non-Protease Inhibitor Antiretrovirals (NRTIs, NNRTIs, and Fusion Inhibitors)." *Psychosomatics* 45:524–535, 2004. Used with permission.

(sometimes referred to as a "cocktail"); these result in side effects, toxicities, and complicated drug interactions, despite great strides in reducing pill burdens with newer "combination" formulations. Strict adherence to the "cocktail" is imperative to limit the development of viral resistance to CART.

At present, CART comprises six broad classes of HIV medications: 1) protease inhibitors, 2) nucleoside analog reverse transcriptase inhibitors (NRTIs), 3) non-nucleoside reverse transcriptase inhibitors (NNRTIs), 4) cell membrane fusion inhibitors, 5) CCR5 antagonists, and 6) integrase inhibitors. The protease inhibitors are selective, competitive inhibitors of protease, an enzyme crucial to viral maturation, infection, and replication. The NRTIs inhibit viral replication by interfering with viral RNA–directed DNA polymerase (reverse transcriptase). Similarly, NNRTIs inhibit viral replication by acting as a specific, noncompetitive reverse transcriptase inhibitor, disrupting that enzyme's catalytic site. Integrase inhibitors disallow insertion of HIV DNA into host cells. The cell membrane fusion inhibitor enfuvirtide (Fuzeon) blocks uptake of the virus by the lymphocyte. CCR5 antagonists currently are used as a last-line treatment for those who are resistant to previous treatments. They block human CCR5 proteins on immune-system cells, which are the protein portal HIV uses to enter and infect the cell.

Protease Inhibitors

Side Effects

Problems that arise from all the protease inhibitors include gastrointestinal disruptions, dysfunctions of lipid and glucose metabolism, sexual dysfunction, hepatic toxicity, and increased risk of bleeding.

Gastrointestinal effects.   Many patients experience a number of gastrointestinal symptoms when either initiating protease inhibitor therapy or changing from one regimen to another. These symptoms frequently cause discontinuation of therapies and can be a major hurdle to effective treatment. d'Arminio Monforte et al. (2000) found that 25% of patients discontinued therapy because of the toxicity of their regimen. Nausea and vomiting are among the most frequent complaints, with some studies as high as 75% (Duran et al. 2001). These symptoms generally abate over 4–6 weeks of therapy, but some patients continue to experience symptoms indefinitely.

Monitoring and treatment of gastrointestinal symptoms can be among the most clinically relevant interventions (Goldsmith and Perry 2003). Tayrouz et al. (2001) found that ritonavir increased the levels of loperamide (Imodium) 3-fold, indicating that less loperamide may be necessary for effect. Patients need to be warned about loperamide toxicity if they are switched from another medication to ritonavir. In patients with severe gastrointestinal distress and wasting with concomitant depression, we find that the tricyclic antidepressants (TCAs) can help greatly with loose stools, poor appetite, and weight loss as well as with mood. Careful blood-level monitoring of the TCA is important, especially with a "pan-cytochrome P450 inhibitor" like ritonavir. Such monitoring is easily accomplished and is an effective form of therapeutic monitoring.

Lipodystrophy, hyperglycemia, and hyperlipidemia. *Lipodystrophy* involves redistribution of fat, with accumulation of neck and abdominal fat (and occasionally a "buffalo hump") and is usually associated with loss of facial, buttocks, and extremity fat. It is resistant to treatment, with the redistributed fat often not reducing even with discontinuation of the offending agent. Glucose intolerance, sometimes leading to diabetes and frank diabetic ketoacidosis, has been increasingly reported (Addy et al. 2003; Hardy et al. 2001). *Hyperlipidemia* is closely associated with both *hyperglycemia* and fat redistribution and may place HIV patients receiving CART at increased risk of cardiovascular disease (Friis-Moller et al. 2003). The combination of protease inhibitors with lipid-lowering "statins" may lower lipid levels and possibly reduce cardiovascular risk but may also lead to rhabdomyolysis if a statin is chosen that is incompatible with the protease inhibitors (see discussion of inhibition in "Drug-Drug Interactions" subsection below).

Sexual dysfunction. Sexual dysfunction occurs very commonly with regimens containing protease inhibitors (Schrooten et al. 2001), more than with most other therapies. Sildenafil (Viagra) must be used with care in combination with ritonavir and saquinavir because of inhibition of sildenafil's 3A4 metabolism (Gilden 1999). Patients are warned not to exceed 25 mg in a 48-hour period to avoid increased risk of side effects of sildenafil (Bartlett 2003). This interaction is likely to occur with all protease inhibitors. Protease inhibitors have not been specifically studied *in vitro* or *in vivo* in combination with the newer erectile dysfunction drugs vardenafil (Levitra) and tadalafil (Cialis), yet the manufacturers of these drugs warn about their use with potent inhibitors of 3A4, which includes all the protease inhibitors (Cialis 2008; Levitra 2007; Reyataz 2008).

Drug-Drug Interactions

Drug interactions account for many of the complications from CART. In the following subsection, we review metabolic inhibition and induction of and by the protease inhibitors, as well as transporter interactions.

Inhibition.   All protease inhibitors are inhibitors of metabolism specifically at the 3A4 enzyme. The same warnings apply to the protease inhibitors as to ciprofloxacin (Cipro), clarithromycin (Biaxin), diltiazem (Cardizem, Tiazac, and others), erythromycin (E-mycin and others), itraconazole (Sporanox), ketoconazole (Nizoral), and nefazodone (Serzone). These drugs all potently inhibit 3A4 metabolism and can greatly affect drugs with narrow margins of safety. For example, a patient taking carbamazepine developed vomiting, vertigo, and elevated liver enzymes with increased serum concentrations of the anticonvulsant within 12 hours of the first dose of ritonavir (Kato et al. 2000). Patients taking antimigraine medications containing ergotamine who are beginning ritonavir regimens risk loss of limbs or death from ergotism (Baldwin and Ceraldi 2003; Tribble et al. 2002).

Other protease inhibitors, although less potent 3A4 inhibitors than ritonavir, have also demonstrated clinically significant inhibition. There have been several case reports of rhabdomyolysis caused by the interaction of a protease inhibitor with HMG-CoA reductase inhibitors or statins. The interaction of simvastatin (Zocor) and nelfinavir (Viracept) caused severe rhabdomyolysis and death (Hare et al. 2002). As a general rule, simvastatin should not be prescribed with a protease inhibitor. Nelfinavir significantly increased levels of atorvastatin (Lipitor) through inhibition of 3A4, while having no interaction with pravastatin (Pravachol) (Armstrong et al. 2002; Fichtenbaum et al. 2002; Lennernas 2003). Liver transplant patients may develop sirolimus and tacrolimus toxicity from protease inhibitor interactions (Jain et al. 2002a, 2002b). Triazolobenzodiazepines—such as midazolam, alprazolam and triazolam—and buspirone are dependent upon 3A4 for metabolism. Coadministration with protease inhibitors may lead to prolonged effects (Clay and Adams 2003; Greenblatt et al. 2000; Merry et al. 1997). (See section

"Hydroxymethylglutaryl–Coenzyme A Reductase Inhibitors (Statins)" in Chapter 13, "Internal Medicine," for further discussion.)

Pimozide, an antipsychotic also used for tic disorders, is dependent upon 3A4 and can cause arrhythmia and seizures. Protease inhibitor use is contraindicated with pimozide (Norvir 2006).

Protease inhibitor metabolism is also affected by other potent 3A4 inhibitors, since all protease inhibitors are metabolized at the 3A4 enzyme. Any drug with a more potent 3A4 inhibition may slow the metabolism of the protease inhibitor. Ritonavir and ketoconazole are considered two of the most potent 3A4 inhibitors in current use. Other potent 3A4 inhibitors that may affect the rest of the protease inhibitors include ciprofloxacin, clarithromycin, diltiazem, erythromycin, itraconazole, nefazodone, and grapefruit juice. Inhibition of protease inhibitor metabolism by these agents may increase protease inhibitor effectiveness or lower the necessary dose (as is seen when ritonavir is used to augment other protease inhibitors or lopinavir). Inhibition of protease inhibitor metabolism may also worsen side effects and toxicity, placing patients at greater risk of headache, nausea, and diarrhea, as well as hepatitis and pancreatitis, and may also lead to nonadherence. More important, the discontinuation of the more potent inhibitor coprescribed with a protease inhibitor would result in a rapid return to the uninhibited state and could quickly reduce circulating levels of the protease inhibitor, placing the patient at risk for developing viral resistance to this class of drug.

Induction. Induction's net effect is usually to decrease parent compound and its effects while enhancing the production of metabolites and increasing the amount of drug ready for elimination. Rifamycins, carbamazepine (Tegretol), phenytoin (Dilantin), ethanol, and barbiturates are "pan-inducers" of multiple cytochrome P450 enzymes. St. John's wort, efavirenz (Sustiva), and nevirapine (Viramune) are all specific 3A4 inducers (Cozza et al. 2003; Piscitelli et al. 2000a). Hamzeh et al. (2003) found that coadministration of indinavir (Crixivan) and rifabutin caused both significant increases in rifabutin levels and decreases in indinavir levels. Induction of protease inhibitor metabolism may reduce circulatory levels of the protease inhibitor, placing the patient at risk for developing viral resistance, which leads to HIV treatment failure (Piscitelli et al. 2000b).

P-glycoproteins. Protease inhibitors are all substrates of P-glycoproteins, which may explain why it is difficult for antiretrovirals to adequately penetrate the blood-brain barrier and other sanctuaries. P-glycoproteins may be responsible for or may enhance many of the drug-drug interactions found with the protease inhibitors (Huang et al. 2001; Lee et al. 1998; Perloff et al. 2000). Ritonavir initially inhibits P-glycoprotein, which may allow for other substrates to then "pass" the barrier and achieve higher serum levels in the circulation, either at the gut wall or at the blood-brain barrier. This P-glycoprotein inhibition in conjunction with 3A4 inhibition would have an additive inhibitory effect and would raise drug levels of dual P-glycoprotein and 3A4 substrates more than expected. *In vitro*, extended exposure to ritonavir then *induces* P-glycoprotein protein expression, reducing the drug's ability to "pass the barrier" and effectively diminishing the bioavailability of the P-glycoprotein substrate (Perloff et al. 2001). Lopinavir exhibits the same pattern of initial inhibition then induction of P-glycoproteins as ritonavir (Vishnuvardhan et al. 2003). There are reports of protease inhibitors apparently inducing the metabolism of drugs such as methadone, and it may be that P-glycoprotein effects are part of the explanation (Bart et al. 2001; McCance-Katz et al. 2003; Phillips et al. 2003).

Specific Protease Inhibitors

The currently available protease inhibitors are presented in Table 14–5, with their most common drug and food interactions, side effects and toxicities, and a list of contraindicated co-medications. Protease inhibitor metabolism is also summarized in Table 14–5. A few of the protease inhibitors deserve special mention.

Atazanavir (Reyataz).　Atazanavir is both metabolized by and inhibits the cytochrome P450 3A4 enzyme. In addition, atazanavir inhibits a phase II glucuronidation enzyme, UGT 1A1. UGT 1A1 is responsible for the metabolism of hemoglobin. According to the Reyataz package insert, inhibition of this phase II enzyme results in inefficient metabolism of hemoglobin and may lead to direct or unconjugated hyperbilirubinemia (Reyataz 2008).

Ritonavir (Norvir).　Ritonavir is the most potent 3A4 inhibitor in the protease inhibitor group. *In vitro* studies have found its inhibitory potency to be slightly less than that of ketoconazole, with amprenavir being an order of magnitude less (von Moltke et al. 2000). Ritonavir also is a potent inhibitor of the 2D6, 2C9, and 2C19 enzymes, and is referred to here as a "pan-inhibitor." *In vitro*, ritonavir is also a moderate inhibitor of the 2B6 enzyme, which is a major metabolic site for bupropion (Wellbutrin) (Hesse et al. 2001). Theoretically, ritonavir may inhibit the metabolism of bupropion, possibly increasing the risk of side effects and toxicity, including seizures. However, the findings of Hesse et al. (2006) indicate that short-term ritonavir dosing has only minimal impact on the pharmacokinetic disposition of a single dose of bupropion in healthy volunteers. A case-series report (Park-Wyllie and Antoniou 2003) found no reported seizures after as long as 2 years of concomitant use of bupropion and ritonavir. While this finding is encouraging, there were no pharmacokinetic data in this report, and none of the patients was receiving high-dose ritonavir. Jover et al. (2002) described a manic HIV patient being treated with ritonavir who became comatose after receiving two risperidone doses of 3 mg each. Ritonavir inhibits risperidone's metabolism at 2D6 and 3A4 (Norvir 2006). Ritonavir with indinavir also resulted in new onset of extrapyramidal symptoms when coadministered with risperidone (Kelly et al. 2002). Greenblatt et al. (2003) conducted a single-dose, blinded, four-way crossover study in healthy volunteers to demonstrate that initial and short-term exposure to ritonavir reduced clearance, prolonged half-life, and increased plasma concentrations of trazodone, as well as increased sedation, fatigue, and worsened performance. Trazodone is metabolized at 3A4, and its active metabolite is metabolized at 2D6 (Rotzinger et al. 1998). Zolpidem is metabolized at 3A4. We have had patients treated with ritonavir who experienced more than 15 hours of sedation with a single 5-mg dose of zolpidem.

Some ritonavir interactions may take longer to become apparent. An HIV patient taking ritonavir was administered the glucocorticoid budesonide (Entocort) for radiation colitis. Within 14 days the patient developed acute hepatitis (ALT=660) (Sagir et al. 2002). Budesonide is used for inflammatory bowel disease because its 90% first-pass clearance allows high dosage within the lumen of the bowel. When 3A4 metabolism of budesonide was inhibited by ritonavir, a large amount of the drug accumulated, which led to hepatic parenchymal damage. There have been case reports of patients taking ritonavir who developed Cushing's syndrome after 5 months of inhaled fluticasone (Flovent) due to ritonavir's inhibition of 3A4 metabolism of the corticosteroid (Clevenbergh et al. 2002; Gupta and Dubé 2002). Although inhibition of the substrates at 3A4 was immediate, the pharmacodynamic effects of the higher concentrations of substrate became apparent over time. Ritonavir, after several weeks, also *induces* 3A4 metabolism. The balance between induction and

TABLE 14–5.   Protease inhibitors (PIs)

Drug	Metabolic site(s)/ Transporter activity	Inhibition[a]	Induction[a]	Common side effects, toxicities, and clinical considerations	Some potential medication interactions
All protease inhibitors	3A4 Transporters	3A4	Varies	Bleeding risk Gastrointestinal disturbance Headaches Hepatitis Lipodystrophy Sexual dysfunction	All 3A4-dependent medications with narrow safety margins, such as: Antiarrhythmics Ergots Lovastatin Pimozide Simvastatin Triazolobenzodiazepines[2] Warfarin Grapefruit juice raises PI serum levels "Pan-inducers[1]" and St. John's wort decrease PI serum levels

TABLE 14–5. Protease inhibitors (PIs) *(continued)*

Drug	Metabolic site(s)/ Transporter activity	Inhibition[a]	Induction[a]	Common side effects, toxicities, and clinical considerations	Some potential medication interactions
Amprenavir (Agenerase) Fosamprenavir (Lexiva)	3A4 Transporters	3A4[b]	3A4 (possibly) Transporters (possibly)	Lactic acidosis Perioral and peripheral paresthesias Stevens-Johnson syndrome Discontinue Vitamin D (high content in formulation) Do not take with high-fat meals	Propafenone
Atazanavir (Reyataz)	3A4 Transporters (possibly)	3A4, 1A2, 2C9, UGT1A1	None known	Direct (unconjugated) hyperbilirubinemia Lactic acidosis Prolonged P-R interval Take with food/ high-fat meal	Irinotecan Proton pump inhibitors

TABLE 14–5.   Protease inhibitors (PIs) *(continued)*

Drug	Metabolic site(s)/ Transporter activity	Inhibition[a]	Induction[a]	Common side effects, toxicities, and clinical considerations	Some potential medication interactions
Indinavir (Crixivan)	3A4 Transporters	3A4	None known	Altered taste Chelitis Dry eyes, skin, mouth Hyperbilirubinemia Nephrolithiasis Neutropenia Paronychia Rash Leukocytoclastic vasculitis Do not take with high-fat meals	

TABLE 14–5. Protease inhibitors (PIs) *(continued)*

Drug	Metabolic site(s)/ Transporter activity	Inhibition[a]	Induction[a]	Common side effects, toxicities, and clinical considerations	Some potential medication interactions
Lopinavir/ritonavir (Kaletra)	3A4 Transporters	*Lopinavir:* 3A4[b], 2D6 *Ritonavir:* **3A4**, **2D6**, **2C9**, **2C19**, **2B6** Transporters (acute)	*Lopinavir:* Glucuronidation *Ritonavir:* **3A4**, 1A2[b], 2C9[b], 2C19 Transporters (chronic)	Diarrhea Pancreatitis Altered taste Take with food	Same as ritonavir Propafenone
Nelfinavir (Viracept)	3A4, 2C19 Transporters	3A4[c], 1A2, 2B6[b], Transporters	2C9 (possibly)	Diarrhea (worst of PIs) Nephrolithiasis	
Ritonavir (Norvir)	3A4, 2D6 Transporters	**3A4**, **2D6**, **2C9**, **2C19**, **2B6** Transporters (acute)	**3A4**, 1A2[b], 2C9[b], 2C19 Transporters (chronic)	Pancreatitis Altered taste	Clozapine Estradiol Meperidine Methadone Propafenone
Saquinavir (Invirase, Fortovase)	3A4 Transporters	3A4[c] Transporters	None known	Altered taste Take with food	

TABLE 14–5. Protease inhibitors (PIs) *(continued)*

Drug	Metabolic site(s)/ Transporter activity	Inhibition[a]	Induction[a]	Common side effects, toxicities, and clinical considerations	Some potential medication interactions
Tipravavir (Aptivus)	3A4 Transporters	3A4	None known	Intracranial hemorrhage Hepatotoxicity Rash Sulfa-based drug Must be taken with ritonavir "boost"	NSAIDs SSRIs (possible bleeding) Vitamin E (bleeding)

Note. Information in this table is based on Albrecht et al. 2002; Engeler et al. 2002; Moyano Calvo et al. 2001; Rachline et al. 2000; Rayner et al. 2001; Rosso et al. 2003; Rotunda et al. 2003; Schupbach et al. 1984; Tosti et al. 1999; Wilde 2000; Wu and Stoller 2000; and all references cited in the "Protease Inhibitors" subsection in text. Additional information taken from the respective manufacturer's package inserts.

NSAID=nonsteroidal anti-inflammatory drug; SSRI=selective serotonin reuptake inhibitor; UGT=uridine 5′-diphosphate glucuronosyltransferase.

[1]Pan-inducers are drugs that induce many if not all cytochrome P450 enzymes and include barbiturates, carbamazepine, ethanol, phenytoin, and rifmycins.

[2]Alprazolam (Xanax), midazolam (Versed), triazolam (Halcion).

[a]**Bold** type indicates potent inhibition or induction.

[b]Moderate inhibition or induction.

[c]Mild inhibition or induction.

inhibition can be quite variable and often unpredictable, ranging from net inhibition to net induction. This necessitates close clinical monitoring for several weeks when one is introducing ritonavir to a patient. Ritonavir has been found to increase or induce the metabolism of meperidine, resulting in increased levels of the neurotoxic metabolite normeperidine, and has been found to reduce the AUC and C_{max} of ethinyl estradiol (Ouellet et al. 1998; Piscitelli et al. 2000b), placing patients at risk for breakthrough bleeding and pregnancy. Ritonavir is also an inducer of the enzymes 1A2, 2C9, and 2C19. In one case study, a patient maintained on a regimen of acenocoumarol (a mixture of R- and S-warfarin, metabolized at 1A2 and 2C9, respectively) was initiated on ritonavir therapy. The patient's INR decreased dramatically even when the dose of acenocoumarol was tripled (Llibre et al. 2002).

Valproic acid has a complicated metabolic profile and may be difficult to use with protease inhibitors and other antiretrovirals. When valproic acid's metabolic enzymes are induced, the production of a toxic metabolite, 4-ene-valproic acid, may be increased. Levels of this toxic metabolite are not measured in routine valproic acid laboratory tests. There have been reports of valproic acid–induced hepatotoxicity with coadministration of P450 inducers such as nevirapine (Cozza et al. 2000). Ritonavir, efavirenz, and lopinavir may also have complicated drug-drug interactions with valproic acid.

As noted earlier, ritonavir may also be a P-glycoprotein inhibitor and inducer. In summary, careful dosing, monitoring, and informed consent are needed with any of the above-mentioned medication combinations with ritonavir.

Lopinavir/ritonavir (Kaletra).   Lopinavir/ritonavir is a combination drug containing ritonavir and lopinavir. Ritonavir, via 3A4 inhibition, increases lopinavir plasma levels for improved clinical results. Interestingly, lopinavir induces glucuronidation. This phase II metabolism induction can greatly reduce levels of the NRTIs zidovudine and abacavir, reducing effectiveness and resulting in viral resistance. Lopinavir has also been found to be a P-glycoprotein inhibitor and inducer (Vishnuvardhan et al. 2003). This complicated pharmacokinetic and P-glycoprotein profile has led to some unexpected clinical findings.

Lopinavir/ritonavir has led to increases in tacrolimus blood concentrations via 3A4 inhibition (Jain et al. 2002b; Schonder et al. 2003) and to decreases in blood concentration (increased clearance) of methadone and ethinyl estradiol. Hogeland et al. (2007) studied healthy volunteers given 2 weeks of lopinavir/ritonavir and bupropion. They report that concurrent use of lopinavir/ritonavir and bupropion resulted in decreased serum levels of bupropion and its active metabolite hydroxybupropion, and they suggest adequate treatment may require as much as a 100% dose increase of bupropion. A probable mechanism for this interaction is the concurrent induction of cytochrome P450 2B6 and UGT enzymes. Lopinavir/ritonavir exposure is unaffected by a single dose of bupropion. *In vitro*, lopinavir/ritonavir also inhibits 2D6 metabolism, but clinical studies are pending (C. Koch, Pharm.D., Medical Information Specialist, and R. Hodd, M.D., Medical Director, Abbott Laboratories, personal communication, December 19, 2000).

Tipranavir/ritonavir.   Tipranavir (Aptivus) is a second-generation protease inhibitor and is useful against HIV strains resistant to other protease inhibitors. It must be "boosted" or used in combination with ritonavir. Tipranavir itself is metabolized by 3A4, but when combined with ritonavir, it also affects cytochromes, glucuronidation, and transporters as ritonavir does (Aptivus 2007; Vourvahis and Kashuba 2007). Tipranavir is a sulfa-based medication and may not be tolerated by patients with sulfonamide allergies. Tipranavir can cause all of the same side effects and toxicities as all the other protease inhibitors. It has also been associated with intracranial hemorrhage, most likely due to direct effect on platelet aggregation. Patients receiving other medications that affect platelets (warfarin, SSRIs, NSAIDs) may be at increased risk (Aptivus 2007; Vourvahis and Kashuba 2007).

Nucleoside and Nucleotide Analogue Reverse Transcriptase Inhibitors

General Principles

The NRTI class of antiretroviral drugs consists of structural analogues of the nucleotide building blocks of RNA and DNA. When incorporated into the viral DNA, these defective nucleotide analogues prevent the formation of a new $3' \rightarrow 5'$ phosphodiesterase bond with the next nucleotide, causing premature termination of strand synthesis and effectively inhibiting viral replication (Hoggard et al. 2000). Metabolism of these drugs depends on intracellular enzymes such as nucleoside kinases, $5'$-nucleotidases, purine and pyrimidine nucleoside monophosphate kinases, and similar enzymes. The NRTIs are not generally metabolized by hepatic cytochrome P450 enzymes (Balzarini 2004).

The metabolism and significant side effects and toxicities of the NRTIs are summarized in Table 14–6.

Nonnucleoside Reverse Transcriptase Inhibitors

General Principles

There are currently four approved NNRTIs: delavirdine (Rescriptor), nevirapine (Viramune), efavirenz (Sustiva), and etravirine (Intelence). Like NRTIs, NNRTIs target HIV reverse transcriptase; however, the mechanism of action is different (De Clercq 1998). Their antiviral potency and tolerability make the NNRTIs a favorable component of CART regimens, and toxicities and viral cross-resistance do not overlap with those of the NRTIs (Balzarini 2004). The most frequently reported adverse effects are mild rash (Bardsley-Elliot and Perry 2000; Scott and Perry 2000), asymptomatic elevation of liver enzymes (Dieterich et al. 2004), and fat redistribution (Adkins and Noble 1998).

Despite different chemical structures and pharmacokinetics, NNRTIs share a similar mechanism of action, and all are metabolized to some degree by cytochrome P450 enzymes. In addition, they act as either inhibitors or inducers of cytochrome P450 enzymes, thus affecting the metabolism of other drugs. The metabolism and significant side effects and toxicities of the NNRTIs are summarized in Table 14–7. For a fully detailed status report on the NNRTIs, including effectiveness, structural and viral mechanisms, dosing schedules, and viral resistance patterns (with excellent tables), please refer to the review by Balzarini (2004). The following sites may also be accessed: http://www.hopkins-hivguide.org and http://www.hiv-druginteractions.org.

Specific NNRTIs

Delavirdine.   Delavirdine (Rescriptor) is associated with a maculopapular rash but rarely as severe or frequent as with nevirapine. Headache, gastrointestinal difficulties, and elevations in liver-associated enzymes also occur. Delavirdine is metabolized by CYP450 3A4 *in vivo,* with *in vitro* data for CYP2D6. Delavirdine potently inhibits CYP450 3A4, thus inhibiting metabolism of other drugs metabolized at that enzyme (e.g., clarithromycin, protease inhibitors,

TABLE 14–6. Nucleoside reverse transcriptase inhibitors (NRTIs)

Drug	Metabolic site(s)/ Transporter activity	Inhibition	Induction	Common side effects, toxicities, and clinical considerations	Some potential interactions
All NRTIs			None known	Hepatomegaly with steatosis Lactic acidosis Lipodystrophy Myopathy Pancreatitis Peripheral neuropathy	
Abacavir (Ziagen)	Alcohol dehydrogenase, Glucuronyl transferase	None known	Unknown	Hypersensitivity reaction Take with or without food	Alcohol

TABLE 14–6. Nucleoside reverse transcriptase inhibitors (NRTIs) *(continued)*

Drug	Metabolic site(s)/ Transporter activity	Inhibition	Induction	Common side effects, toxicities, and clinical considerations	Some potential interactions
Didanosine (ddl, Videx, Videx EC)	Purine nucleoside phosphorylase	Unknown	None known	Optic neuritis Retinal depigmentation Pancreatitis Peripheral neuropathy Take on empty stomach Do not crush or chew EC tablets	Allopurinol Dapsone[1] Delavirdine Ganciclovir Itraconazole[1] Ketoconazole[1] Methadone Quinolones[2] Ribavirin Stavudine Tenofovir Tetracyclines[2] Trimethoprim/ sulfamethoxazole
Emtricitabine (Emtriva)	Full recovery in urine and feces	None known	None known	Discontinuation with hepatitis B virus; may exacerbate hepatitis	

TABLE 14–6. Nucleoside reverse transcriptase inhibitors (NRTIs) *(continued)*

Drug	Metabolic site(s)/ Transporter activity	Inhibition	Induction	Common side effects, toxicities, and clinical considerations	Some potential interactions
Lamivudine (3TC, Epivir)	Minimal metabolism Renal transporters	None known	None known	Generally well tolerated	Ribavirin Trimethoprim/ sulfameth-oxazole[3] Zalcitabine[3]
Stavudine (d4T, Zerit)	Not yet known	None known	None known	Peripheral neuropathy (increased risk with didanosine)	Didanosine Ribavirin Trimethoprim/ sulfamethoxazole[4] Zidovudine[4]
Tenofovir disoproxil fumarate (Viread)	Renal transporters	1A2	None known	Nausea	Atazanavir Didanosine[5]
Zalcitabine (ddC, Hivid)	Renal transporters	Unknown	None known	Pancreatitis Peripheral neuropathy Stomatitis	Antacids[6] Didanosine Doxorubicin Lamivudine Metoclopramide[6] Ribavirin Stavudine

TABLE 14–6. Nucleoside reverse transcriptase inhibitors (NRTIs) *(continued)*

Drug	Metabolic site(s)/ Transporter activity			Common side effects, toxicities, and clinical considerations	Some potential interactions
		Inhibition	Induction		
Zidovudine (AZT, Retrovir)	UGT2B7 Transporters	Unknown		Anemia Granulocytopenia Headache Gastrointestinal cytopenia	Atovaquone[7] Fluconazole[7] Ganciclovir[8] Methadone Rifampin[9] Ritonavir[9] Valproic acid[7]

[1]Drug that requires an acidic pH for absorption and thus should not be coadministered with didanosine, which is buffered with an antacid.

[2]Antibiotic that would be chelated by didanosine's buffered formulation.

[3]Lamivudine and zalcitabine may inhibit each other's intracellular phosphorylation, worsening toxicity, and should not be coadministered.

[4]Zidovudine may inhibit the intracellular phosphorylation of stavudine, worsening toxicity, so these two drugs should not be coadministered.

[5]Take tenofovir 2 hours before or 1 hour after didanosine.

[6]Decrease absorption of zalcitabine.

[7]Increase zidovudine levels and toxicity.

[8]Increase hematologic toxicity.

[9]Decrease zidovudine levels.

TABLE 14–7. Non-nucleoside reverse transcriptase inhibitors (NNRTIs)

Drug	Metabolic site(s)/ Transporter activity	Inhibition[a]	Induction[a]	Common side effects, toxicities, and clinical considerations	Some potential medication interactions
All NNRTIs				Rash Asymptomatic elevation of liver-associated enzymes	
Delavirdine (Rescriptor)	3A4, 2D6, 2C9, 2C19	3A4, 2C9, 2D6, 2C19	None known		Atorvastatin Calcium channel blockers Ergots Lovastatin Phenobarbital Phenytoin Pimozide Rifampin St. John's wort Triazolobenzo-diazepines[1]

TABLE 14–7. Non-nucleoside reverse transcriptase inhibitors (NNRTIs) *(continued)*

Drug	Metabolic site(s)/ Transporter activity	Inhibition[a]	Induction[a]	Common side effects, toxicities, and clinical considerations	Some potential medication interactions
Etravirine (Intelence)	Methyl hydroxylation 3A4, 2C19, 2C9 glucuronidation	2C9[c], 2C19[c]	3A4, glucuronidation	Rash (severe) Peripheral neuropathy	NNRTs Oral contraceptives "Pan-inducers" Protease inhibitors St. John's wort Triazolobenzo-diazepines
Efavirenz (Sustiva)	3A4, 2B6	3A4, 2C9, 2C19, 2D6, 1A2	3A4[b], 2B6[c]	CNS symptoms: Confusion Depression Euphoria Insomnia Vivid dreams Agitation Altered cognition Amnesia Hallucinations Stupor	Atorvastatin Carbamazepine Ergots Lovastatin Pimozide Phenobarbital Phenytoin Rifampin St. John's wort

TABLE 14–7. Non-nucleoside reverse transcriptase inhibitors (NNRTIs) *(continued)*

Drug	Metabolic site(s)/ Transporter activity	Inhibition[a]	Induction[a]	Common side effects, toxicities, and clinical considerations	Some potential medication interactions
Nevirapine (Viramune)	3A4, 2B6	None known	3A4[b], 2B6[b]	Hepatotoxicity	Antiarrhythmics Beta-blockers Cyclosporine Protease inhibitors St. John's wort Tacrolimus

[1]Alprazolam (Xanax), midazolam (Versed), triazolam (Halcion).
[a]**Bold** type indicates potent inhibition or induction.
[b]Moderate induction.
[c]Mild inhibition or induction.

HMG-CoA reductase inhibitors [statins], sildenafil, triazolobenzodiazepines) (Scott and Perry 2000; Tran et al. 2001) and raises the serum levels of these drugs. *In vitro* studies show that delavirdine also inhibits CYP2C9 and CYP2C19 (von Moltke et al. 2001). The metabolism of delavirdine can be inhibited as well, leading to greater serum concentrations of the drug as well as potentially intensifying noxious side effects such as nausea, vomiting, and diarrhea. Ketoconazole is a potent CYP3A4 inhibitor and has increased concentrations of delavirdine (Tseng and Foisy 1997). Most protease inhibitors are potent inhibitors of CYP3A4 and may also raise levels of delavirdine when coadministered (Smith et al. 2001). Conversely, medications such as rifabutin and rifampin, St. John's wort, ritonavir (Norvir), and modafinil (Provigil) all induce CYP3A4, which may result in greater metabolic efficiency and lower serum levels of delavirdine (Cozza et al. 2003). Finally, it should be noted that antacids decrease delavirdine absorption, so dosing of NNRTIs and antacids should be at least an hour apart (Tran et al. 2001).

Efavirenz.   Efavirenz (Sustiva) is associated with greater side effects and toxicities than the other NNRTIs and has a more complicated drug interaction profile. Fontas et al. (2004) reported that their patients receiving NNRTIs had higher levels of total cholesterol and low-density lipoprotein cholesterol than did antiretroviral-naïve patients. The patients receiving efavirenz had higher levels of total cholesterol and triglycerides than those receiving nevirapine.

Neuropsychiatric side effects have also been seen with efavirenz. Lochet et al. (2003) evaluated neuropsychiatric symptoms in subjects taking efavirenz in a prospective, multicenter survey study. They found that sleep disturbances were the most common complaint in their 174 subjects. Neuropsychiatric complaints also included impaired concentration (18.9%) and memory (23.0%), anxiety (15,5%), sadness (19.3%), and suicidal ideations (9.2%). Twenty-three percent of patients rated their global neuropsychiatric discomfort as moderate to severe after 3 months of treatment.

Fumaz et al. (2002), in a randomized, prospective, two-arm controlled study, compared quality of life and neuropsychiatric side effects in patients receiving an efavirenz-containing regimen versus a group whose treatment did not include efavirenz. They found that the group receiving an efavirenz-containing regimen reported a better quality of life, particularly because their regimen was easier. They found, however, that 13% of their patients reported "character change" (mood swings, melancholy, and irritability) at week 48. In our clinic, patients with a previous history of mood disorders often have a worsening of their symptoms at the onset of efavirenz treatment, a minority of them unable to tolerate the medication because of neuropsychiatric side effects. We provide a psychiatric assessment of all patients with a previous history of psychiatric disorder before initiation of efavirenz, and most efavirenz-treated patients are followed for at least 6 months after initiation of efavirenz to monitor for neuropsychiatric symptoms that might worsen their adherence and their quality of life. Patients with a previous history of mood disorder are informed of the risk of recurrence with efavirenz and often are cotreated with therapy and antidepressants if efavirenz is absolutely warranted.

Efavirenz is metabolized at CYP3A4 and 2B6. *In vivo*, efavirenz is an inducer of 3A4; *in vitro*, it also inhibits 3A4, 2C9, 2C19, 2D6, and 1A2 (the last two only weakly) (Sustiva 2002; von Moltke et al. 2001). Concomitant use with 3A4-dependent drugs with narrow therapeutic windows warrants caution. Drugs such as tricyclic antidepressants, antiarrhythmics, and triazolobenzodiazepines may initially require lower dosing when taken with efavirenz because of efavirenz's *inhibition* of 3A4. Subsequently, however, efavirenz's *induction* of 3A4 may lead to decreased levels of the 3A4-dependent medications, requiring increased dosages of them at a later date. In one study it was demonstrated that 3A4 induction by efavirenz reached full magnitude on day 10 of administration (Mouly et al. 2002).

Efavirenz is often used in combination with protease inhibitors and can induce or increase their metabolism, bringing plasma levels below therapeutic ranges. Efavirenz can decrease amprenavir levels 1–2 weeks after efavirenz initiation (Duval et al. 2000). It is fortunate that this effect can be reversed with the addition of protease inhibitors that inhibit 3A4. Efavirenz induces metabolism of all protease inhibitors except nelfinavir and ritonavir. In a study of 11 patients receiving methadone therapy, efavirenz resulted in more than a 50% decrease in methadone AUC after only 24 hours of efavirenz administration. This led to nine patients complaining of withdrawal symptoms beginning on day 8 of efavirenz treatment (Clarke et al. 2001). Efavirenz can also be induced by the usual potent CYP 3A4 inducers: phenytoin, carbamazepine, alcohol, rifamycins, St. John's wort, and barbiturates. Rifampin and rifapentine induce the metabolism of protease inhibitors and NNRTIs and can decrease serum concentrations, leading to viral resistance (decreased sensitivity of the virus).

Nevirapine. Seventeen percent of patients treated with nevirapine (Viramune) develop a maculopapular rash within the first 6 weeks, and several cases of Stevens-Johnson syndrome have been reported (Balzarini 2004). Other, less frequently reported side effects include hepatitis, gastrointestinal complaints, fatigue, depression, and headache (Vandamme et al. 1998). Nevirapine is an inducer of 3A4 as well. Because this enzyme metabolizes the protease inhibitors, the concomitant use of nevirapine and protease inhibitors typically requires dose modification. For example, studies show significant reductions of plasma saquinavir and indinavir concentrations when either of these agents is coadministered with nevirapine (Murphy et al. 1999; Sahai et al. 1997). However, this induction is compensated for by coadministration of ritonavir (Skowron et al. 1998). Some other commonly prescribed 3A4 substrates that have the potential for drug-drug interactions with nevirapine include methadone (nevirapine effects a decreased plasma concentration of methadone [Altice et al. 1999]), certain HMG-CoA reductase inhibitors (excluding pravastatin and fluvastatin, which are metabolized by alternate cytochrome enzymes [Hsyu et al. 2001]), and oral contraceptives (oral contraceptive concentrations are reduced in the presence of nevirapine [Back et al. 2003]). Because nevirapine is also a substrate of CYP3A4, it induces its own metabolism as well. Therefore, nevirapine doses are escalated upon initiation of therapy (the current recommended schedule is 200 mg/day for 14 days then 200 mg twice per day). Nevirapine serum levels may also be reduced by other medications that induce 3A4, including St. John's wort, the rifamycins, and troglitazone (de Maat et al. 2001; Maldonado et al. 1999).

CCR5 Antagonists

The first CCR5 antagonist to be introduced in the United States is maraviroc (Selzentry). Maraviroc works by blocking the protein CCR5 on human immune system cells, which HIV uses as a portal to enter and infect the cell. Its mechanism is to prevent HIV from entering target cells, preventing the initiation of HIV's replication cycle. This sets it apart from the currently available oral HIV/AIDS antiretroviral drugs (such as protease inhibitors, nucleoside reverse transcriptase inhibitors, and nonnucleoside reverse transcriptase inhibitors), which work by inhibiting HIV/AIDS replication intracellularly, and also from the intramuscular agent enfuvirtide. *In vitro,* maraviroc seems to be effective against HIV/AIDS strains that are resistant to the current classes of HIV/AIDS antiretroviral agents. Maraviroc is a substrate of 3A4 and P-gp trasporters and has potential interactions with protease inhibitors. For example, levels of maraviroc are increased in patients also taking atazanavir, ritonavir-boosted lopinavir, and ritonavir-boosted saquinavir (Muirhead et al. 2004). Potent inducers of 3A4 will lower serum levels of maraviroc, and dose escalation may be necessary. A complete listing of interactions with maraviroc can be found at http://www.hiv-druginteractions.org/frames.asp?new/ Content.asp?ID=340&TDM=False (accessed October 8, 2007). (See Table 14–8.)

Cell Membrane Fusion Inhibitors

Fusion inhibitors such as enfuvirtide (Fuzeon) are a novel class of antiretroviral drugs. Fusion inhibitors prevent the conformational changes that are necessary for the fusion of virions to host cells (Cervia and Smith 2003; Hardy and Skolnick 2004). Enfuvirtide is a synthetic peptide derived from a naturally occurring amino acid sequence in viral membrane proteins. Because of the complicated production, cost, and need for subcutaneous injections, it is currently used in treatment-experienced patients, generally those who have failed traditional antiretroviral combinations. The most frequent and disturbing complication with this treatment is injection-site reactions, most of which resolve on their own and do not interfere with continuation for most patients. Patients are encouraged to rotate injection sites, but the need for injections may affect adherence (Maggi et al. 2004).

It is expected that because enfuvirtide is a peptide, it is metabolized in the liver and kidneys by peptidases, with recycling of the amino acids. *In vitro* studies indicate that there are no cytochrome P450 interactions, and clinical trials with potent P450 inhibitors and inducers such as the protease inhibitors, NNRTIs like efavirenz and nevirapine, and rifampin revealed no clinically significant interactions (Hardy and Skolnick 2004). Zhang et al. (2004) conducted an open-label, one-sequence crossover study in 12 HIV-infected patients. They assessed clinical interactions between five probe drugs and enfuvirtide and found no clinically significant P450 interactions. (See Table 14–8.)

TABLE 14–8. CCR5 antagonists, fusion inhibitors, integrase inhibitors

Drug	Metabolic site(s)/ Transporter activity	P450 and transporter inhibition/ induction	Common side effects, toxicities, and clinical considerations	Some potential medication interactions
CCR5 antagonist				
Maraviroc (Selzentry)	3A4 Transporters	None	Allergic reactions (systemic) Hepatotoxicity Rash	Clarithromycin Efavirenz Inhibitors of 3A4, e.g., grapefruit juice Inducers of 3A4, e.g., St. John's wort Protease inhibitors Rifampin
Fusion inhibitor				
Enfuvirtide (Fuzeon)	Peptidases in liver and kidneys	None	Subcutaneous nodules	None known
Integrase inhibitor				
Raltegravir (Isentress)	UGT1A1	None	Gastrointestinal Headache	Inhibitors of glucuronidation, e.g., atazanavir Inducers of glucuronidation, e.g., rifampin

Note. UGT = uridine 5′-diphosphate glucuronosyltransferase.

Integrase Inhibitor

Raltegravir (Isentress) is an HIV integrase strand transfer inhibitor that blocks the action of HIV integrase, thus preventing HIV from inserting the HIV DNA into the host DNA. Common side effects include gastrointestinal disturbances and headaches. Raltegravir has less impact on lipids than do efavirenz and other antiretrovirals (Grinsztejn et al. 2007). Raltegravir is mainly metabolized by UGT1A1 via glucuronidation. Potent inducers of glucuronidation/UGT1A1, such as rifampin and tipranavir/ritonavir, reduce plasma concentrations of raltegravir. Potent inhibitors of glucuronidation, such as atazanavir, may increase serum levels of raltegravir and potentially worsen side effects and toxicity. Raltegravir neither utilizes nor affects the P450 system (Isentress 2007). (See Table 14–8.)

Summary: Antiretrovirals

Antiretroviral drugs have dramatically changed the survival rate for persons with HIV/AIDS. However, these drugs are a complicated group of medications with significant side effects, toxicities, and drug interactions. Non-HIV specialists will encounter HIV/AIDS patients taking antiretrovirals and will be faced with assisting with the life-changing complications these drugs bring.

The protease inhibitors are the group with the most difficult drug-drug interactions to monitor. In brief, all protease inhibitors may cause nausea, diarrhea, headache, sexual dysfunction, hepatic toxicity, and risk of bleeding. Lipodystrophy, glucose intolerance, and hyperlipidemia result in grave morbidity. All protease inhibitors are susceptible to significant drug-drug interactions. All protease inhibitors are inhibitors of the cytochrome P450 3A4 enzyme, and ritonavir is a "pan-inhibitor" of multiple P450 enzymes. This pattern of enzyme inhibition may slow the metabolism of coadministered drugs and may cause frank toxicity, including death, with drugs that have narrow therapeutic indices and must be metabolized via the enzymes that protease inhibitors inhibit. Ritonavir and lopinavir/ritonavir combination drugs also *induce* cytochrome P450 enzymes and may reduce the effectiveness of coadministered drugs such as oral contraceptives or immunosuppressants. The protease inhibitors also have provided clinical evidence of the importance of P-glycoproteins in the maintenance of effective drug concentrations. All protease inhibitors are substrates of P-glycoproteins, which limits oral availability and blood-brain barrier penetration of these agents. Ritonavir and lopinavir seem to be dual inhibitors and inducers of P-glycoproteins, making predictions about potential drug interactions difficult. A working knowledge of the protease inhibitors—including the pill burdens, side effects, toxicities, and drug interactions of these therapies—is of benefit for all clinicians.

NRTIs have no CYP 450 interactions but are subject to interactions of absorption and glucuronidation. NRTIs have serious side effects and toxicities, which are additive with those of other similar drugs.

NNRTIs are metabolized by P450 enzymes and are subject to P450 interactions, being substrates as well as inhibitors and inducers of those metabolic enzymes themselves. NNRTIs have their own significant side-effect and toxicity profiles.

The integrase inhibitor raltegravir is metabolized via glucuronidation at UGT1A1. Plasma concentration may be affected by potent inhibitors and inducers of glucuronidation.

The CCR5 antagonist maraviroc is dependent on CYP3A4 for metabolism and is susceptible to inhibition and induction at this enzyme.

References

Abdel-Rahman SM, Gotschall RR, Kauffman RE, et al: Investigation of terbinafine as a CYP2D6 inhibitor in vivo. Clin Pharmacol Ther 65:465–472, 1999

Addy CL, Gavrila A, Tsiodras S, et al: Hypoadiponectinemia is associated with insulin resistance, hypertriglyceridemia, and fat redistribution in human immunodeficiency virus–infected patients treated with highly active antiretroviral therapy. J Clin Endocrinol Metab 88:627–636, 2003

Adkins JC, Noble S: Efavirenz. Drugs 56:1055–1066, 1998

Akdag I, Ersoy A, Kahvecioglu S, et al: Acute colchicines intoxication during clarithromycin administration in patients with chronic renal failure. J Nephrol 19:515–517, 2006

Albengres E, LeLouet H, Tillement JP: Systemic antifungal agents: drug interactions of clinical significance. Drug Saf 18:83–97, 1998

Albrecht D, Vieler T, Horst HA: Rash-associated severe neutropenia as a side-effect of indinavir in HIV postexposure prophylaxis. AIDS 16:2098–2099, 2002

Allen A, Bygate E, Vousden M, et al: Multiple-dose pharmacokinetics and tolerability of gemifloxacin administered orally to healthy volunteers. Antimicrob Agents Chemother 45:540–545, 2001

Allington DR, Rivey MP: Quinupristin/Dalfopristin: a therapeutic review. Clin Ther 23:24–44, 2001

Altice FL, Friedland GH, Cooney EL: Nevirapine induced opiate withdrawal among injection drug users with HIV infection receiving methadone. AIDS 13:957–962, 1999

Ammassari A, Antinori A, Cozzi-Lepri A, et al; AdICoNA Study Group: Relationship between HAART adherence and adipose tissue alterations. J Acquir Immune Defic Syndr 31 (suppl 3):S140–S144, 2002

Antoniou T, Gough K, Yoong D, et al: Severe anemia secondary to a probable drug interaction between zidovudine and valproic acid. Clin Infect Dis 38:e38–e40, 2004

Aptivus (package insert). Ridgefield, CT, Boehringer Ingelheim, 2007

Armstrong SC, Cozza KL, Oesterheld JR: Med-psych drug-drug interactions update: statins. Psychosomatics 43:77–81, 2002

Ausband SC, Goodman PE: An unusual case of clarithromycin associated ergotism. J Emerg Med 21:411–413, 2001

Back D, Gibbons S, Khoo S: Pharmacokinetic drug interactions with nevirapine. J Acquir Immune Defic Syndr 34 (suppl 1):S8–S14, 2003

Backman JT, Kivisto KT, Olkkola KT, et al: The area under the plasma concentration-time curve for oral midazolam is 400-fold larger during treatment with itraconazole than with rifampicin. Eur J Clin Pharmacol 54:53–58, 1998

Balfour JA, Figgitt DP: Telithromycin. Drugs 61:815–829, 2001

Balfour JA, Wiseman LR: Moxifloxacin. Drugs 57:363–373, 1999

Baldwin ZK, Ceraldi CC: Ergotism associated with HIV antiviral protease inhibitor therapy. J Vasc Surg 37:676–678, 2003

Ball P, Stahlmann R, Kubin R, et al: Safety Profile of Oral and Intravenous Moxifloxacin: Cumulative Data from Clinical Trials and Postmarketing Studies. Clin Ther 26:940–950, 2004

Balzarini J: Current status of the non-nucleoside reverse transcriptase inhibitors of human immunodeficiency virus type 1. Curr Top Med Chem 4:921–944, 2004

Bang LM, Scott LJ: Emtricitabine: an antiretroviral agent for HIV infection. Drugs 63:2413–2424, 2003

Barbier O, Turgeon D, Girard C, et al: 3′-Azido-3′deoxythymidine (AZT) is glucuronidated by human UDP-glucuronosyltransferase 2B7 (UGT2B7). Drug Metab Dispos 28:497–502, 2000

Bardsley-Elliot A, Perry CM: Nevirapine: a review of its use in the prevention and treatment of paediatric HIV infection. Paediatr Drugs 2:373–407, 2000

Bart PA, Rizzardi PG, Gallant S, et al: Methadone blood concentrations are decreased by the administration of abacavir plus amprenavir. Ther Drug Monit 23:553–555, 2001

Bartlett JG: The Johns Hopkins Hospital 2003 Guide to Medical Care of Patients With HIV Infection, 10th Edition. Philadelphia, PA, Lippincott Williams & Wilkins, 2003, pp 101–111

Bearden DT, Neuhauser MM, Garey KW: Telithromycin: an oral ketolide for respiratory infections. Pharmacotherapy 21:1204–1222, 2001

Bebchuk JM, Stewart DE: Drug interaction between rifampin and nortriptyline: a case report. Int J Psychiatry Med 21:183–187, 1991

Bhargava V, Leroy B, Shi J, et al: Effect of telithromycin on pharmacokinetics of theophylline in healthy volunteers. Poster presented at 42nd Interscience Conference on Antimicrobial Agents and Chemotherapy, San Diego, CA, September 27–30, 2002

Biaxin (package insert). North Chicago, IL, Abbott Laboratories, 2007

Blanchard JN, Wohlfciler M, Canas A, et al: Pancreatitis with didanosine and tenofovir disoproxil fumarate. Clin Infect Dis 37:e57–e62; correction, 37:995, 2003

Boelaert JR, Dom GM, Huitema AD, et al: The boosting of didanosine by allopurinol permits a halving of the didanosine dosage (letter). AIDS 16:2221–2223, 2002

Brinkman K, Smeitink JA, Romijn JA, et al: Mitochondrial toxicity induced by nucleoside-analogue reverse-transcriptase inhibitors is a key factor in the pathogenesis of antiretroviral-therapy–related lipodystrophy. Lancet 354:1112–1115, 1999

Brockmeyer NH, Tillmann I, Mertins L, et al: Pharmacokinetic interaction of fluconazole and zidovudine in HIV-positive patients. Eur J Med Res 2:377–383, 1997

Butt AA: Fatal lactic acidosis and pancreatitis associated with ribavirin and didanosine therapy. AIDS Read 13:344–348, 2003

Caraco Y, Sheller J, Wood AJ: Pharmacogenetic determinants of codeine induction by rifampin: the impact on codeine's respiratory, psychomotor and miotic effects. J Pharmacol Exp Ther 281:330–336, 1997

Carr A: Lactic acidemia in infection with human immunodeficiency virus. Clin Infect Dis 36 (suppl 2):S96–S100, 2003

Carr A, Miller J, Law M, et al: A syndrome of lipoatrophy, lactic acidemia and liver dysfunction associated with HIV nucleoside analogue therapy: contribution to protease inhibitor–related lipodystrophy syndrome. AIDS 14:F25–F32, 2000

Carranco E, Kareus J, Co S, et al: Carbamazepine toxicity induced by concurrent erythromycin therapy. Arch Neurol 42:187–188, 1985

Cervia JS, Smith MA: Enfuvirtide (T-20): a novel human immunodeficiency virus type 1 fusion inhibitor. Clin Infect Dis 37:1102–1106, 2003

Chien JY, Peter RM, Nolan CM, et al: Influence of polymorphic N-acetyltransferase phenotype on the inhibition and induction of acetaminophen bioactivation with long-term isoniazid. Clin Pharmacol Ther 61:24–34, 1997

Cialis (package insert). Indianapolis, IN, Eli Lilly and Company, 2008

Cipro (package insert). Kenilworth, NJ, Schering-Plough, 2007

Claessens YE, Cariou A, Monchi M, et al: Detecting life-threatening lactic acidosis related to nucleoside-analog treatment of human immunodeficiency virus–infected patients, and treatment with l-carnitine. Crit Care Med 31:1042–1047, 2003

Clarke SM, Mulcahy FM, Tjia J, et al: The pharmacokinetics of methadone in HIV-positive patients receiving the non-nucleoside reverse transcriptase inhibitor efavirenz. Br J Clin Pharmacol 51:213–217, 2001

Clay PG: The abacavir hypersensitivity reaction: a review. Clin Ther 24:1502–1514, 2002

Clay PG, Adams MM: Pseudo-Parkinson disease secondary to ritonavir–buspirone interaction. Ann Pharmacother 37:202–205, 2003

Clevenbergh P, Corcostegui M, Gerard D, et al: Iatrogenic Cushing's syndrome in an HIV-infected patient with inhaled corticosteroids (fluticasone propionate) and low dose ritonavir enhanced PI containing regimen. J Infect 44:194–195, 2002

Cobo J, Ruiz MF, Figueroa MS, et al: Retinal toxicity associated with didanosine in HIV-infected adults (letter). AIDS 10:1297–3000, 1996

Cohen LG, Chesley S, Eugenio L, et al: Erythromycin-induced clozapine toxic reaction. Arch Intern Med 156:675–677, 1996

Court MH, Krishnaswamy S, Hao Q, et al: Evaluation of 3′-azido-3′deoxythymidine, morphine, and codeine as probe substrates for UDP-glucuronosyltransferase 2B7 (UGT2B7) in human liver microsomes: specificity and influence of the UGT2B7*2 polymorphism. Drug Metab Dispos 31:1125–1133, 2003

Cozza KL, Swanton EJ, Humphreys CW: Hepatotoxicity with combination of valproic acid, ritonavir, and nevirapine: a case report. Psychosomatics 41:452–453, 2000

Cozza KL, Armstrong SC, Oesterheld JR: Drug Interaction Principles for Medical Practice: Cytochrome P450s, UGTs, P-Glycoproteins, 2nd Edition. Washington, DC, American Psychiatric Publishing, 2003

d'Amati G, Kwan W, Lewis W: Dilated cardiomyopathy in a zidovudine-treated AIDS patient. Cardiovasc Pathol 1:317–320, 1992

d'Arminio Monforte A, Lepri AC, Rezza G: Insights into the reasons for discontinuation of the first highly active antiretroviral therapy (HAART) regimen in a cohort of antiretroviral naive patients. I.CO.N.A. Study Group: Italian Cohort of Antiretroviral-Naive Patients. AIDS 14:499–507, 2000

Davis R, Markham A, Balfour JA: Ciprofloxacin: an updated review of its pharmacology, therapeutic efficacy and tolerability. Drugs 51:1019–1074, 1996

Davy M, Allen A, Bird N, et al: Lack of effect of gemifloxacin on the steady-state pharmacodynamics of theophylline in healthy volunteers. Chemotherapy 45:478–484, 1999a

Davy M, Bird N, Rost KL: Lack of effect of gemifloxacin on the steady-state pharmacodynamics of warfarin in healthy volunteers. Chemotherapy 45:491–495, 1999b

De Clercq E: The role of non-nucleoside reverse transcriptase inhibitors (NNRTIs) in the therapy of HIV-1 infection. Antiviral Res 38:153–179, 1998

de la Asuncion JG, del Olmo ML, Sastre J, et al: AZT treatment induces molecular and ultrastructural oxidative damage to muscle mitochondria: prevention by antioxidant vitamins. J Clin Invest 102:4–9, 1998

de Maat MMR, Mathot R, Hoetelmans R, et al: A population pharmacokinetic model of nevirapine reveals drug interaction with St John's wort in a cohort of HIV-1 infected patients, in Abstracts of the Second International Workshop on Clinical Pharmacology of HIV Therapy, Nordwijk, the Netherlands, April 2001, abstract 1.2

De Sarro A, De Sarro G: Adverse reactions to fluoroquinolones: an overview on mechanistic aspects. Curr Med Chem 8:371–384, 2001

Desta Z, Kerbusch T, Soukhova N, et al: Identification and characterization of human cytochrome P450 isoforms interacting with pimozide. J Pharmacol Exp Ther 285:428–437, 1998

Desta Z, Kerbusch T, Flockhart DA: Effect of clarithromycin on the pharmacokinetics and pharmacodynamics of pimozide in healthy poor and extensive metabolizers of cytochrome P450 2D6 (CYP2D6). Clin Pharmacol Ther 65:10–20, 1999

Dieterich DT, Robinson PA, Love J, et al: Drug-induced liver injury associated with the use of nonnucleoside reverse-transcriptase inhibitors. Clin Infect Dis 38 (suppl 2):S80–S89, 2004

Douglas JG, McLeod M: Pharmacokinetic factors in modern drug treatment of tuberculosis. Clin Pharmacokinet 37:127–146, 1999

Duran S, Spire B, Raffi F, et al: Self-reported symptoms after initiation of a protease inhibitor in HIV-infected patients and their impact on adherence to HAART. APROCO Cohort Study Group. HIV Clin Trials 2:38–45, 2001

Duval X, Le Moing V, Longuet C, et al: Efavirenz-induced decrease in plasma amprenavir levels in human immunodeficiency virus-infected patients and correction by ritonavir (letter). Antimicrob Agents Chemother 44:2593, 2000

Efthymiopoulos C: Pharmacokinetics of grepafloxacin. J Antimicrob Chemother 40 (suppl A):35–43, 1997

Emtriva (package insert). Foster City, CA, Gilead Sciences, 2004

Engeler DS, John H, Rentsch KM, et al: Nelfinavir urinary stones. J Urol 167:1384–1385, 2002

Eron J, Benoit S, Jemsek J, et al: Treatment with lamivudine, zidovudine, or both in HIV-positive patients with 200 to 500 CD4+ cells per cubic millimeter. North American HIV Working Party. N Engl J Med 333:1662–1669, 1995

Ethell BT, Anderson GD, Burchell B: The effect of valproic acid on drug and steroid glucuronidation by expressed human UDP-glucuronosyltransferases. Biochem Pharmacol 65:1441–1449, 2003

Factive (package insert). Waltham, MA, Oscient, 2005

Fayz S, Inaba T: Zidovudine azido-reductase in human liver microsomes: activation by ethacrynic acid, dipyridamole, and indomethacin and inhibition by human immunodeficiency virus protease inhibitors. Antimicrob Agents Chemother 42:1654–1648, 1998

Fichtenbaum CJ, Gerber JG, Rosenkranz SL: Pharmacokinetic interactions between protease inhibitors and statins in HIV seronegative volunteers: ACTG Study A5047. AIDS 16:569–577, 2002

Finch CK, Chrisman CR, Baciewicz AM, et al: Rifampin and rifabutin drug interactions. Arch Intern Med 162:985–992, 2002

Fish DN: Fluoroquinolone adverse effects and drug interactions. Pharmacotherapy 21:253S–272S, 2001

Flockhart DA, Drici M, Kerbusch T, et al: Studies on the mechanism of a fatal clarithromycin-pimozide interaction in a patient with Tourette syndrome. J Clin Psychopharmacol 20:317–324, 2000

Fontas E, van Leth F, Sabin CA, et al: Lipid profiles in HIV-infected patients receiving combination antiretroviral therapy: are different antiretroviral drugs associated with different lipid profiles? D:A:D Study Group. J Infect Dis 189:1056–1074, 2004

Freund YR, Dabbs J, Creek MR, et al: Synergistic bone marrow toxicity of pyrimethamine and zidovudine in murine in vivo and in vitro models: mechanism of toxicity. Toxicol Appl Pharmacol 181:16–26, 2002

Friis-Moller N, Weber R, Reiss P, et al; DAD Study Group: Cardiovascular disease risk factors in HIV patients—association with antiretroviral therapy: results from the DAD Study. AIDS 17:1179–1193, 2003

Fromenty B, Pessayre D: Impaired mitochondrial function in microvesicular steatosis: effects of drugs, ethanol, hormones and cytokines. J Hepatol 26 (suppl 2):43–53, 1997

Fumaz CR, Tuldra A, Ferrer MJ, et al: Quality of life, emotional status, and adherence of HIV-1-infected patients treated with efavirenz versus protease inhibitor-containing regimens. J Acquir Immune Defic Syndr 29:244–253, 2002

Fung HB, Kirschenbaum HL, Ojofeitimi BO: Linezolid: an oxazolidinone antimicrobial agent. Clin Ther 23:356–391, 2001

Gallicano KD, Sahai J, Shukla VK, et al: Induction of zidovudine glucuronidation and amination pathways by rifampicin in HIV-infected patients. Br J Clin Pharmacol 48:168–179, 1999

Gallicchio VS, Hughes NK, Tse KF: Comparison of dideoxynucleoside (DDI and zidovudine) and induction of hematopoietic toxicity using normal human bone marrow cells in vitro. Int J Immunopharmacol 15:263–268, 1993

Gilden D: Protease inhibitors, sexual dysfunction and Viagra. GMHC Treat Issues 13:12, 1999

Gold R, Meurers B, Reichmann H: Mitochondrial myopathy caused by long-term zidovudine therapy (letter). N Engl J Med 323:994, 1990

Goldschmidt N, Azaz-Livshits T, Gotsman [I], et al: Compound cardiac toxicity of oral erythromycin and verapamil. Ann Pharmacother 35:1396–1399, 2001

Goldsmith DR, Perry CM: Atazanavir. Drugs 63:1679–1693, 2003

Gorski JC, Jones DR, Haehner-Daniels BD, et al: The contribution of intestinal and hepatic CYP3A to the interaction between midazolam and clarithromycin. Clin Pharmacol Ther 64:133–143, 1998

Grasela TH, Walawander CA, Beltangady M, et al: Analysis of potential risk factors associated with the development of pancreatitis in phase I patients with AIDS or AIDS-related complex receiving didanosine. J Infect Dis 169:1250–1255, 1994

Greenblatt DJ, von Moltke LL, Harmatz JS, et al: Inhibition of triazolam clearance by macrolide antimicrobial agents: in vitro correlates and dynamic consequences. Clin Pharmacol Ther 64:278–285, 1998

Greenblatt DJ, von Moltke LL, Harmatz JS, et al: Alprazolam–ritonavir interaction: implications for product labeling. Clin Pharmacol Ther 67:335–341, 2000

Greenblatt DJ, von Moltke LL, Harmatz JS, et al. Short-term exposure to low-dose ritonavir impairs clearance and enhances adverse effects of trazodone. J Clin Pharmacol 43:414–422, 2003

Greiner B, Eichelbaum M, Fritz P, et al: The role of intestinal P-glycoprotein in the interaction of digoxin and rifampin. J Clin Invest 104:147–153, 1999

Griffith DE, Brown BA, Girard WM, et al: Adverse events associated with high-dose rifabutin in macrolide-containing regimens for the treatment of *Mycobacterium avium* complex lung disease. Clin Infect Dis 21:594–598, 1995

Grim SA, Romanelli F: Tenofovir disoproxil fumarate. Ann Pharmacother 37:849–859, 2003

Grinsztejn B, Nguyen BY, Katlama C, et al: Safety and efficacy of the HIV-1 integrase inhibitor raltegravir (MK-0518) in treatment-experienced patients with multi-drug-resistant virus: a phase II randomised controlled trial. Lancet 369:1261–1269, 2007

Gupta SK, Dubé MP: Exogenous Cushing syndrome mimicking human immunodeficiency virus lipodystrophy. Clin Infect Dis 35(6):E69–E71 (Epub), 2002

Gustavson LE, Shi H, Palmer R, et al: Drug interaction between clarithromycin and fluconazole in healthy subjects (abstract). Presented at 97th annual meeting, American Society for Clinical Pharmacology and Therapeutics, Orlando, FL, March 20–22, 1996

Hagg S, Spigset O, Mjorndal T, et al: Absence of interaction between erythromycin and a single dose of clozapine. Eur J Clin Pharmacol 55:221–226, 1999

Hamzeh FM, Benson C, Gerber J, et al: Steady-state pharmacokinetic interaction of modified-dose indinavir and rifabutin. AIDS Clinical Trials Group 365 Study Team. Clin Pharmacol Ther 73:159–169, 2003

Hardy H, Skolnick PR: Enfuvirtide, a new fusion inhibitor for therapy of human immunodeficiency virus infection. Pharmacotherapy 24:198–211, 2004

Hardy H, Esch LDF, Morse GD: Glucose disorders associated with HIV and its drug therapy. Ann Pharmacother 35:343–351, 2001

Hare CB, Vu MP, Grunfeld C: Simvastatin-nelfinavir interaction implicated in rhabdomyolysis and death. Clin Infect Dis 35:E111–E112 (Epub), 2002

Harris S, Leroy B: Effects of telithromycin on the pharmacokinetics of cisapride, simvastatin, and midazolam (abstract). Presented at American College of Clinical Pharmacy Annual Meeting, Tampa, FL, October 21–24, 2001

Hendershot PE, Antal EJ, Welshman IR, et al: Linezolid: pharmacokinetic and pharmacodynamic evaluation of coadministration with pseudoephedrine HCl, phenylpropanolamine HCl, and dextromethorphan H. Br J Clin Pharmacol 41:563–572, 2001

Hennessy M, Kelleher D, Spiers JP, et al: St John's wort increases expression of P-glycoprotein: implications for drug interactions. Br J Clin Pharmacol 53:75–82, 2002

Herrlin K, Segerdahl M, Gustafsson LL, et al: Methadone, ciprofloxacin, and adverse drug reactions. Lancet 356:2069–2070, 2000

Hesse LM, von Moltke LL, Shader RI, et al: Ritonavir, efavirenz, and nelfinavir inhibit CYP2B6 activity in vitro: potential drug interactions with bupropion. Drug Metab Dispos 29:100–102, 2001

Hesse LM, Greenblatt DJ, von Moltke LL, et al: Ritonavir has minimal impact on the pharmacokinetic disposition of a single dose of bupropion administered to human volunteers. J Clin Pharmacol 46:567–576, 2006

Hiller A, Olkkola KT, Isohanni P, et al: Unconsciousness associated with midazolam and erythromycin. Br J Anaesth 65:826–828, 1990

Hogeland GW, Swindells S, McNabb JC, et al: Lopinavir/ritonavir reduces bupropion plasma concentrations in healthy subjects. Clin Pharmacol Ther 81:69–75, 2007

Hoggard PG, Sales SD, Kewn S, et al: Correlation between intracellular pharmacological activation of nucleoside analogues and HIV suppression in vitro. Antivir Chem Chemother 11:353–358, 2000

Holmes VF: Rifampin-induced methadone withdrawal in AIDS (letter). J Clin Psychopharmacol 10:443–444, 1990

Horowitz RS, Dart RC, Gomez HF: Clinical ergotism with lingual ischemia induced by clarithromycin-ergotamine interaction. Arch Intern Med 156:456–458, 1996

Hsyu PH, Schultz-Smith MD, Lillibridge JH, et al: Pharmacokinetic interactions between nelfinavir and 3-hydroxy-3-methylglutaryl coenzyme A reductase inhibitors atorvastatin and simvastatin. Antimicrob Agents Chemother 45:3445–3450, 2001

Huang L, Wring SA, Woolley JL: Induction of P-glycoprotein and cytochrome P450 3A by HIV protease inhibitors. Drug Metab Dispos 29:754–760, 2001

Hung I, Wu A, Cheng V, et al: Fatal interaction between clarithromycin and colchicine in patients with renal insufficiency: a retrospective study. Clin Infect Dis 41:291–300, 2005

Hyland R, Jones BC, Smith DA: Identification of the Cytochrome P450 Enzymes Involved in the N-Oxidation of Voriconazole. Drug Metab Dispos 31:540–547, 2003

Intelence (package insert). Raritan, NJ, Tibotec, January 2008

Isentress (package insert). Whitehouse Station, NJ, Merck & Co., October 2007

Ito K, Ogihara K, Kanamitsu S: Prediction of the in vivo interactions between midazolam and macrolides based on in vitro studies using human liver microsomes. Drug Metab Dispos 31:945–954, 2003

Jain AK, Venkataramanan R, Fridell JA: Nelfinavir, a protease inhibitor, increases sirolimus levels in a liver transplantation patient: a case report. Liver Transpl 8:838–840, 2002a

Jain AK, Venkataramanan R, Shapiro R: The interaction between antiretroviral agents and tacrolimus in liver and kidney transplant patients. Liver Transpl 8:841–845, 2002b

Jayasagar G, Dixit AA, Kishan V, et al: Effect of clarithromycin on the pharmacokinetics of tolbutamide. Drug Metabol Drug Interact 16:207–215, 2000

John M, McKinnon EJ, James IR: Randomized, controlled, 48-week study of switching stavudine and/or protease inhibitors to Combivir/abacavir to prevent or reverse lipoatrophy in HIV-infected patients. J Acquir Immune Defic Syndr 33:29–33, 2003

Jover F, Cuadrado JM, Andreu L, et al: Reversible coma caused by risperidone-ritonavir interaction. Clin Neuropharmacol 25:251–253, 2002

Kakuda TN: Pharmacology of nucleoside and nucleotide reverse transcriptase inhibitor–induced mitochondrial toxicity. Clin Ther 22:685–708, 2000

Kamali F, Thomas SHL, Edwards C: The influence of steady-state ciprofloxacin on the pharmacokinetics and pharmacodynamics of a single dose of diazepam in healthy volunteers. Eur J Clin Pharmacol 44:365–367, 1993

Kantola T, Kivisto KT, Neuvonen PJ: Erythromycin and verapamil considerably increase serum simvastatin and simvastatin acid concentrations. Clin Pharmacol Ther 64:177–182, 1998

Kantola T, Backman JT, Niemi M, et al: Effect of fluconazole on plasma fluvastatin and pravastatin. Eur J Clin Pharmacol 56:225–229, 2000

Kato Y, Fujii T, Mizoguchi N, et al: Potential interaction between ritonavir and carbamazepine. Pharmacotherapy 20:851–854, 2000

Kay L, Kampmann JP, Svendsen T, et al: Influence of rifampicin and isoniazid on the kinetics of phenytoin. Br J Clin Pharmacol 20:323–326, 1985

Kearney BP, Flaherty JF, Shah J: Tenofovir disoproxil fumarate: clinical pharmacology and pharmacokinetics. Clin Pharmacokinet 43:595–612, 2004

Kelly DV, Beique LC, Bowmer MI: Extrapyramidal symptoms with ritonavir/indinavir plus risperidone. Ann Pharmacother 36:827–830, 2002

Ketek (package insert). Bridgewater, NJ, Sanofi-Aventis, 2007

Kim YH, Cha IJ, Shim JC, et al: Effect of rifampin on the plasma concentration and the clinical effect of haloperidol concomitantly administered to schizophrenic patients. J Clin Psychopharmacol 16:247–252, 1996

Kinzig-Schippers M, Fuhr U, Zaigler M, et al: Interaction of pefloxacin and enoxacin with the human cytochrome P450 enzyme CYP1A2. Clin Pharmacol Ther 65:262–274, 1999

Kivisto KT, Lamberg TS, Kantola T, et al: Plasma buspirone concentrations are greatly increased by erythromycin and itraconazole. Clin Pharmacol Ther 62:348–354, 1997

Kivisto KT, Kantola T, Neuvonen PJ: Different effects of itraconazole on the pharmacokinetics of fluvastatin and lovastatin. Br J Clin Pharmacol 46:49–53, 1998

Kivisto KT, Lamberg TS, Neuvonen PJ: Interactions of buspirone with itraconazole and rifampicin: effects on the pharmacokinetics of the active 1-(2-pyrimidinyl)-piperazine metabolite of buspirone. Pharmacol Toxicol 84:94–97, 1999

Kreek MJ, Garfield JW, Gutjahr CL, et al: Rifampin induced methadone withdrawal. N Engl J Med 294:1104–1106, 1976

Lamberg TS, Kivisto KT, Neuvonen PJ: Concentrations and effects of buspirone are considerably reduced by rifampin. Br J Clin Pharmacol 45:381–385, 1998

Lamisil (package insert). East Hanover, NJ, Novartis, November 2005

Lana R, Nunez M, Mendoza JL, et al: Rate and risk factors of liver toxicity in patients receiving antiretroviral therapy. Med Clin (Barc) 117:607–610, 2001

Lazar JD, Wilner KD: Drug interactions with fluconazole. Rev Infect Dis 12 (suppl 3):S327–S333, 1990

Lea AP, Faulds D: Stavudine: a review of its pharmacodynamic and pharmacokinetic properties and clinical potential in HIV infection. Drugs 51:846–864, 1996

Lee CG, Gottesman MM, Cardarelli CO, et al: HIV-1 protease inhibitors are substrates for the MDR1 multidrug transporter. Biochemistry 37:3594–3601, 1998

Lennernas H: Clinical pharmacokinetics of atorvastatin. Clin Pharmacokinet 42:1141–1160, 2003

Levitra (package insert). West Haven, CT, Bayer Pharmaceuticals, 2007

Llibre JM, Romeu J, Lopez E, et al: Severe interaction between ritonavir and acenocoumarol. Ann Pharmacother 36:621–623, 2002

Lochet P, Peyriere H, Lotthe A, et al: Long-term assessment of neuropsychiatric adverse reactions associated with efavirenz. HIV Med 4:62–66, 2003

Luurila H, Olkkola KT, Neuvonen PJ: Interaction between erythromycin and the benzodiazepines diazepam and flunitrazepam. Pharmacol Toxicol 78:117–122, 1996

Maggi P, Ladisa N, Cinori E, et al: Cutaneous injection site reactions to long-term therapy with enfuvirtide. J Antimicrob Chemother 53:678–681, 2004

Maldonado S, Lamson M, Gigliotti M, et al: Pharmacokinetic interaction between nevirapine and rifabutin, in Abstracts of the 39th International Conference on Antimicrobial Agents and Chemotherapy. Washington, DC, American Society for Microbiology, 1999, abstract 341

Markowitz JS, DeVane CL: Suspected ciprofloxacin inhibition of olanzapine resulting in increased plasma concentration. J Clin Psychopharmacol 19:289–291, 1999

Markowitz JS, DeVane CL: Rifampin-induced selective serotonin reuptake inhibitor withdrawal syndrome in a patient treated with sertraline (letter). J Clin Psychopharmacol 20:109–110, 2000

Markowitz JS, Gill HS, DeVane CL, et al: Fluoroquinolone inhibition of clozapine metabolism (letter). Am J Psychiatry 154:881, 1997

Martin AM, Hammond E, Nolan D, et al: Accumulation of mitochondrial DNA mutations in human immunodeficiency virus–infected patients treated with nucleoside-analogue reverse-transcriptase inhibitors. Am J Hum Genet 72:549–560, 2003; correction, 72:1358, 2003

Maxa JL, Melton LB, Ogu CC, et al: Rhabdomyolysis after concomitant use of cyclosporine, simvastatin, gemfibrozil, and itraconazole. Ann Pharmacother 36:820–823, 2002

McCance-Katz EF, Rainey PM, Jatlow P, et al: AIDS Clinical Trials Group 262: Methadone effects on zidovudine disposition. J Acquir Immune Defic Syndr Hum Retrovirol 18:435–443, 1998

McCance-Katz EF, Rainey PM, Friedland G, et al: The protease inhibitor lopinavir-ritonavir may produce opiate withdrawal in methadone-maintained patients. Clin Infect Dis 37:476–482, 2003

McNeely MC, Yarchoan R, Broder S, et al: Dermatologic complications associated with administration of 2′,3′-dideoxycytidine in patients with human immunodeficiency virus infection. J Am Acad Dermatol 21:1213–1217, 1989

Merry C, Mulcahy F, Barry M, et al: Saquinavir interaction with midazolam: pharmacokinetic considerations when prescribing protease inhibitors for patients with HIV disease. AIDS 11:268–269, 1997

Mizuki Y, Fujiwara I, Yamaguchi T: Pharmacokinetic interactions related to the chemical structures of fluoroquinolones. J Antimicrob Chemother 37 (suppl A):41–55, 1996

Montay G, Shi J, Leroy B, et al: Effects of telithromycin on the pharmacokinetics of digoxin in healthy men (abstract no. A-1834), in Abstracts of the 42nd International Conference on Antimicrobial Agents and Chemotherapy. San Diego, CA, American Society for Microbiology, 2002, p 28

Mouly S, Lown KS, Kornhauser D, et al: Hepatic but not intestinal CYP3A4 displays dose-dependent induction by efavirenz in humans. Clin Pharmacol Ther 72:1–9, 2002

Moyano Calvo JL, Huesa Martinez I, Cruz Navarro N, et al: Urinary lithiasis secondary to indinavir in an HIV-positive patient. Arch Esp Urol 54:1117–1120, 2001

Moyle G: Clinical manifestations and management of antiretroviral nucleoside analog-related mitochondrial toxicity. Clin Ther 22:911–939, 2000

Moyle GJ, Sadler M: Peripheral neuropathy with nucleoside antiretrovirals: risk factors, incidence and management. Drug Saf 6:481–494, 1998

Moyle GJ, Baldwin C, Langroudi B, et al: A 48-week, randomized, open-label comparison of three abacavir-based substitution approaches in the management of dyslipidemia and peripheral lipoatrophy. J Acquir Immune Defic Syndr 33:22–28, 2003

Muirhead GJ, Faulkner S, Harness JA, et al: The effects of steady-state erythromycin and azithromycin on the pharmacokinetics of sildenafil in healthy volunteers. Br J Clin Pharmacol 53 (suppl 1):375–435, 2002

Muirhead G, Ridgway C, Leahy D, et al: A study to investigate the combined coadministration of P450 CYP3A4 inhibitors and inducers on the pharmacokinetics of the novel CCR5 inhibitor UK-427,857. Seventh International Congress on Drug Therapy in HIV Infection, Glasgow, abstract P284, 2004

Murphy RL, Sommadossi JP, Lamson M, et al: Antiviral effect and pharmacokinetic interaction between nevirapine and indinavir in persons infected with human immunodeficiency virus type 1. J Infect Dis 179:1116–1123, 1999

Mycamine (package insert). Deerfield, IL, Astellas Pharma US, 2008

Negredo E, Molto J, Burger D, et al: Unexpected CD4 count decline in patients receiving didanosine and tenofovir-based regimens despite undetectable viral load. AIDS 18:459–463, 2004

Neuvonen PJ, Kantola T, Kivisto KT: Simvastatin but not pravastatin is very susceptible to interaction with the CYP3A4 inhibitor itraconazole. Clin Pharmacol Ther 63:332–341, 1998

Nishimura Y, Kurata N, Sakurai E, et al: Inhibitory effect of antituberculosis drugs on human cytochrome P450-mediated activities. J Pharmacol Sci 96:293–300, 2004

Norvir (package insert). North Chicago, IL, Abbott Laboratories, January 2006

Olkkola KT, Aranko K, Luurila H, et al: A potentially hazardous interaction between erythromycin and midazolam. Clin Pharmacol Ther 53:298–305, 1993

Ouellet D, Hsu A, Qian J, et al: Effect of ritonavir on the pharmacokinetics of ethinyl oestradiol in healthy female volunteers. Br J Clin Pharmacol 46:111–116, 1998

Pan-Zhou XR, Cretton-Scott E, Zhou ZJ, et al: Role of human liver P450s and cytochrome b5 in the reductive metabolism of 3'-azido-3'deoxythymidine (AZT) to 3'-amino-3'deoxythymidine. Biochem Pharmacol 55:757–766, 1998

Park-Wyllie LY, Antoniou T: Concurrent use of bupropion with CYP2B6 inhibitors, nelfinavir, ritonavir and efavirenz: a case series. AIDS 17:638–640, 2003

Park-Wyllie, L, Juurlink D, Kopp A, et al: Outpatient gatifloxacin therapy and dysglycemia in older adults. N Engl J Med 354:1352–1361, 2006

Perloff MD, von Moltke LL, Fahey JM, et al: Induction of P-glycoprotein expression by HIV protease inhibitors in cell culture. AIDS 14:1287–1289, 2000

Perloff MD, von Moltke LL, Marchand JE, et al: Ritonavir induces P-glycoproteins expression, multidrug resistance-associated protein (MRP1) expression, and drug transporter–mediated activity in a human intestinal cell line. J Pharm Sci 90:1829–1837, 2001

Peyriere H, Reynes J, Rouanet I, et al: Renal tubular dysfunction associated with tenofovir therapy: report of 7 cases. J Acquir Immune Defic Syndr 35:269–273, 2004

Phillips EJ, Rachlis AR, Ito S: Digoxin toxicity and ritonavir: a drug interaction mediated through P-glycoprotein? AIDS 17:1577–1578, 2003

Piscitelli SC, Burstein AH, Chaitt D, et al: Indinavir concentrations and St John's wort. Lancet 355:547–548, 2000a

Piscitelli SC, Kress DR, Bertz RJ, et al: The effect of ritonavir on the pharmacokinetics of meperidine and normeperidine. Pharmacotherapy 20:549–553, 2000b

Placidi L, Cretton EM, Placidi M, et al: Reduction of 3'-azido-3'deoxythymidine to 3'-amino-3'deoxythymidine in human liver microsomes and its relationship to cytochrome P450. Clin Pharmacol Ther 54:168–176, 1993

Pollak PT, Sketris IS, MacKenzie SL, et al: Delirium probably induced by clarithromycin in a patient receiving fluoxetine. Ann Pharmacother 29:486–488, 1995

Quinupristin/Dalfopristin. Med Lett Drugs Ther 41:109–110, 1999

Raaska K, Neuvonen PJ: Ciprofloxacin increases serum clozapine and N-desmethylclozapine: a study in patients with schizophrenia. Eur J Clin Pharmacol 56:585–589, 2000

Rachline A, Lariven S, Descamps V, et al: Leucocytoclastic vasculitis and indinavir. Br J Dermatol 143:1112–1113, 2000

Randinitis E, Alvey C, Koup J: Drug interactions with clinafloxacin. Antimicrob Agents Chemother 45:2543–2552, 2001

Rapp RP: Pharmacokinetics and pharmacodynamics of intravenous and oral azithromycin: enhanced tissue activity and minimal drug interactions. Ann Pharmacother 32:785–793, 1998

Ravnan SL, Locke C: Levofloxacin and warfarin interaction. Pharmacotherapy 21:884–885, 2001

Ray AS, Olson L, Fridland A: Role of purine nucleoside phosphorylase in interactions between 2'3'-dideoxyinosine and allopurinol, gancyclovir, or tenofovir. Antimicrob Agents Chemother 48:1089–1095, 2004

Rayner CR, Esch LD, Wynn HE: Symptomatic hyperbilirubinemia with indinavir/ritonavir-containing regimen. Ann Pharmacother 35:1391–1395, 2001

Reyataz (package insert). Princeton, NJ, Bristol-Myers Squibb, February 2008

Robson RA: The effects of quinolones on xanthine pharmacokinetics. Am J Med 92:22S–25S, 1992

Roffey SJ, Cole S, Comby P, et al: The disposition of voriconazole in mouse, rat, rabbit, guinea pig, dog and human. Drug Metab Dispos 31:731–741, 2003

Rossero R, Asmuth DM, Grady JJ, et al: Hydroxyurea in combination with didanosine and stavudine in antiretroviral-experienced HIV-infected subjects with a review of the literature. Int J STD AIDS 14:350–355, 2003

Rosso R, Di Biagio A, Ferrazin A, et al: Fatal lactic acidosis and mimicking Guillain-Barre syndrome in an adolescent with human immunodeficiency virus infection. Pediatr Infect Dis J 22:668–670, 2003

Rotunda A, Hirsch RJ, Scheinfeld N, et al: Severe cutaneous reactions associated with the use of human immunodeficiency virus medications. Acta Derm Venereol 83:1–9, 2003

Rotzinger S, Fang J, Baker GB: Trazodone is metabolized to m-chlorophenylpiperazine by CYP3A4 from human sources. Drug Metab Dispos 26:572–575

Ruane PJ, Parenti DM, Margolis DM, et al: Compact quadruple therapy with the lamivudine/zidovudine combination tablet plus abacavir and efavirenz, followed by the lamivudine/zidovudine/abacavir triple nucleoside tablet plus efavirenz in treatment-naive HIV-infected adults. COL30336 Study Team. HIV Clin Trials 4:231–243, 2003

Rubenstein E, Prokocimer P, Talbot GH: Safety and tolerability of quinupristin/dalfopristin: administration guidelines. J Antimicrob Chemother 44 (suppl A):37–46, 1999

Sagir A, Wettstein M, Oette M, et al: Budesonide-induced acute hepatitis in an HIV-positive patient with ritonavir as a co-medication. AIDS 16:1191–1192, 2002

Sahai J, Gallicano K, Pakuts A, et al: Effect of fluconazole on zidovudine pharmacokinetics in patients infected with human immunodeficiency virus. J Infect Dis 169:1103–1107, 1994

Sahai J, Cameron W, Salgo M, et al: Drug interaction study between saquinavir (SQV) and nevirapine (NVP), in Abstracts of the 4th Conference on Retroviruses and Opportunistic Infections, Chicago, IL, February 1997, abstract 496. Available at http://www.retroconference.org/1997/Abstracts/A496_abstract.htm. Accessed July 30, 2008.

Scholtz HE, Pretorius SG, Wessels DH, et al: Telithromycin (HMR 3647), a new ketolide antimicrobial, does not affect the pharmacodynamics or pharmacokinetics of warfarin in healthy adult males. Poster presented at 5th International Conference on Macrolides, Azalides, Streptogramins, Ketolides and Oxazolidinones, Seville, Spain, January 2000

Schonder KS, Shullo MA, Okusanya O: Tacrolimus and lopinavir/ritonavir interaction in liver transplantation. Ann Pharmacother 37:1793–1796, 2003

Schrooten W, Colebunders R, Youle M, et al: Sexual dysfunction associated with protease inhibitor containing highly active antiretroviral treatment. Eurosupport Study Group. AIDS 15:1019–1023, 2001

Schupbach J, Popovic M, Gilden RV, et al: Serological analysis of a subgroup of human T-lymphotropic retroviruses (HTLV-III) associated with AIDS. Science 224:503–505, 1984

Scott LJ, Perry CM: Delavirdine: a review of its use in HIV infection. Drugs 60:1411–1444, 2000

Seithel A, Ebert S, Singer K, et al: The influence of macrolide antibiotics on the uptake of organic anions and drugs mediated by OATP1B1 and OATP1B3. Drug Metab Dispos 35:779–786, 2007

Self T, Corley CR, Nabhan S, et al: Case report: interaction of rifampin and nortriptyline. Am J Med Sci 311:80–81, 1996

Self TH, Chrisman CR, Baciewicz AM, et al: Isoniazid drug and food interactions. Am J Med Sci 317:304–311, 1999

Seminari E, Castagna A, Lazzarin A: Etravirine for the treatment of HIV infection.. Expert Rev Anti Infect Ther 6:427–433, 2008

Shi J, Montay G, Bhargava VO: Clinical pharmacokinetics of telithromycin, the first ketolide antibacterial. Clin Pharmacokinet 44:915–934, 2005

Simdon J, Watters D, Bartlett S, et al: Ototoxicity associated with use of nucleoside analog reverse transcriptase inhibitors: a report of 3 possible cases and review of the literature. Clin Infect Dis 32:1623–1627, 2001

Simpson DM, Citak KA, Godfrey E, et al: Myopathies associated with human immunodeficiency virus and zidovudine: can their effects be distinguished? Neurology 43:971–976, 1993

Simpson DM, Tagliati M: Nucleoside analogue–associated peripheral neuropathy in human immunodeficiency virus infection. J Acquir Immune Defic Syndr Hum Retrovirol 9:153–161, 1995

Skowron G, Leoung G, Kerr B, et al: Lack of pharmacokinetic interaction between nelfinavir and nevirapine (editorial). AIDS 12:1243–1244, 1998

Smith PF, DiCenzo R, Morse GD: Clinical pharmacokinetics of non-nucleoside reverse transcriptase inhibitors. Clin Pharmacokinet 40:893–905, 2001

Spicer ST, Liddle C, Chapman JR, et al: The mechanism of cyclosporine toxicity induced by clarithromycin. Br J Clin Pharmacol 43:194–196, 1997

Spina E, Pisani F, Perucca E: Clinically significant pharmacokinetic drug interactions with carbamazepine: an update. Clin Pharmacokinet 31:198–214, 1996

Sporanox (package insert). Titusville, NJ, Janssen Pharmaceutica, 2000

Spruance SL, Pavia AT, Mellors JW, et al: Clinical efficacy of monotherapy with stavudine compared with zidovudine in HIV-infected, zidovudine experienced patients: a randomized, double-blind, controlled trial. Bristol-Myers Squibb Stavudine 019 Study Group. Ann Intern Med 126:355–363, 1997

Staas H, Kubitza D: Lack of pharmacokinetic interaction between moxifloxacin, a novel 8-methoxyfluoroquinolone, and theophylline. Clin Pharmacokinet 40 (suppl 1):63–70, 2001a

Staas H, Kubitza D: Profile of moxifloxacin drug interactions. Clin Infect Dis 32 (suppl 1):S47–S50, 2001b

Strayhorn VA, Baciewicz AM, Self TH: Update on rifampin drug interactions, III. Arch Intern Med 157:2453–2458, 1997

Sulkowski MS: Hepatotoxicity associated with antiretroviral therapy containing HIV-1 protease inhibitors. Semin Liver Dis 23:183–194, 2003

Sustiva (package insert). Princeton, NJ, Bristol-Myers Squibb, 2002

Synercid (package insert). Greenville, NC, DSM Pharmaceuticals, July 2003

Taburet AM, Singlas E: Drug interactions with antiretroviral drugs. Clin Pharmacokinet 30:385–401, 1996

Takeda M, Nishinuma K, Yamashita S, et al: Serum haloperidol levels of schizophrenics receiving treatment for tuberculosis. Clin Neuropharmacol 9:386–397, 1986

Tayrouz Y, Ganssmann B, Ding R: Ritonavir increases loperamide plasma concentrations without evidence for P-glycoprotein involvement. Clin Pharmacol Ther 70:405–414, 2001

Temple ME, Nahata MC: Rifapentine: its role in the treatment of tuberculosis. Ann Pharmacother 33:1203–1210, 1999

Theuretzbacher U, Ihle F, Derendorf H: Pharmacokinetics/pharmacodynamic profile of voriconazole. Clin Pharmacokinet 45:649–663, 2006

Tokinaga N, Kondo T, Kaneko S, et al: Hallucinations after a therapeutic dose of benzodiazepine hypnotics with co-administration of erythromycin. Psychiatry Clin Neurosci 50:337–339, 1996

Tosti A, Piraccini BM, D'Antuono A, et al: Paronychia associated with antiretroviral therapy. Br J Dermatol 140:1165–1168, 1999

Tran JQ, Gerber JG, Kerr BM: Delavirdine: clinical pharmacokinetics and drug interactions. Clin Pharmacokinet 40:207–226, 2001

Trapnell CB, Klecker RW, Jamis-Dow C, et al: Glucuronidation of 3'-azido-3'-deoxythymidine (zidovudine) by human liver microsomes: relevance to clinical pharmacokinetic interactions with atovaquone, fluconazole, methadone, and valproic acid. Antimicrob Agents Chemother 42:1592–1596, 1998

Tribble MA, Gregg CR, Margolis DM, et al: Fatal ergotism induced by an HIV protease inhibitor. Headache 42:694–695, 2002

Tseng AL, Foisy MM: Management of drug interactions in patients with HIV. Ann Pharmacother 31:1040–1058, 1997

Vandamme AM, Van Vaerenbergh K, DeClercq E: Anti-human immunodeficiency virus drug combination strategies. Chem Chemother 9:187–203, 1998

Varis T, Kivisto KT, Backman JT, et al: The cytochrome P450 3A4 inhibitor itraconazole markedly increases the plasma concentrations of dexamethasone and enhances its adrenal-suppressant effect. Clin Pharmacol Ther 68:487–494, 2000

Venkatakrishnan K, von Moltke LL, Greenblatt DJ: Effects of the antifungal agents on oxidative drug metabolism: clinical relevance. Clin Pharmacokinet 38:111–180, 2000

Viagra (package insert). New York, NY, Pfizer Inc., 2000

Videx (package insert). Princeton, NJ, Bristol-Myers Squibb, 2003

Villikka K, Kivisto KT, Backman JT, et al: Triazolam is ineffective in patients taking rifampin. Clin Pharmacol Ther 61:8–14, 1997a

Villikka K, Kivisto KT, Luurila H, et al: Rifampin reduces plasma concentrations and effects of zolpidem. Clin Pharmacol Ther 62:629–634, 1997b

Villikka K, Kivisto KT, Neuvonen PJ: The effect of rifampin on the pharmacokinetics of oral and intravenous ondansetron. Clin Pharmacol Ther 65:377–381, 1999

Viread (package insert). Foster City, CA, Gilead Sciences, October 26, 2001

Vishnuvardhan D, von Moltke LL, Richert C, et al: Lopinavir: acute exposure inhibits P-glycoprotein: extended exposure induces P-glycoprotein. AIDS 17:1092–1094, 2003

von Moltke LL, Durol AL, Duan SX, et al: Potent mechanism-based inhibition of human CYP3A in vitro by amprenavir and ritonavir: comparison with ketoconazole. Eur J Clin Pharmacol 56:259–261, 2000

von Moltke LL, Greenblatt DJ, Granda BW, et al: Inhibition of human cytochrome P450 isoforms by nonnucleoside reverse transcriptase inhibitors. J Clin Pharmacol 41:85–91, 2001

Vourvahis M, Kashuba AD: Mechanisms of pharmacokinetic and pharmacodynamic drug interactions associated with ritonavir-enhanced tipranavir. Pharmacotherapy 27:888–908, 2007

Vousden M, Allen A, Lewis A, et al: Lack of pharmacokinetic interaction between gemifloxacin and digoxin in healthy elderly volunteers. Chemotherapy 45:485–90, 1999

Wang EJ, Lew K, Casciano CN, et al: Interaction of common azole antifungals with P-glycoprotein. Antimicrob Agents Chemother 46:160–165, 2002

Watkins JS, Polk RE, Stotka JL: Drug interactions of macrolides: emphasis on dirithromycin. Ann Pharmacother 31:349–356, 1997

Wilde JT: Protease inhibitor therapy and bleeding. Haemophilia 6:487–490, 2000

Wu DS, Stoller ML: Indinavir urolithiasis. Curr Opin Urol 10:557–561, 2000

Wynn GH, Zapor MJ, Smith BH, et al: Antiretrovirals, part 1: overview, history, and focus on protease inhibitors. Psychosomatics 45:262–270, 2004

Yasui N, Otani K, Kaneko S, et al: A kinetic and dynamic study of oral alprazolam with and without erythromycin in humans: in vivo evidence for the involvement of CYP3A4 in alprazolam metabolism. Clin Pharmacol Ther 59:514–519, 1996

Yasui N, Otani K, Kaneko S, et al: Carbamazepine toxicity induced by clarithromycin coadministration in psychiatric patients. Int Clin Psychopharmacol 12:225–229, 1997

Yeates RA, Laufen H, Zimmermann T, et al: Pharmacokinetic and pharmacodynamic interaction study between midazolam and the macrolide antibiotics, erythromycin, clarithromycin, and the azalide azithromycin. Int J Clin Pharmacol Ther 35:577–599, 1997

Zapor MJ, Cozza KL, Wynn GH, et al: Antiretrovirals, part II: focus on non-protease inhibitor antiretrovirals (NRTIs, NNRTIs, and fusion inhibitors). Psychosomatics 45:524–535, 2004

Zhang X, Lalezari JP, Badley AD, et al: Assessment of drug-drug interaction potential of enfuvirtide in human immunodeficiency virus type 1–infected patients. Clin Pharmacol Ther 75:558–568, 2004

Zhao XJ, Koyama E, Ishizaki T: An in vitro study on the metabolism and possible drug interactions of rokitamycin, a macrolide antibiotic, using human liver microsomes. Drug Metab Dispos 27:776–785, 1999

Ziagen (package insert). Research Triangle Park, NC, GlaxoSmithKline, 2006

Zyvox (package insert). New York, NY, Pharmacia & Upjohn Company, 2007

15

Neurology

Gary H. Wynn, M.D.

Scott C. Armstrong, M.D., D.F.A.P.A., F.A.P.M.

Reminder: This chapter is dedicated primarily to metabolic and transporter interactions. Interactions due to displaced protein-binding, alterations in absorption or excretion, and pharmacodynamics are not covered.

Neurology and psychiatry share many pharmacological treatments, and they often share patients as well. The use of tricyclic antidepressants (TCAs) and selective serotonin reuptake inhibitors for poststroke depression, TCAs for migraine prophylaxis, cholinesterase inhibitors for dementia, and the overlapping psychiatric and neurological uses of antiepileptic drugs are just a few examples. The antiepileptics have significant potential for phase I and phase II drug interactions. These agents are substrates of phase I and phase II enzymes, but they interact to both induce and inhibit these enzymes as well. In this chapter, we cover antiepileptic agents, antiparkinsonian drugs, cognitive enhancers, triptans, and ergots for migraine headaches (see Table 15–1).

325

TABLE 15–1.   Antiepileptic drugs

Drug	Metabolism	Enzyme(s) inhibited	Enzyme(s) induced[a]
Carbamazepine (Tegretol)	3A4, 2B6, 2C8, 2E1,2C9, 1A2, UGT2B7, ABCB1	?2C19	3A4, 1A2, 2B6, 2C8, 2C9, UGT1A4
Ethosuximide (Zarontin)	3A4, phase II	None known	?Pan-inducer
Felbamate (Felbatol)	3A4, 2E1, ABCB1	2C19	3A4
Gabapentin (Neurontin)	Excreted in urine unchanged	None known	None known
Lamotrigine (Lamictal)	UGT1A4; excreted in urine unchanged	None known	UGT1A4 (mild, autoinduction)
Levetiracetam (Keppra)	Non–P450 phase I hydrolysis	None known	None known
Methsuximide (Celontin)	3A4, phase II	None known	?Pan-inducer
Oxcarbazepine (Trileptal)	3A4, ABCB1	2C19	3A4[b], UGT1A4[b]
Phenobarbital	2C9, 2C19, 2E1; 25% excreted in urine unchanged	3A4, ?phase II	UGTs, 3A4, 2C9, 2C19, 1A2, ?others
Phenytoin (Dilantin)	2C9, 2C19, UGT1A family, ABCB1	None known	2B6, 2C9, 2C19, 3A4, UGT1A1, UGT1A4
Primidone (Mysoline)	2C9, 2C19, 2E1; 25% excreted in urine unchanged	3A4, ?phase II	UGTs, 3A4, 2C9, 2C19, 1A2, ?others
Tiagabine (Gabitril)	3A4, UGTs	None known	None known
Topiramate (Topamax)	70% excreted in urine unchanged; phase I, phase II, ABCB1	2C19[b]	None known; decreases ethinyl estradiol levels

TABLE 15–1. Antiepileptic drugs *(continued)*

Drug	Metabolism	Enzyme(s) inhibited	Enzyme(s) induced[a]
Valproic acid (Depakote/ Depakene)	Complex: 2C9, 2C19, 2A6, UGT1A6, 1A9, 2B7, β-oxidation	2D6, 2C9, UGTs 1A4, 1A9, 2B7, 2B15, epoxide hydroxylase	None known, though some evidence of 3A4, ABCB1
Vigabatrin (Sabril)	Primarily excreted in urine unchanged	None known	Decreases phenytoin levels by unknown mechanism
Zonisamide (Zonegran)	3A4, acetylation, sulfonation	None known	None known

Note.   UGT=uridine 5'-diphosphate glucuronosyltransferase.
[a]**Bold** type indicates potent induction.
[b]Moderate inhibition or induction.
[c]Mild inhibition or induction.

Antiepileptic Drugs

Phenobarbital was first noted to be efficacious for seizure control in 1912; phenytoin (Dilantin) was marketed in 1938; and valproic acid (valproate; Depakote) was marketed in 1968. The metabolism of these three drugs was, and in some ways remains, a mystery. Drugs developed after the late 1980s have undergone human liver microsome studies and some *in vivo* testing with probes. Unfortunately, many older drugs, being off patent, have not been reevaluated with this new technology. The hepatic enzymes involved in the metabolism of phenobarbital and valproic acid have been deduced from their *in vivo* drug interactions, and the older literature is sometimes confusing. Phenytoin has been studied more extensively, having become a probe drug for 2C9 and 2C19. In this section, we briefly discuss the numerous antiepileptic drugs and their P450 and phase II profiles.

General Drug Interactions

Barbiturates

Twenty-five percent of primidone (Mysoline) is metabolized to phenobarbital; primidone is also metabolized to phenylethylmalonamide (a weak anticonvulsant). Therefore, primidone's pharmacokinetic effects on other drugs may be similar to those of phenobarbital.

Phenobarbital is metabolized at 2C9, 2C19, and 2E1 (Tanaka 1999). 2C9 is the drug's major enzyme of metabolism. Metabolism of phenobarbital may be inhibited by potent inhibitors of these enzymes, including other anticonvulsants such as valproic acid (Perucca 2006). It would be expected that potent 2C9 and 2C19 inhibitors such as fluvoxamine and the "pan-inhibitor" ritonavir would also inhibit the metabolism of phenobarbital and other barbiturates, resulting in significant consequences, but currently there is scant evidence of this interaction. Phenobarbital causes potent induction of CYP isozymes, a "pan-inducer," (Turnheim 2004) resulting in an overall decrease in plasma levels of CYP substrates and potential decreased efficacy. Many psychotropics are affected by phenobarbital's pan-induction, including nortriptyline (von Bahr et al. 1998), desipramine (Spina et al. 1996a), clozapine, haloperidol, and chlorpromazine (Besag and Berry 2006), and others.

Carbamazepine

Carbamazepine is metabolized primarily by 3A4, with minor contribution from 1A2, 2B6, 2C8/9, 2E1, and glucuronidation via UGT2B7 (Pearce et al. 2002; Spina et al. 1996b; Staines et al. 2004). Carbamazepine is also a substrate and inhibitor for ABCB1 transport, although the clinical relevance of inhibition is unknown (Pavek et al. 2001; Potschka et al. 2001; Weiss et al. 2003). It is also probably an inhibitor of 2C19, as evidenced by carbamazepine's ability to increase blood levels of both phenytoin and clomipramine (Lakehal et al. 2002; Zielinski and Haidukewych 1987). Carbamazepine is a potent inducer of 3A4 (Arana et al. 1988; Ucar et al. 2004), and it also induces 1A2, 2B6, 2C8/9, and UGT1A4 (Bottiger et al. 1999; Faucette et al. 2004; Parker et al. 1998).

Oxcarbazepine

Oxcarbazepine is quickly metabolized to an active monohydroxyoxcarbazepine (MHD) metabolite by the action of arylketone reductase. Both oxcarbazepine and MHD are metabolized in part through phase II glucuronidation. Additionally, oxcarbazepine is a substrate of ABCB1 (Marchi et al. 2005). Oxcarbazepine is a mild inducer of 3A4 (Fattore et al. 1999) and a moderate inducer of UGT1A4 (O'Neill and de Leon 2007). The metabolite MHD is an inhibitor of 2C19 (Lakehal et al. 2002).

Felbamate

Because of its potential for hepatotoxicity and for producing aplastic anemia, felbamate (Felbatol) is reserved for treatment of refractory seizures. The drug is metabolized by 3A4 and 2E1 and is also a substrate for ABCB1 transport (Bialecka et al. 2005; Harden et al. 2006), and its metabolism has been induced by carbamazepine, phenytoin, and phenobarbital. The manufacturer recommends a 33% reduction in phenytoin dose when felbamate is added to a regimen (Felbatol 2002). The presumed mechanism is 2C19 inhibition by felbamate. Felbamate also induces 3A4 (Turnheim 2004), reducing stable carbamazepine levels (Glue et al. 1997) and concentrations of the progestogenic component of oral contraceptives (Harden et al. 2006).

Gabapentin

Gabapentin (Neurontin) is distributed throughout the body, largely unbound (<3% bound) with proteins, and is excreted unchanged in the urine (Neurontin 2007). In clinical studies, there were no significant interactions with phenytoin, carbamazepine, valproic acid, phenobarbital, cimetidine, or oral contraceptives.

Lamotrigine

Lamotrigine is metabolized primarily by UGT1A4 (Hiller et al. 1999; Linnet et al. 2002), although one or more P450 enzymes, not yet fully understood, serve as a secondary pathway. However, this P450 pathway leads to the generation of toxic metabolites (Maggs et al. 2000). In the presence of a UGT1A4 inhibitor such as valproic acid, a greater proportion of lamotrigine is metabolized through this P450 metabolic pathway, leading to production of these toxic metabolites. This effect helps to explain why the combination of valproic acid and lamotrigine is associated with a greater incidence of both Stevens-Johnson syndrome and toxic epidermal necrolysis, even when low dosages of lamotrigine are used. Some weak autoinduction (at UGT1A4) has been noted (Lamictal 2007).

Levetiracetam

Levetiracetam (Keppra) was developed in the late 1990s as an agent for add-on therapy in patients with refractory complex partial seizures. The agent has a limited metabolism in humans, with 66% of the dose excreted in the urine unchanged (Keppra 2008). Twenty-four percent is metabolized through enzymatic hydrolysis by the acetamide group, resulting in a carboxylic acid metabolite. Levetiracetam is not metabolized by P450 enzymes, nor does the drug inhibit or induce these enzymes (Patsalos 2004). Studies by the manufacturer indicated no significant interactions with phenytoin, carbamazepine, valproic acid, phenobarbital, lamotrigine, gabapentin, primidone, oral contraceptives, digoxin, or warfarin (Keppra 2008).

Phenytoin

Phenytoin is primarily a substrate of 2C9 and 2C19 (Lakehal et al. 2002; Mamiya et al. 1998), UGT1A enzymes, and ABCB1 transport (Zhou et al. 2004). Potent 2C9 inhibitors include fluvoxamine (Luvox) and fluconazole (Diflucan); fluoxetine (Prozac) and modafinil (Provigil) are moderate inhibitors. Potent inhibitors of 2C19 include fluvoxamine, ticlopidine (Ticlid), and omeprazole (Prilosec); again, fluoxetine is a moderate inhibitor. Phenytoin induces multiple enzymes, including 2B6, 2C9/19, 3A4, and UGTs 1A1 and 1A4 (Chetty et al. 1998; Faucette et al. 2004; Gibson et al. 2002; Raucy 2003; Ritter et al. 1999).

Tiagabine

Tiagabine (Gabitril) is metabolized at 3A4, and its metabolism is susceptible to induction by other antiepileptic drugs that induce this enzyme (Gabitril 2005); use of tiagabine with carbamazepine, phenytoin, or phenobarbital caused a 60% increase in the clearance of tiagabine. Tiagabine administration in patients taking valproic acid has been associated with a mild decrease in valproic acid levels (10%), the mechanism of which is unclear (Brodie 1995; Gustavson et al. 1998); in general, however, tiagabine is not considered a P450 inhibitor or inducer (Kalviainen 1998).

Topiramate

Seventy percent of topiramate (Topamax) is excreted unchanged in the urine, with six other metabolites making up the rest of the clearance through various enzymes of phase I and phase II (Topamax 2007). Topiramate is also a substrate for ABCB1 transport (Bialecka et al. 2005). Nevertheless, topiramate's metabolism appears to be inducible, and topiramate concentrations can be decreased by 40%–48% when the drug is used with phenytoin or carbamazepine. Topiramate modestly inhibits 2C19 (Anderson 1998; Sachdeo et al. 2002). It is unclear whether topiramate induces any enzymes; however, when it was used with oral contraceptives containing ethinyl estradiol (EE), the mean total exposure to EE decreased by 18%, 21%, and 30% at daily doses of 200, 400, and 800 mg of topiramate, respectively. Therefore, efficacy of oral contraceptives may be compromised by topiramate through an unknown mechanism.

Valproic Acid

The metabolism of valproic acid is exceedingly complex, involving multiple phase I and II pathways: 2A6 and 2C9 (Sadeque et al. 1997); UGTs 1A6, 1A9, and 2B7 (Ethell et al. 2003); and β-oxidation (Pisani 1992). Valproic acid is a moderate inhibitor of 2C9 (Wen et al. 2001). It also inhibits multiple UGTs, including 1A4, 1A9, 2B7, and 2B15 (Ethell et al. 2003), as well as epoxide hydrolase, the enzyme that metabolizes the principal metabolite of carbamazepine (carbamazepine-10,11-epoxide) (Bernus et al. 1994; Bourgeois 1988). There is evidence that valproic acid is an inducer of 3A4 and ABCB1 (Eyal et al. 2006), but the clinical impact of this effect has yet to be fully elucidated.

Vigabatrin

Vigabatrin (Sabril) is excreted almost entirely unchanged in the urine, so pharmacokinetic drug interactions are not expected. However, through an unknown mechanism, vigabatrin decreases phenytoin levels by 20% (Richens 1995).

Zonisamide

Zonisamide (Zonegran) undergoes acetylation to form N-acetylzonisamide and reduction to form the open-ring metabolite 2-sulfamoylacetylphenol (SMAP). Of the excreted dose, 35% was recovered as zonisamide, 15% as N-acetylzonisamide, and 50% as the glucuronide of SMAP. Reduction of zonisamide to SMAP is mediated by 3A4. Zonisamide does not induce its own metabolism (Zonegran 2007). Zonisamide has a long half-life (63 hours), but because of induction (probably of 3A4), its half-life is significantly shorter when the drug is used with phenytoin (27 hours), phenobarbital (38 hours), or valproic acid (46 hours).

Psychotropic Drug Interactions

Induction

Carbamazepine, phenytoin, phenobarbital, and felbamate may all decrease drug levels by induction, especially if the other drug is metabolized primarily by 3A4. Therefore psychotropics such as TCAs and triazolobenzodiazepines may need dosage adjustments when administered along with these 3A4 inducers (Facciola et al. 1998). Carbamazepine administration with most antipsychotics—including olanzapine, clozapine, risperidone, paliperidone, ziprasidone, and haloperidol— results in decreases in plasma concentration of the antipsychotic. Carbamazepine has been shown to induce 1A2 *in vitro* and *in vivo* (Lucas et al. 1998).

In a controlled clinical trial, Hesslinger et al. (1999) studied the effects of carbamazepine and valproic acid on the pharmacokinetics of haloperidol (Haldol). Subjects with schizophrenia or schizoaffective disorder received haloperidol alone, haloperidol and carbamazepine, or haloperidol and valproic acid. The haloperidol dose remained stable, and the antiepileptic agents were adjusted to therapeutic levels. Carbamazepine significantly decreased plasma haloperidol levels, producing worsened clinical symptoms compared with symptoms in patients taking haloperidol only. Valproic acid had no effect on haloperidol levels or clinical outcome. The authors suggested that haloperidol taken with carbamazepine may result in treatment failure if haloperidol doses are not increased. Patients who discontinue carbamazepine will undergo a process of uninduction, which can take 1–2 weeks and may put them at risk for haloperidol toxicity.

Inhibition

Fluvoxamine and fluoxetine are inhibitors of 2C9 and 2C19, and phenytoin levels have increased with the addition of fluoxetine. Shad and Preskorn (1999) reported a decrease in phenytoin levels and effect when fluoxetine was discontinued.

Valproic acid is a moderate inhibitor. In an open-label, sequential, two-period study involving healthy volunteers, valproic acid increased amitriptyline serum levels by 31% and increased combined TCA levels 19% (Wong et al. 1996).

Summary: Antiepileptic Drugs

Antiepileptic drugs are generally inducers of P450 enzymes and can therefore reduce the efficacy of coadministered drugs, particularly oral contraceptives (Guberman 1999). Carbamazepine and many other agents are metabolized at 3A4 and may reach toxic levels when administered with potent inhibitors such as erythromycin, ritonavir, or grapefruit juice (intestinal 3A4 inhibition). Phenytoin is metabolized at 2C9 and 2C19, so administration with drugs such as fluvoxamine and ticlopidine may lead to phenytoin toxicity. Valproic acid is a drug with a complicated metabolism and the potential for hepatotoxicity; therefore, serum levels of all coadministered medications with potential for toxicity should be carefully monitored, along with liver-associated enzymes.

Antiparkinsonian Drugs

Bromocriptine

Bromocriptine (Parlodel), an ergot alkaloid, was first released in the United States in 1978 for the treatment of amenorrhea and galactorrhea secondary to hyperprolactinemia. In 1982, the agent was approved for use in patients with Parkinson's disease. Because it is an older drug, little is known about its metabolism or its potential to inhibit or induce liver enzymes. There is evidence some of bromocriptine's metabolism is through 3A4 (Peyronneau et al. 1994) and the drug inhibits 3A4 (Wynalda and Wienkers 1997) (Table 15–2).

Carbidopa-Levodopa

Carbidopa-levodopa (Sinemet) is a mainstay in the treatment of symptoms of Parkinson's disease and has been available in the United States since 1982. Levodopa is easily metabolized to dopamine in the central nervous system (CNS) and peripherally, and carbidopa is added to inhibit peripheral destruction of the levodopa so that more levodopa can enter the CNS. Carbidopa does not cross the blood-brain barrier and is not appreciably metabolized but rather excreted unchanged (Sinemet 2006). Although pharmacokinetic drug interactions are not a problem with carbidopa-levodopa, significant pharmacodynamic drug interactions do occur. Carbidopa-levodopa should not be administered with monoamine oxidase inhibitors (MAOIs), and problems may arise with concomitant administration of carbidopa-levodopa and antipsychotics. Further research is necessary to elucidate the potential complications associated with these interactions.

Entacapone

Entacapone (Comtan) is a reversible catecholamine O-methyltransferase (COMT) inhibitor used as an adjunct in the treatment of Parkinson's disease. It is metabolized primarily through glucuronidation via the UGT 1A family (Luukkanen et al. 2005). However, the glucuronidated metabolite is primarily (90%) excreted through the biliary system, so caution should be used in patients with the potential for biliary obstruction (Comtan 2000). Although *in vitro* studies by the manufacturer revealed that entacapone is a very mild inhibitor of nearly all P450 enzymes, this inhibition is considered clinically in-

TABLE 15–2. Antiparkinsonian drugs

Drug	Metabolism	Enzyme(s) inhibited[a]	Enzyme(s) induced[a]
COMT inhibitors			
Entacapone (Comtan)	UGT1A family	P450[c]	None known
Tolcapone (Tasmar)	UGT1A9, COMT, 3A4, 2A6	2C9[c]	None known
Dopamine agonists			
Bromocriptine (Parlodel)	3A4, ?others	3A4[b]	None known
Carbidopa-levodopa (Sinemet)	Carbidopa: excreted unchanged; Levodopa: aromatic amino acid decarboxylase	None known	None known
Pergolide (Permax)	Phase II, 3A4	3A4	None known
Pramipexole (Mirapex)	Excreted in urine unchanged	None known	None known
Ropinirole (Requip)	1A2, 3A4	1A2[c]	None known
Monoamine oxidase B inhibitors			
Selegiline (Eldepryl)	2B6, 3A4, 2A6	2C19[1]	None known
Rasagiline (Azilect)	1A2	None known	None known

Note. COMT=catechol *O*-methyltransferase; UGT=uridine 5′-diphosphate glucuronosyltransferase.
[1]Significance unknown.
[a]**Bold** type indicates potent inhibition or induction.
[b]Moderate inhibition.
[c]Mild inhibition.

significant. There are no reports to date of significant drug-drug interactions involving entacapone. However, because most of entacapone's excretion is through the biliary system, as noted above, caution should be exercised when entacapone is administered with drugs known to interfere with biliary excretion or glucuronidation. These drugs include probenecid, cholestyramine, and some antibiotics (e.g., erythromycin, rifampicin, ampicillin, and chloramphenicol).

Pergolide

In 1988, pergolide (Permax), an ergot alkaloid, received approval in the United States for use in patients with Parkinson's disease. Pergolide and bromocriptine work in a similar way, but pergolide is much more potent. Metabolism occurs through multiple mechanisms and there are 10 known metabolites, including both oxidative metabolites and conjugates from various phase II enzymes. Pergolide appears to inhibit 3A4 (Wynalda and Wienkers 1997), but no case reports exist to elucidate the clinical significance of this inhibition. In March 2007, pergolide was voluntarily withdrawn from the U.S. market (Permax 2003).

Pramipexole

When pramipexole (Mirapex) was released in the United States in 1997, it represented the first new agent for the treatment of Parkinson's disease in nearly 10 years. Because of its chemical structure, pramipexole is classified as an aminobenzothiazole (Hubble 2000). It is excreted unchanged in the urine, and there are no known metabolites (Mirapex 2007). Drug interactions involving pramipexole have not been reported.

Rasagiline

Rasagiline (Azilect) is a selective monoamine oxidase type B (MAO-B) inhibitor used in the treatment of Parkinson's disease. Rasagiline undergoes oxidative metabolism via CYP1A2 with subsequent conjugation and renal excretion (Guay 2006). Inhibitors of 1A2, such as fluvoxamine and ciprofloxacin, may significantly increase the plasma levels of rasagiline, resulting in increased rates of side effects and potential toxicities. Similarly, inducers of 1A2, such as rifampin and caffeine, may decrease rasagiline levels, resulting in ineffective therapy. Rasagiline is not known to cause inhibition or induction.

Ropinirole

Ropinirole (Requip) was released in the United States soon after pramipexole, in late 1997. This nonergot dopamine agonist is primarily metabolized by 1A2 (Kaye and Nicholls 2000; Requip 2006), with minor contribution by 3A4. There are no known active metabolites. Most of the 1A2 metabolite is excreted unchanged, with a smaller amount glucuronidated before excretion.

Because 1A2 represents the major means of clearance for ropinirole, clinicians must remember, when using ropinirole, that 1A2 can be induced (e.g., caffeine, esometrazole, and griseofulvin) or inhibited (e.g., ciprofloxacin and fluvoxamine). Few reports of drug interactions involving ropinirole exist. However, the manufacturer states that when ropinirole is used with ciprofloxacin (a potent inhibitor of 1A2), ropinirole's AUC can increase by up to 84% and its maximum concentration (C_{max}) by as much as 60% (Requip 2006). The AUC and C_{max} of theophylline, a drug dependent on 1A2 for clearance with a narrow therapeutic or safety window, did not change significantly when the agent was used with ropinirole, indicating that ropinirole probably has minimal 1A2 inhibition (Thalamas et al. 1999).

Tolcapone

Tolcapone (Tasmar), like entacapone, is a reversible COMT inhibitor, but tolcapone is relegated as a second-line agent for the treatment of Parkinson's disease because of its potential for liver toxicity (Tasmar 2006). Tolcapone is metabolized primarily through glucuronidation via UGT1A9 (Martignoni et al. 2005). Other pathways for metabolism include COMT, 3A4, and 2A6. Premarketing *in vitro* studies by the manufacturer revealed that tolcapone may inhibit 2C9 (Tasmar 2006), but when tolbutamide (which is metabolized by 2C9) was used with tolcapone in an *in vivo* study, the pharmacokinetics of tolbutamide remained unchanged.

Selegiline

Selegiline (Eldepryl) is an irreversible monoamine oxidase B (MAO B) inhibitor. Because of its selectivity, it is presumed to have less potential for side effects and pharmacodynamic interactions (e.g., hypertensive crisis) typically associated with other MAOIs. Selegiline is used in treatment for Parkinson's disease because MAO B is responsible for most of dopamine's metabolism in the brain, and inhibition of MAO B thus increases dopamine levels but has no effect on CNS serotonin or epinephrine levels. Selegilene is metabolized primarily via 2B6 with minor involvement of 3A4 and 2A6 (Benetton et al. 2007). It appears that selegiline moderately inhibits 2C19, but no case reports have yet indicated that this effect is clinically relevant.

Cognitive Enhancers

Since the advent of the first cholinesterase inhibitor, tacrine, in 1993, cognitive enhancers have grown in use and popularity. Their greater use is due in part to increasing awareness of dementive processes, the aging populus, and further understanding of effective treatments of dementia. Because they are prescribed to elderly patients, who are often taking multiple drugs, the potential for pharmacokinetic drug interactions is high (see Table 15–3).

Tacrine

Tacrine (Cognex) was the first cholinesterase inhibitor to be released; it was made available in 1993. Tacrine is rarely used now, primarily because of its association with liver toxicity. In addition, the agent must be taken four times a day, which makes it the least convenient of the four available cholinesterase inhibitors (the others can be prescribed for once or twice a day). Use of tacrine all but ceased after donepezil (Aricept) became available in 1997.

Tacrine is metabolized by 1A2 and, to a smaller extent, 2D6 (Cognex 2006). Tacrine also inhibits 1A2. Because theophylline's metabolism is highly dependent on 1A2, theophylline levels are increased 2-fold when that drug is used with tacrine. Pseudo-parkinsonism resulting from tacrine coadministration with haloperidol (partially 1A2 metabolized) has been noted in case reports (Maany 1996; Strain et al. 2004) and may be the result of 1A2 inhibition. Concentrations of caffeine and pentoxifylline, which are also dependent on 1A2 for metabolic clearance, can be expected to increase with tacrine coadministration. Tacrine levels increase significantly when tacrine is administered with cimetidine (Tagamet), a pan-inhibitor of the P450 enzymes.

TABLE 15–3. Cognitive enhancers

Drug	Metabolism	Enzyme(s) inhibited	Enzyme(s) induced
Donepezil (Aricept)	2D6, 3A4, UGTs	None known	None known
Galantamine (Razadyne)	2D6, 3A4, UGTs; 50% excreted in urine unchanged	None known	None known
Memantine (Namenda)	75% excreted in urine unchanged; oxidation via unknown enzyme	2B6	None known
Rivastigmine (Exelon)	Local cholinesterases; excreted in urine	None known	None known
Tacrine (Cognex)	1A2,[1] 2D6[2]	1A2	None known

Note. UGT=uridine 5′-diphosphate glucuronosyltransferase.
[1]Major metabolizer.
[2]Minor metabolizer.

Donepezil

Donepezil (Aricept), a reversible selective acetylcholinesterase inhibitor, was introduced in 1997. Donepezil is metabolized primarily by 2D6 and 3A4 (Aricept 2006) with minor contribution from glucuronidation. Donepezil metabolism results in two major metabolites; an active metabolite via hydrolysis (6-O-desmethyldonepezil) and an inactive metabolite via oxidation (donepezil-*cis*-N-oxide). Donepezil and the active metabolite appear to have similar pharmacologic activity. Because of the multiple pathways for its metabolism and the lack of a narrow therapeutic margin, donepezil poses a low risk for serious drug interactions, although the potential for minor complications of therapy remain. In addition, unlike tacrine, donepezil does not inhibit or induce any metabolic enzymes that could affect clearance of other drugs. More specifically, healthy volunteer studies have shown no clinical or pharmacokinetic impact when donepezil is coadministered with theophylline, warfarin, memantine, risperidone, thioridazine, levodopa/carbidopa, cimetidine, or digoxin (Jann et al. 2002; Okereke et al. 2004; Periclou et al. 2004; Zhao et al. 2003).

Rivastigmine

Rivastigmine (Exelon) is a pseudo-irreversible cholinesterase inhibitor for mild to moderate dementia that was introduced in the United States in 2000. It has a unique metabolic profile compared with other cholinesterase inhibitors. Rivastigmine is metabolized at its site of action—the cholinesterases—and this product is then cleared almost entirely by the kidneys (Exelon 2007). Grossberg et al. (2000) found few clinically significant pharmacokinetic and pharmacodynamic drug interactions between rivastigmine and 22 classes of medications.

Memantine

Memantine (Namenda) is a moderate affinity noncompetitive NMDA antagonist used in the treatment of mild to moderate dementia (Namenda 2007). Memantine is primarily excreted unchanged in urine (75%–90%). The metabolized portion of memantine undergoes hydroxylation and N-oxidation—implying P450 involvement, although this has not yet been confirmed. *In vitro* studies have shown 2B6 inhibition by memantine (Micuda et al.

2004). Memantine's lack of significant metabolism and its inhibition of a relatively minor enzyme (2B6) translate into a low likelihood of drug-drug interactions.

Galantamine

Galantamine (Razadyne; formerly Reminyl) is the newest cholinesterase inhibitor on the U.S. market, introduced in 2001. The manufacturer studied many facets of potential pharmacokinetic drug interactions before releasing galantamine (Razadyne 2007). Fifty percent of galantamine is metabolized by 2D6, 3A4, and UGTs; the other 50% is excreted in the urine unchanged. The C_{max} in poor metabolizers at 2D6 (7% of Caucasians) is similar to that in normal metabolizers at 2D6. Galantamine does not inhibit or induce any metabolic enzymes. Because no single metabolic or clearance pathway predominates, galantamine is less likely to interact with drugs that inhibit or induce metabolic enzymes. Known inhibitors of 2D6 (fluoxetine, paroxetine, quinidine, and cimetidine) and 3A4 (erythromycin and ketoconazole) increase area under the plasma concentration-time curve (AUC) by 10%–40%, but the clinical significance of this effect appears to be small.

Triptans

Triptans are potent serotonin type 1B/1D ($5\text{-HT}_{1B/1D}$) receptor agonists used to abort and treat migraine headaches. They are believed to cause vasoconstriction of cranial blood vessels at arteriovenous anastomoses (sites of many $5\text{-HT}_{1B/1D}$ receptors, whereas 5-HT_2 receptors are found in peripheral arteries) and reduce inflammation. Sumatriptan (Imitrex) is a first-generation triptan but has many limitations, including poor oral bioavailability in oral formulations, a short half-life, and an inability to cross the blood-brain barrier. Newer and better-tolerated triptans have been developed. All triptans have very similar pharmacodynamic (receptor) characteristics but differ pharmacokinetically.

All triptans go through phase I (oxidative) metabolism. Unlike most drug classes, however, triptans undergo oxidative metabolism through both the P450 system enzymes and monoamine oxidase (MAO). MAO exists to metabolize endogenous biogenic amines. The two MAO enzymes are MAO A and MAO B. A particular triptan may be metabolized by the P450 system,

MAO, or both. None of the triptans appear to actively inhibit or induce P450 metabolism themselves (see Table 15–4).

Triptan metabolism may be affected by competing drugs that use, inhibit, or induce the P450 system, particularly if the triptan in question is dependent on a specific P450 enzyme for its metabolism or is coadministered with drugs that inhibit MAO activity. Triptan toxicities and side effects include dizziness, chest or neck tightness, palpitations, shortness of breath, and acute anxiety. Myocardial ischemia has been reported with triptan use, especially in patients with coronary artery disease (Jhee et al. 2001). MAOIs—particularly MAO A inhibitors such as moclobemide—are contraindicated with triptans whose metabolism is solely dependent on MAO (rizatriptan [Maxalt] and sumatriptan). Gardner and Lynd (1998) found no reports of adverse events with simultaneous administration of sumatriptan and MAOIs, but the manufacturer has listed this combination as absolutely contraindicated (Imitrex 2007). Propranolol (Inderal) may be an inhibitor of MAO A and has been found to increase plasma concentrations of rizatriptan, so dose reduction of MAO A–dependent triptans is recommended with concomitant use of propranolol. The other beta-blockers do not seem to have this interaction (Goldberg et al. 2001).

P450 interactions with triptans are predictable, based on which P450 enzyme is the site of metabolism for a given triptan. Eletriptan (Relpax) and almotriptan (Axert) are primarily metabolized at 3A4. One would predict that potent inhibitors of 3A4 such as nefazodone, clarithromycin (Biaxin), erythromycin, ketoconazole, itraconazole (Sporanox), ritonavir (Norvir), ciprofloxacin (Cipro), and grapefruit juice might increase plasma levels of the triptans and worsen side effects or toxicity. In healthy volunteers, verapamil (a moderate inhibitor of 3A4) and fluoxetine (a moderate inhibitor of both 3A4 and 2D6) caused a moderate increase in the C_{max} and AUC of almotriptan, which is metabolized by MAO, 3A4, and 2D6 (Fleishaker et al. 2000, 2001). In both of these studies, no significant clinical events occurred, and the authors suggest that no dose adjustment is necessary. These modest findings reflect almotriptan's multiple avenues of metabolism, which allow the drug to be biotransformed despite roadblocks at some of its metabolic sites. At the time of writing, there were no reports on the use of triptans with potent 3A4 inhibitors. However, caution is advised, particularly with eletriptan, which is predominantly metabolized at 3A4.

TABLE 15–4. Triptans

Drug	Metabolism	Enzyme(s) inhibited	Enzyme(s) induced
Almotriptan (Axert)	3A4, 2D6, MAO A	None known	None known
Eletriptan (Relpax)	3A4, ABCB1	None known	None known
Frovatriptan (Frova)	1A2	None known	None known
Naratriptan (Amerge)	P450, MAO A, excreted in urine	None known	None known
Rizatriptan (Maxalt)	MAO A	None known	None known
Sumatriptan (Imitrex)	MAO A	None known	None known
Zolmitriptan (Zomig)	1A2, MAO A	None known	None known

Note.   MAO A=monoamine oxidase A.

Triptans dependent on 1A2 (frovatriptan [Frova] and zolmitriptan [Zomig]) may become toxic when administered with potent 1A2 inhibitors such as fluvoxamine and ciprofloxacin (Millson et al. 2000). Buchan et al. (2002) reviewed results of *in vitro* studies, healthy-volunteer studies, and a retrospective analysis of phase I clinical data concerning triptans and drugs commonly coadministered with frovatriptan. In addition to noting inhibition by the aforementioned potent 1A2 inhibitors, they found the AUC and C_{max} of frovatriptan to be lower in tobacco smokers (tobacco smoke is a potent inducer of 1A2). Eletriptan requires higher dosing than other second-generation triptans due to the activity of ABCB1 (Millson et al. 2000).

Ergotamines

Ergot derivatives are used as vasodilators in migraine treatment, and more recently they have been used in the treatment of dementia. Ergotamines are metabolized at 3A4 and are mild to moderate inhibitors of 3A4 as well. Potent 3A4 inhibitors, particularly the macrolide antibiotics, have caused frank ergotism in patients. Horowitz et al. (1996) reported that a patient receiving clarithromycin therapy developed lingual ischemia after taking 2 mg of ergotamine tartrate. Nicergoline, an ergot not available in the United States, has been used for the treatment of dementia in Europe. This agent appears to be metabolized via 2D6, as determined by studies involving healthy volunteers (Bottiger et al. 1996). Studies of this drug in combination with potent 2D6 inhibitors have not yet been conducted.

References

Anderson GD: A mechanistic approach to antiepileptic drug interactions. Ann Pharmacother 32:554–563, 1998

Anderson GD, Yua MK, Gidal BE, et al: Bidirectional interaction of valproate and lamotrigine in healthy subjects. Clin Pharmacol Ther 60:145–156, 1996

Arana GW, Epstein S, Molloy M, et al: Carbamazepine-induced reduction of plasma alprazolam concentrations: a clinical case report. J Clin Psychiatry 49:448–449, 1988

Aricept (package insert). Woodcliff Lake, NJ, Eisai Inc., 2006

Benetton SA, Fang C, Yang YO, et al: P450 phenotyping of the metabolism of selegilene to desmethylselegilene and methamphetamine. Drug Metab Pharmacokinet 22:78–87, 2007

Bernus I, Dickinson RG, Hooper WD, et al: Inhibition of phenobarbitone N-glucosidation by valproate. Br J Clin Pharmacol 38:411–416, 1994

Besag FM, Berry D: Interactions between antiepileptic and antipsychotic drugs. Drug Saf 29:95–118, 2006

Bialecka M, Hnatyszyn G, Bielicka-Cymerman J, et al: The effect of MDR1 gene polymorphism in the pathogenesis and the treatment of drug-resistant epilepsy. Neurol Neurochir Pol 39:476–481, 2005

Bottiger Y, Dostert P, Benedetti MS, et al: Involvement of CYP2D6 but not CYP2C19 in nicergoline metabolism in humans. Br J Clin Pharmacol 42:707–711, 1996

Bottiger Y, Svensson JO, Stahle L: Lamotrigine drug interactions in a TDM material. Ther Drug Monit 21:171–174, 1999

Bourgeois BF: Pharmacologic interactions between valproate and other drugs. Am J Med 84(1A):29–33, 1988

Brodie JM: Tiagabine pharmacology in profile. Epilepsia 36 (suppl 6):S7–S9, 1995

Buchan P, Wade A, Ward C, et al: Frovatriptan: a review of drug-drug interactions. Headache 42 (suppl 2):63–73, 2002

Chetty M, Miller R, Seymour MA: Phenytoin auto-induction. Ther Drug Monit 1998; 20:60–62

Citrome L, Macher JP, Salazar DE, et al: Pharmacokinetics of aripiprazole and concomitant carbamazepine. J Clin Psychopharmacol 27:279–283, 2007

Cognex (package insert). Atlanta, GA, Sciele Pharma, Inc., 2006

Comtan (package insert). East Hanover, NJ, Novartis Pharmaceuticals Corp., 2000

Deleu D, Hanssens Y: Current and emerging second-generation triptans in acute migraine therapy: a comparative review. J Clin Pharmacol 40:687–700, 2000

Ethell BT, Anderson GD, Burchell B: The effect of valproic acid on drug and steroid glucuronidation by expressed human UDP-glucuronosyltransferases. Biochem Pharmacol 65:1441–1449, 2003

Exelon (package insert). East Hanover, NJ, Novartis Pharmaceuticals Corp., 2007

Eyal S, Lamb JG, Smith-Yockman M, et al: The antiepileptic and anticancer agent valproic acid, induces P-glycoprotein in human tumour cell lines and in rat liver. Br J Pharmacol 149:250–260, 2006

Facciola G, Avenoso A, Spina E, et al: Inducing effect of phenobarbital on clozapine metabolism in patients with chronic schizophrenia. Ther Drug Monit 20:628–630, 1998

Fattore C, Cipolla G, Gatti G, et al: Induction of ethinylestradiol and levonorgestrel metabolism by oxcarbazepine in healthy women. Epilepsia 40:783–787, 1999

Faucette SR, Wang H, Hamilton GA, et al: Regulation of CYP2B6 in primary human hepatocytes by prototypical inducers. Drug Metab Dispos 32:348–358, 2004

Felbatol (package insert). Cranbury, NJ, Wallace Laboratories, 2002

Fleishaker JC, Sisson TA, Carel BJ, et al: Pharmacokinetic interaction between verapamil and almotriptan in healthy volunteers. Clin Pharmacol Ther 67:498–503, 2000

Fleishaker JC, Ryan KK, Carel BJ, et al: Evaluation of the potential pharmacokinetic interaction between almotriptan and fluoxetine in healthy volunteers. J Clin Pharmacol 41:217–223, 2001

Forgue ST, Reece PA, Sedman AJ, et al: Inhibition of tacrine metabolism by cimetidine. Clin Pharmacol Ther 59:444–449, 1996

Gabitril (package insert). West Chester, PA, Cephalon Inc., 2005

Gardner DM, Lynd LD: Sumatriptan contraindications and the serotonin syndrome. Ann Pharmacother 32:33–38, 1998

Gibson GG, el-Sankary W, Plant NJ: Receptor-dependent regulation of the CYP3A4 gene. Toxicology 181–182:199–202, 2002

Glue P, Banfield CR, Perhach JL, et al: Pharmacokinetic interactions with felbamate: in vitro–in vivo correlation. Clin Pharmacokinet 33:214–224, 1997

Goldberg MR, Sciberras D, De Smet M, et al: Influence of beta-adrenoceptor antagonists on the pharmacokinetics of rizatriptan, a 5-HT1B/1D agonist: differential effects of propranolol, nadolol and metoprolol. Br J Clin Pharmacol 52:69–76, 2001

Grossberg GT, Stahelin HB, Messina JC, et al: Lack of adverse pharmacodynamic drug interactions with rivastigmine and twenty-two classes of medications. Int J Geriatr Psychiatry 15:242–247, 2000

Guay DR: Rasagiline (TVP-1012): a new selective monoamine oxidase inhibitor for Parkinson's disease. Am J Geriatr Pharmacother 4:330–346, 2006

Guberman A: Hormonal contraception with epilepsy. Neurology 53:S38–S40, 1999

Gustavson LE, Sommerville KW, Boellner SW, et al: Lack of a clinically significant pharmacokinetic drug interaction between tiagabine and valproate. Am J Ther 5:73–79, 1998

Hachad H, Ragueneau-Majlessi I, Levy RH: New antiepileptic drugs: review on drug interactions. Ther Drug Monit 24:91–103, 2002

Harden CL, Leppik I: Optimizing therapy of seizures in women who use oral contraceptives. Neurology 67:S56–S58, 2006

Hesslinger B, Normann C, Langosch JM, et al: Effects of carbamazepine and valproate on haloperidol plasma levels and on psychopathologic outcome in schizophrenic patients. J Clin Psychopharmacol 19:310–315, 1999

Hidestrand M, Oscarson M, Salonen JS, et al: CYP2B6 and CYP2C19 as the major enzymes responsible for the metabolism of selegiline, a drug used in the treatment of Parkinson's disease, as revealed from experiments with recombinant enzymes. Drug Metab Dispos 29:1480–1484, 2001

Hiller A, Nguyen N, Strassburg CP, et al: Retigabine N-glucuronidation and its potential role in enterohepatic circulation. Drug Metab Dispos 27:605–612, 1999

Horowitz RS, Dart RC, Gomez HF: Clinical ergotism with lingual ischemia induced by clarithromycin-ergotamine interaction. Arch Intern Med 156:456–458, 1996

Hubble JP: Pre-clinical studies of pramipexole: clinical significance. Eur J Neurol 7 (suppl 1):15–20, 2000

Imitrex (package insert). Research Triangle Park, NC, GlaxoSmithKline, 2007

Jann M, Shirley K, Small G: Clinical pharmacokinetics and pharmacodynamics of cholinesterase inhibitors. Clin Pharmacokinet 41:719–739, 2002

Jhee SS, Shiovitz T, Crawford AW, et al: Pharmacokinetics and pharmacodynamics of the triptan antimigraine agents. Clin Pharmacokinet 40:189–205, 2001

Kalviainen R: Tiagabine: a new therapeutic option for people with intellectual disability and partial epilepsy. J Intellect Disabil Res 42 (suppl 1):63–67, 1998

Kaye CM, Nicholls B: Clinical pharmacokinetics of ropinirole. Clin Pharmacokinet 39:243–254, 2000

Keppra (package insert). Smyrna, GA, UCB Pharmaceuticals, Inc., 2008

Lakehal F, Wurden CJ, Kalhorn TF, et al: Carbamazepine and oxcarbazepine decrease phenytoin metabolism through inhibition of CYP2C19. Epilepsy Res 52:79–83, 2002

Lamictal (package insert). Greenville, NC, GlaxoSmithKline, 2007

Linnet K: Glucuronidation of olanzapine by cDNA-expressed human UDP-glucuronosyltransferases and human liver microsomes. Hum Psychopharmacol 17:233–238, 2002

Lucas RA, Gilfillan DJ, Bergstrom RF: A pharmacokinetic interaction between carbamazepine and olanzapine: observations on possible mechanism. Eur J Clin Pharmacol 54:639–643, 1998

Luukkanen L, Taskinen J, Kurkela M, et al: Kinetic characterization of the 1A subfamily of recombinant human UDP-glucuronosyltransferases. Drug Metab Dispos 33:1017–1026, 2005

Maany I: Adverse interaction of tacrine and haloperidol (letter). Am J Psychol 153:1504, 1996

Madden S, Spaldin V, Park BK: Clinical pharmacokinetics of tacrine. Clin Pharmacokinet 28:449–457, 1995

Maggs JL, Naisbitt DJ, Tettey JN, et al: Metabolism of lamotrigine to a reactive arene oxide intermediate. Chem Res Toxicol 13:1075–1081, 2000

Mamiya K, Ieiri I, Shimamoto J, et al: The effects of genetic polymorphisms of CYP2C9 and CYP2C19 on phenytoin metabolism in Japanese adult patients with epilepsy: studies in stereoselective hydroxylation and population pharmacokinetics. Epilepsia 1998; 39:1317–1323

Marchi N, Guiso G, Rizzi M, et al: A pilot study on brain-to-plasma partition of 10,11-dihydro-10-hydroxy-5H-dibenzo(b,f)azepine-5—carboxamide and MDR1 brain expression in epilepsy patients not responding to oxcarbazepine. 46:1613–1619, 2005

Martignoni E, Cosentino M, Ferrari M, et al: Two patients with COMT inhibitor-induced hepatic dysfunction and UGT1A9 genetic polymorphism. Neurology 65:1820–1822, 2005

Matsuo F: Lamotrigine. Epilepsia 40 (suppl 5):S30–S36, 1999

Micuda S, Mundlova L, Anzenbacherova E et al. Inhibitory effects of memantine on human cytochrome P450 activities: prediction of in vivo drug interactions. Eur J Clin Pharmacol 60: 583–589, 2004

Millson DS, Tepper SJ, Rapoport AM: Migraine pharmacotherapy with oral triptans: a rational approach to clinical management. Expert Opin Pharmacother 1:391–404, 2000

Mirapex (package insert). Ridgefield, CT, Boehringer Ingelheim Pharmaceuticals, Inc., 2007

Nair DR, Morris HH: Potential fluconazole-induced carbamazepine toxicity. Ann Pharmacother 33:790–792, 1999

Namenda (package insert). St. Louis, MO, Forest Pharmaceuticals Inc., 2007

Neurontin (package insert). Morris Plains, NJ, Parke-Davis, 2007

Okereke C, Kirby L, Kumar D, et al: Concurrent administration of donepezil HCl and levodopa/carbidopa in patients with Parkinson's disease: assessment of pharmacokinetic changes and safety following multiple oral doses. Br J Clin Pharmacol 58:S41–S49, 2004

O'Neill A, de Leon J: Two case reports of oral ulcers with lamotrigine several weeks after oxcarbazepine withdrawal. Bipolar Disord 9:310–313, 2007

Owen A, Goldring C, Morgan P, et al: Induction of P-glycoprotein in lymphocytes by carbamazepine and rifampicin: the role of nuclear hormone response elements. Br J Clin Pharmacol 62:237–242, 2006

Parker AC, Pritchard P, Preston T, et al: Induction of CYP1A2 activity by carbamazepine in children using the caffeine breath test. Br J Clin Pharmacol 45:176–178, 1998

Patsalos PN: Clinical pharmacokinetics of levetiracetam. Clin Pharmacokinet 43:707–724, 2004

Pavek P, Fendrich Z, Staud F, et al: Influence of P-glycoprotein on the transplacental passage of cyclosporine. J Pharm Sci 90:1583–1592, 2001

Pearce RE, Vakkalagadda GR, Leeder JS: Pathways of carbamazepine bioactivation in vitro, 1: characterization of human cytochrome P450 responsible for the formation of 2- and 3-hydroxylated metabolites. Drug Metab Dispos 30:1170–1179, 2002

Periclou A, Ventura D, Sherman T, et al: Lack of pharmacokinetic interaction between memantine and donepezil. Ann Pharmacother 38:1389–1394, 2004

Permax (pergolide mesylate) 2003 safety alert. U.S. Food and Drug Administration Medwatch, 2003. Available at http://www.fda.gov/medwatch/SAFetY/2003/permax.htm.

Perucca E: Clinically relevant drug interactions with antiepileptic drugs. Br J Clin Pharmacol 61:246–255, 2006

Peyronneau MA, Delaforge M, Riviere R, et al: High affinity of ergopeptides for cytochromes P450 3A: importance of their peptide moiety for P450 recognition and hydroxylation of bromocriptine. Eur J Biochem 223:947–956, 1994

Pisani F: Influence of co-medication on the metabolism of valproate. Pharm Weekbl Sci 14:108–113, 1992

Potschka H, Fedrowitz M, Löscher W: P-glycoprotein and multi-drug resistance-associated protein are involved in the regulation of extracellular levels of the major antiepileptic drug carbamazepine in the brain. Neuroreport 12:3557–3560, 2001

Raucy JL: Regulation of CYP3A4 expression in human hepatocytes by pharmaceuticals and natural products. Drug Metab Dispos 2003; 31:533–539

Razadyne (package insert). Titusville, NJ, Janssen Pharmaceutica Products, LP, 2007

Requip (package insert). Research Triangle Park, NC, GlaxoSmithKline, 2001

Richens A: Pharmacokinetic and pharmacodynamic drug interactions during treatment with vigabatrin. Acta Neurol Scand Suppl 162:43–46, 1995

Ritter JK, Kessler FK, Thompson MT, et al: Expression and inducibility of the human bilirubin UDP-glucuronosyltransferase UGT1A1 in liver and cultured primary hepatocytes: evidence for both genetic and environmental influences. Hepatology 30:476–484, 1999

Sachdeo RC, Sachdeo SK, Levy RH, et al: Topiramate and phenytoin pharmacokinetics during repetitive monotherapy and combination therapy to epileptic patients. Epilepsia 43:691–696, 2002

Sadeque AJM, Fisher MB, Korzekwa KR, et al: Human CYP2C9 and CYP2A6 mediate formation of the hepatotoxin 4-ene-valproic acid. J Pharmacol Exp Ther 283:698–703, 1997

Shad MU, Preskorn SH: Drug-drug interaction in reverse: possible loss of phenytoin efficacy as a result of fluoxetine discontinuation. J Clin Psychopharmacol 19:471–472, 1999

Sinemet (package insert). Princeton, NJ, Bristol-Myers Squibb, 2006

Spina E, Avenoso A, Campo GM, et al: Phenobarbital induces the 2-hydroxylation of desipramine. Ther Drug Monit 18:60–64, 1996a

Spina E, Pisani F, Perucca E: Clinically significant pharmacokinetic drug interactions with carbamazepine. Clin Pharmacokinet 31:198–214, 1996b

Spina E, Arena D, Scordo MG, et al: Elevation of plasma carbamazepine concentrations by ketoconazole in patients with epilepsy. Ther Drug Monit 19:535–538, 1997

Staines AG, Coughtrie MW, Burchell B: N-glucuronidation of carbamazepine in human tissues is mediated by UGT2B7. J Pharmacol Exp Ther 311:1131–1137, 2004

Strain J, Niem C, Kaiser S et al: Psychotropic drug versus psychotropic drug: update. Gen Hosp Psychiatry 26:85–105, 2004

Tanaka E: Clinically significant pharmacokinetic drug interactions between antiepileptic drugs. J Clin Pharm Ther 24:87–92, 1999

Tasmar (package insert). Montvale, NJ, Roche Laboratories Inc., 2006

Thalamas C, Taylor A, Brefel-Courbon C, et al: Lack of pharmacokinetic interaction between ropinirole and theophylline in patients with Parkinson's disease. Eur J Pharmacol 55:299–303, 1999

Tiseo PJ, Perdomo CA, Friedhoff LT: Concurrent administration of donepezil HCl and cimetidine: assessment of pharmacokinetic changes following single and multiple doses. Br J Clin Pharmacol 46 (suppl 1):25–29, 1998

Topamax (package insert). Raritan NJ, Ortho-McNeil Pharmaceutical, Inc., 2007

Trileptal (package insert). East Hanover, NJ, Novartis Pharmaceutical Corp., 2007

Turnheim K: Drug interactions with antiepileptic agents. Wien Klin Wochenschr 116:112–118, 2004

Ucar M, Neuvonen M, Luurila H, et al: Carbamazepine markedly reduces serum concentrations of simvastatin and simvastatin acid. Eur J Clin Pharmacol 59:879–882, 2004

von Bahr C, Steiner E, Koike Y, et al: Time course of enzyme induction in humans: effect of pentobarbital on nortriptyline metabolism. Clin Pharmacol Ther 64:18–26, 1998

Weiss J, Kerpen CJ, Lindenmaier H, et al: Interaction of antiepileptic drugs with human P-glycoprotein in vitro. J Pharmacol Exp Ther 307:262–267, 2003

Wen X, Wang JS, Kivisto KT, et al: In vitro evaluation of valproic acid as an inhibitor of human cytochrome P450 isoforms: preferential inhibition of cytochrome P450 2C9 (CYP2C9). Br J Clin Pharmacol 52:547–553, 2001

Wong SL, Cavanaugh J, Shi H, et al: Effects of divalproex sodium on amitriptyline and nortriptyline pharmacokinetics. Clin Pharmacol Ther 60:48–53, 1996

Wynalda MA, Wienkers LC: Assessment of potential interactions between dopamine receptor agonists and various human cytochrome P450 enzymes using a simple in vitro inhibition screen. Drug Metab Dispos 25:1211–1214, 1997

Zhao Q, Xie C, Pesco-Koplowitz L, et al: Pharmacokinetic and safety assessment of concurrent administration of risperidone and donepezil. J Clin Pharmacol 43:180–186, 2003

Zhou S, Lim LY, Chowbay B: Herbal modulation of P-glycoprotein. Drug Metab Rev 36:57–104, 2004

Zielinski JJ, Haidukewych D: Dual effects of carbamazepine-phenytoin interaction. Ther Drug Monit 9:21–23, 1987

Zonegran (package insert). South San Francisco, CA, Elan Pharmaceuticals, Inc., 2007

16

Oncology

Scott G. Williams, M.D.

Justin M. Curley, M.D.

Gary H. Wynn, M.D.

***Reminder:** This chapter is dedicated primarily to metabolic and ABCB1 (P-gp) interactions. Interactions due to displaced protein-binding, alterations in absorption or excretion, and pharmacodynamics may be mentioned but should not be considered a comprehensive review.*

The pharmacology of oncology is exceptionally complex. Accelerated approval of new agents by the U.S. Food and Drug Administration means medications are hitting the market nearly continuously. Polypharmacy is the norm, so attention needs to be paid to the interactions between drugs. The P450 effects of oncology drugs are in varying stages of investigation.

The elucidation of oncology drug interactions is hampered for several reasons. Many of the medications used are old and have not been studied comprehensively. The toxicity of these medications makes *in vivo* pharmaco-kinetic studies both difficult for and unattractive to primary investigators. Also, many oncology agents are used together, making identification of some interactions and effects difficult. Additionally, many cancer patients receive a wide variety of medications for disease- and pain-related complications; these complications range from mucositis to cytopenias to depression. Given these issues, much of the knowledge of oncology drug metabolism comes from *in vitro* cell studies, tissue culture studies, and even animal models.

In pediatric patients, isoforms of enzymes are known to develop at different rates, with some enzymes not being present at all and others not present in amounts equivalent to those in adults. For example, fetal liver microsomes have approximately 1% of adult activity levels, and these levels increase to an average of 70% by day 7 of extrauterine life. The majority of the data in this chapter can be extrapolated to apply to children older than 10–12 years, but consulting a more thorough review of developmental pharmacokinetics in the pediatric subpopulation is warranted. A good review can be found in a report by Leeder and Kearns (1997).

This chapter is arranged by major classes of agents, and drugs in each class are discussed individually. These drugs are then evaluated in light of current available research. We also provide an overview of the classes and drugs routinely used in clinical oncology practice—including, but in less detail, new and investigational agents.

Antineoplastic Drugs

Alkylating Agents

Alkylating agents are some of the more frequently used drugs in chemotherapy. The main effect of these drugs is the cross-linking of strands of DNA, resulting in breakage and cell death. Although the cross-linking of DNA strands works most effectively in cells that are rapidly proliferating, alkylating agents can damage cells during any phase of the cell cycle (see Table 16–1).

Busulfan

The primary site of metabolism of busulfan (Myleran) is 3A4. Buggia and colleagues (1996) studied the effects of itraconazole (Sporanox) and fluconazole

TABLE 16–1. Alkylating agents

Drug	Metabolism site(s)	Enzyme(s) inhibited	Enzyme(s) induced[a]
Busulfan (Myleran)	3A4	None known	None known
Carmustine (BCNU)	Unknown, possibly 2C9	None known	None known
Chlorambucil (Leukeran)	Unknown	None known	None known
Cisplatin (Platinol)	Unknown	None known	None known
Cyclophosphamide (Cytoxan)[1]	3A4, 2B6	None known	**3A4, 2B6**
Dacarbazine (DTIC)	1A2, 1A1, 2E1	None known	None known
Ifosfamide (Ifex)	3A4, 2B6	None known	**3A4**[b], 2C9
Lomustine (Belustine, CCNU)	2D6	Unknown, possibly 2D6	None known
Melphalan (Alkeran)	Unknown	None known	None known
Streptozocin (STZ, Zanosar)	Unknown	None known	None known
Temozolomide (Temodar)	Unknown	None known	None known
Thiotepa (Testamine)	3A4, 2B6	2B6	None known
Trofosfamide (Ixoten)	3A4, 2B6	None known	None known

[1]N-Dealkylation to a neurotoxic agent, chloroacetaldehyde
[a]**Bold** type indicates potent induction.
[b]Moderate induction.
[c]Mild inhibition or induction.

(Diflucan) on busulfan levels. Itraconazole and fluconazole are known to be potent inhibitors of 3A4 and 2C9, respectively. The subjects who received coadministered itraconazole had a roughly 20% decrease in busulfan clearance compared with those receiving fluconazole and busulfan. An additional study by Hassan et al. (1993) showed increased clearance of busulfan with coadministration of phenytoin (Dilantin), a known 3A4 inducer. Busulfan is not known to induce or inhibit any hepatic enzymes.

Carmustine and Lomustine

Delineating the metabolism of carmustine (BCNU) and lomustine (Belustine) has proven difficult. In 1975, Hill et al. demonstrated that carmustine is metabolized by mice via microsomal enzymes in the liver and lungs. In 1979, Levin et al. showed that pretreatment with phenobarbital in a rat model increases the clearance of carmustine and lomustine, whereas pretreatment with phenytoin or dexamethasone (Decadron), both inducers of 3A4, does not affect clearance rates. Weber and Waxman (1993) demonstrated that phenytoin and dexamethasone induce metabolism of carmustine and lomustine when the glutathione-S-transferase system is inhibited. Both the P450 system (only 2%–3% of overall metabolism) and the glutathione-S-transferase system play roles in the metabolism of these drugs. This is because of the need for both denitrosation and NADPH (nicotinamide adenine dinucleotide phosphate, reduced form)–dependent P450 metabolism. Le Guellec et al. (1993) also proposed that lomustine may be metabolized by and possibly inhibit 2D6, but data are limited. More research, including studies beyond animal models, needs to be done to determine the clinical implications of this complicated system of metabolism.

Cisplatin and Other Platinum Agents

Platinum agents are the only heavy metal compounds used for antineoplastic chemotherapy. Administration of these drugs results in covalent cross-linking strands of DNA, leading to an inability to replicate the DNA strands. There may be a pharmacokinetic interaction between cisplatin (Platinol) and the antiemetic ondansetron (Zofran) (Cagnoni et al. 1999), resulting in decreased area under the plasma concentration-time curve (AUC) for cisplatin, which would imply possible metabolism via P450, but the mechanism remains uncertain. There are numerous reports that the use of cisplatin may decrease serum levels of anticonvulsants, although none of these reports includes comments on a specific mechanism. One possibility might be moderate induction of a transporter by cisplatin (Masuyama et al. 2005). Although platinum agents are predominantly cleared renally, there is a need for further study of these pharmacokinetic effects.

Ando and colleagues (1998) reported that satraplatin (JM216), an oral platinum agent, inhibited several P450 enzymes, including 3A4, 2C9, 1A1, 1A2, 2A6, 2E1, and 2D6. These authors provided no *in vivo* data and suggested that further studies need to be done.

Cyclophosphamide, Ifosfamide, and Trofosfamide

Cyclophosphamide (Cytoxan), ifosfamide (Ifex), and trofosfamide (Ixoten) are alkylating pro-drugs that require metabolism by P450 enzymes for antitumor activity. The primary mechanism of bioactivation of cyclophosphamide and ifosfamide is via 4-hydroxylation at 2B6 and 3A4 (Brain et al. 1998; Huang et al. 2000b). Inactivation of these metabolites occurs via *N*-dechloroethylation to a potent neurotoxin, chloroacetaldehyde. Chloroacetaldehyde's neurotoxic effect, commonly known as "ifosfamide encephalopathy," may present cerebellar ataxia, confusion, complex visual hallucinations, extrapyramidal signs, mutism, and seizures (Primavera et al. 2002). Cyclophosphamide is *N*-dechloroethylated—or inactivated—at 3A4. Ifosfamide is *N*-dechloroethylated (and also inactivated) through 3A4 and 2B6. Because these agents have somewhat different inactivation sites, the production of the potent neurotoxin chloroacetaldehyde may vary under different circumstances. As with cisplatin, ondansetron may reduce the AUC of cyclophosphamide (Cagnoni et al. 1999). Ifosfamide-treated patients with increased metabolism at 2B6 because of genetic predisposition or induction of 2B6 by other factors may have an increased ability to form chloroacetaldehyde, resulting in increased levels of chloroacetaldehyde; in ifosfamide-treated patients who have lower capacity at 2B6 and 3A4, or who are not taking inducing comedications, this risk is less. According to Huang et al. (2000a), there may be clinical benefit to medically inhibiting 2B6 in patients known to have higher capacity at 2B6, given the risk of neurotoxic metabolite formation.

Schmidt et al. (2001) showed that in the hepatic microsomes of females treated with ifosfamide, more chloroacetaldehyde was produced than in male-counterpart microsomes. Further research is needed on the possible clinical relevance of gender difference in chloroacetaldehyde formation.

Clinical anecdotes suggest that cyclophosphamide may induce its own metabolism. These anecdotes are supported by an *in vitro* study using human liver microsomes that showed administration of either cyclophosphamide or ifosfamide results in increased production of several P450 enzymes, including 3A4 and 2C9 (Chang et al. 1997). In a study by Baumhakel et al. (2001), human liver microsomes of 3A4 were given a stable substrate of dihydropyridine denitronifedipine with cyclophosphamide and ifosfamide. Both ifosfamide and cyclophosphamide significantly inhibited oxidation of dihydropyridine denitronifedipine, indicating that treatment with either drug may have clinical implications, given the drugs' effective inhibition of 3A4. Because of conflicting evidence on the activity of these medications, and the lack of *in vivo* studies confirming activity, further research into the autoinduction of cyclophosphamide and inhibition at 3A4 is needed.

Trofosfamide is a newer alkylating agent that also requires metabolism for activation. Trofosfamide is bioactivated by being metabolized to ifosfamide and cyclophosphamide (Hempel et al. 1997) through 3A4 and 2B6 (May-Manke et al. 1999). Brinker et al. (2002) found low levels of ifosfamide and cyclophosphamide present during treatment, which would suggest there is a component of direct 4-hydroxylation. Given that trofosfamide has several pathways of metabolism to include metabolism to cyclophosphamide and ifosfamide as well as direct 4-hydroxylation, the findings by May-Manke et al. (1999) demonstrating minimal induction or inhibition are likely due to the minimal availability of inhibiting metabolites. The amount of available neurotoxic metabolite from the conversion of cyclophosphamide and ifosfamide is also lower because of the variety of available metabolic pathways.

Dacarbazine

Dacarbazine (DTIC), a pro-drug like cyclophosphamide, was found to be metabolized to active forms by 1A1, 1A2, and 2E1 (Patterson and Murray 2002). Inhibitors of these enzymes significantly decreased *N*-demethylation of dacarbazine *in vitro* (Reid et al. 1999).

Streptozocin

Chang et al. (1976) reported decreased clearance of doxorubicin (Adriamycin) in patients who were also receiving streptozocin (Zanosar). The increased severity of side effects of doxorubicin in that study suggests hepatic dysfunction due to streptozocin, but no specific mention was made of P450 involvement.

Thiotepa

Thiotepa (Testamine) is an alkylating agent that requires metabolism to its pharmacologically active metabolite, TEPA. Jacobson et al. (2002) showed that thiotepa's metabolism is primarily by 3A4, with a minor portion performed by 2B6. Huitema et al. (2001) and Rae et al. (2002) showed reversible inhibition of 2B6 by thiotepa through increased levels of cyclophosphamide when the latter is coadministered with thiotepa. These studies suggest that clinicians using chemotherapeutic regimens including thiotepa should be aware of its inhibitory effects on drugs metabolized by 2B6.

Antimetabolites

Antimetabolites come in many forms, with multiple mechanisms of action. These drugs exert their cytotoxicity by acting as false substrates in multiple biochemical pathways. Nucleoside analogs such as fludarabine (Fludara), mercaptopurine (Purinethol), pentostatin (Nipent), fluorouracil (Adrucil, Efudex), gemcitabine (Gemzar), capecitabine (Xeloda) and cytarabine (Cytosar) are incorporated into newly formed DNA and result in termination of DNA growth. Other antimetabolite agents include enzyme inhibitors such as methotrexate (Trexall). All antimetabolites ultimately result in faulty DNA synthesis. Most of the antimetabolites are not processed by the P450 system, and only those that have relevant clinical data will be listed here.

Capecitabine

Capecitabine is a pyrimidine analog that does not appear to be metabolized by the P450 system. Although the metabolism is incompletely understood, there have been reports of clinically significant interactions with wafarin (Coumadin). Janney and Waterbury (2005) postulate that capecitabine might downregulate 2C9, leading to elevated international normalized ratio (INR) values. In at least one case, the dosage of warfarin was reduced by more than 85% during a 3-week course of capecitabine.

Cytarabine

Cytarabine is a pyrimidine analog metabolized exclusively by 3A4. Colburn et al. (2004) performed *in vitro* analysis and showed cytarabine metabolism was significantly decreased when combined with caspofungin (Cancidas) or itraconazole. Cytarabine also inhibits 3A4 and has the potential to increase serum levels of drugs metabolized at this enzyme. At present, however, there are limited data regarding these potential interactions.

Fluorouracil

van Meerten et al. (1995) reviewed the then-available literature on drug interactions and antineoplastic agents. They noted that both cimetidine and metronidazole (Flagyl) inhibit metabolism of fluorouracil, which suggests metabolism by P450 enzymes (3A4, 2D6). Interferon-α appears to decrease clearance of fluorouracil (Adrucil, Efudex), a finding that is consistent with studies showing interferon's effect on the downregulation of some P450 enzymes. Through an *in vivo* study, Gunes et al. (2006) demonstrated that flurouracil inhibited 2C9 activity and altered the metabolism of losartan (Cozaar).

Methotrexate

Methotrexate is metabolized by oxidation, but not via P450 enzymes (Chladek et al. 1997). More recent studies have reconfirmed that the P450 system plays no role in methotrexate metabolism (Baumhakel et al. 2001; Rozman 2002). A study using human liver microsomes showed that ethanol and acetaminophen, both inducers of 2E1, also increase levels of tumor necrosis factor-a, interleukin (IL)-6, and IL-8 (Neuman et al. 1999). These increases led to reduced mitochondrial and cytosolic glutathione levels and an increase in the oxidative stress on cell cultures. These same cells were then exposed to methotrexate and showed increased methotrexate-induced cytotoxicity and an increased rate of programmed cell death (apoptosis). This phenomenon is not specifically related to P450 enzymes, but it is important to consider, given the known effects of cytokines on P450 activity. (See "Immunomodulators" later in this chapter.) Elimination interactions do exist. Penicillins (Dean et al. 1992) and NSAIDs (Frenia and Long 1992) may inhibit renal tubular secretion of methotrexate resulting in elevated plasma levels.

Mitosis Inhibitors

Mitosis inhibitors interfere with the mitotic phase of tumor cell replication (and also with normal host tissue). Various methods of inhibiting mitosis include stabilizing microtubules and preventing their disassembly (with taxanes), binding to tubulin polymers and preventing their growth (with vinca alkaloids), and inhibiting the DNA replication enzyme topoisomerase. In addition to P450 interactions, mitosis inhibitors are also substrates of transporters such as ABCB1 (P-glycoprotein) (see Table 16–2).

Taxanes

Docetaxel (Taxotere) and paclitaxel (Taxol) belong to a group of compounds isolated from plants, specifically yews. Both docetaxol and paclitaxel are metabolized to their inactive metabolites by liver microsomal enzymes and are further eliminated through the biliary system.

Cresteil et al. (2002) showed that docetaxel is metabolized by 3A4 oxidation of the *tert*-butyl group on the lateral chain to its inactive form before removal through the biliary system. Royer et al. (1996) demonstrated that ketoconazole and erythromycin (known 3A4 inhibitors) significantly inhibit oxidation of docetaxel. Malingre et al. (2001) determined that coadministration of cyclosporine (Neoral) (a known 3A4 and ABCB1 inhibitor) and docetaxel increases the bioavailability of docetaxel. Given the presence of ABCB1 in the intestine, as well as first-pass P450 metabolism by 3A4, coadministration will likely result in increased levels of docetaxel.

Cresteil and colleagues (2002) showed that paclitaxel is metabolized by 2C8 6-hydroxylation of the taxane ring before further removal. Paclitaxel is also metabolized by 3A4 (Spratlin and Sawyer 2007) and a substrate of ABCB1. Dai et al. (2001) demonstrated that metabolism of paclitaxel by 2C8 is affected by the allele coding the 2C8 enzyme. 2C8 alleles vary according to the coding allele, and these alleles are present in various amounts in different ethnic groups. In patients homozygous for the allele that shows decreased ability to metabolize paclitaxel, the ability to metabolize the drug is significantly decreased, and thus the incidence of toxicity is higher.

Britten et al. (2000) studied the effect of oral paclitaxel and cyclosporine coadministration. Use of oral paclitaxel typically does not result in therapeutic levels of the agent, but with coadministration of cyclosporine, effective blood concentrations of paclitaxel can be achieved. Given that cyclosporine is a potent 3A4 and ABCB1 inhibitor, this result may be due to cyclosporine's effect on ABCB1 or hepatic metabolism. Schwartz et al. (1999) report a possible P450 interaction with saquinavir (a known 3A4 and ABCB1 inhibitor) and delavirdine (a 3A4 inhibitor) causing severe paclitaxel toxicity.

It is well established that paclitaxel and docetaxel are metabolized by the P450 system, and that the transporter system seems to have a significant effect

TABLE 16–2. Mitosis inhibitors

Drug	Metabolism site(s)	Enzyme(s) inhibited[a]	Enzyme(s) induced[a]
Taxanes			
Docetaxel (Taxotere)	3A4, ABCB1	3A4	None known
Paclitaxel (Taxol)	3A4, 2C8, ABCB1	None known	2C8[b], 3A4[b]
Vinca alkaloids			
Vinblastine (Velban)	3A4, ABCB1	None known	None known
Vincristine (Oncovin)	3A4, ABCB1	None known	None known
Vindesine (Eldesine)	3A4, ABCB1	None known	None known
Vinorelbine (Navelbine)	3A4, ABCB1	None known	None known
Topoisomerase inhibitors			
Etoposide (Toposar)	3A4, 1A2, 2E1, ABCB1	None known	3A4
Irinotecan (Camptosar)	3A4, UGT1A1, ABCB1	3A4	None known
Teniposide (Vumon)	3A4, ABCB1	3A4	None known
Topotecan (Hycamtin)	3A4, ABCB1	None known	None known

Note. ABCB1 = P-glycoprotein.
[a]**Bold** type indicates potent inhibition or induction.
[b]Moderate induction.
[c]Mild inhibition or induction.

on these drugs as well. Data indicate that inhibition of ABCB1 activity in normal tissues by effective modulators, and the physiological and pharmacological consequences of this treatment, cannot be predicted solely by monitoring plasma drug levels. More research is needed on the effect of various transporter mechanism on docetaxel and paclitaxel (see Chapter 4, "Transporters.")

Vinca Alkaloids

Vinca alkaloids are another group of plant-derived drugs. Chan (1998) conducted a thorough review of case reports on and clinical studies of vinca alkaloids and their pharmacokinetics. All inhibitors of 3A4 reduce the metabolism of vincristine *in vitro*, and several case reports indicate increased vincristine toxicity with concomitant use of 3A4 inhibitors such as itraconazole and cyclosporine.

Drugs such as carbamazepine (Tegretol) and phenytoin, known inducers of 3A4, have been shown to increase the metabolism of vincristine (used in combination with lomustine and procarbazine) in human volunteer subjects (Villikka et al. 1999). No studies have evaluated the clinical impact of these findings.

Kajita et al. (2000) used human liver microsome preparations to demonstrate that 3A4 is the main P450 enzyme responsible for metabolism of vinorelbine. The same investigators showed that high concentrations (100 μM) also inhibit 3A4 activity without inhibiting other P450 enzymes. This inhibitory effect, however, seems to occur at plasma vinorelbine concentrations greater than those found in humans during standard therapy.

Vinca alkaloids are also substrates of the transporter system, and concomitant use of transport inhibitors such as amiodarone, nifedipine, and verapamil have been shown in animal models to increase the concentration of vinblastine in the liver and the kidney but not the brain and testes (Arboix et al. 1997). This finding likely represents the various type of transporter present at each organ site in conjunction with the relevant interaction.

Topoisomerase Inhibitors

Replication of DNA is a complex process that is mediated by several enzymes. Tension is created in the DNA molecule during normal replication. Topoisomerase enzymes are crucial to making the DNA molecule flexible during replication. If topoisomerase enzymes are inhibited, this tension is not relieved and thus DNA replication and cell production cannot take place.

Etoposide (Toposar; VP-16) and teniposide (Vumon) are metabolized mainly by 3A4 (Kawashiro et al. 1998; McLeod 1998; Relling et al. 1994). 3A4 inducers such as glucocorticoids and anticonvulsants increase clearance of etoposide and teniposide (Bagniewski et al. 1996; Relling et al. 2000). Cyclosporine and quinidine have been shown to decrease the clearance of etoposide both *in vitro* and clinically, but this is likely due to inhibition of a transporter rather than a P450-mediated effect (Yu 1999).

Irinotecan (Camptosar) is a topoisomerase inhibitor used in treating patients with lung or colon cancer (Berkery et al. 1997). Santos et al. (2000) used human liver microsome preparations to demonstrate that irinotecan is metabolized predominantly via 3A4. More recent studies have shown irinotecan metabolism is vulnerable to both inhibition (by potent 3A4 inhibitors such as ketoconazole) and induction (by carbamazepine, phenobarbital, phenytoin, and St. John's wort) (Kehrer et al. 2002; Mathijssen et al. 2002). The decrease in irinotecan concentration that occurs with phenytoin coadministration is likely also mediated by the uridine 5'-diphosphate glucuronosyltransferase (UGT) 1A1 enzyme. Additionally, transporter mediated biliary excretion was inhibited by cyclosporine (Innocenti et al. 2004). Given the multiple avenues of irinotecan metabolism—including 3A4 and UGT1A1 as well as disposition via transporter, there is a need for further study of irinotecan regarding these potential interactions.

Antitumor Antibiotics

Often referred to as anthracyclines, antitumor antibiotics are cytotoxic by a variety of mechanisms, but the underlying mechanism of these drugs is DNA damage. Some antitumor antibiotics lead to creation of free radicals that make double- and single-strand breaks in the DNA, whereas others cross-link DNA strands, preventing replication. Antitumor antibiotics are processed via a variety of mechanisms. Those with substantial P450 metabolism will be discussed in detail (see Table 16–3).

Doxorubicin and Daunorubicin

Balis (1986) showed that P450 enzymes are involved in anthracycline metabolism, although no specific enzymes were delineated. Use of phenytoin or

TABLE 16–3. Antitumor antibiotics

Drug	Metabolism site(s)	Enzyme(s) inhibited[a]	Enzyme(s) induced[a]
Bleomycin	Renal	None known	None known
Daunorubicin (Cerubidine)	ABCB1	None known	ABCB1
Doxorubicin (Adriamycin)	3A4, ABCB1	None known	ABCB1
Epirubicin (Ellence)	UGT2B7	None known	None known
Idarubicin (Idamycin)	2D6, 2C9	2D6[b]	None known

Note.   ABCB1 = P-glycoprotein.
[a]**Bold** type indicates potent inhibition or induction.
[b]Moderate induction.
[c]Mild inhibition or induction.

phenobarbital has been shown to induce the metabolism of doxorubicin. These data suggest doxorubicin is partially metabolized at 3A4. Baumhakel et al. (2001) showed that doxorubicin has an inhibitory effect on 3A4 *in vitro*, but there are no *in vivo* studies to support this observation. There are multiple reports of *in vitro* studies regarding doxorubicin's effects on taxane metabolism, but no *in vivo* data could be found to demonstrate clinical relevance. Along with daunorubicin, doxorubicin is both a transporter substrate and inducer.

Idarubicin

Based on *in vitro* studies, idarubicin (Idamycin) appears to be extensively metabolized via 2D6 and 2C9. Though there are no *in vivo* studies to demonstrate clinically significant interactions, Colburn et al. (2004) did show that idarubicin inhibits 2D6. This finding suggests that idarubicin has the potential for interactions with 2D6 substrates including TCAs, antipsychotics, some selective serotonin reuptake inhibitors (SSRIs), beta-blockers, and some opiates.

Antiandrogens and Antiestrogens

Because androgens (e.g., testosterone) and estrogens have such dramatic effects on some carcinomas (e.g., breast and prostate cancer), modulation of aromatase, the enzyme responsible for conversion of androgens to estrogens in many human tissues, is a useful therapeutic modality (see Table 16–4).

Aminoglutethimide

Aminoglutethimide (Cytadren) is a potent aromatase inhibitor used in the treatment of metastatic breast cancer. Santner et al. (1984) studied preparations of human placental microsomes with inhibitors of aromatase and noted that coadministration of two aromatase inhibitors, aminoglutethimide and testolactone, produces an additive effect, evidenced by a decrease in aromatase activity greater than that seen with either agent alone. These investigators suggested that a reduction in aminoglutethimide dose might be possible with coadministration of aminoglutethimide and testolactone. This reduction would decrease possible side effects such as sedation, rash, and orthostatic hypotension.

TABLE 16–4. Antiandrogens and antiestrogens

Drug	Metabolism site(s)	Enzyme(s) inhibited	Enzyme(s) induced
Antiandrogens			
Bicalutamide (Casodex)	3A4 (*R*), glucuronidation (*S*)	3A4 (*R*), 2D6, 2I9, 2C9, 3A4	None known
Flutamide (Eulexin)	1A2, 3A4	1A2[1]	None known
Antiestrogens			
Aminoglutethimide (Cytadren)	Unknown	None known	None known
Anastrozole (Arimidex)	3A4	1A2, 2C9, 3A4, 2C8	None known
Exemestane (Aromasin)	3A4	None known	None known
Letrozole (Femara)	2A6, 3A4	2A6, 2C19	None known
Tamoxifen (Nolvadex)	2D6, 3A4, 2C9, 1A2, 2A6, 2B6, 2E1	3A4, 2C9	3A4
Toremifene (Fareston)	3A4, 1A2	None known	None known

[1]Flutamide's primary metabolite, 2-hydroxyflutamide, is a potent 1A2 inhibitor.

Anastrozole

Anastrozole (Arimidex) is a third-generation aromatase inhibitor used in the treatment of breast cancer (usually advanced-stage disease in postmenopausal women). A potent inhibitor of aromatase, anastrozole decreases the progrowth effect of estrogen on breast tissue—specifically, cancerous breast tissue. Metabolism occurs via 3A4. Grimm and Dyroff (1997) used *in vitro* studies to show that anastrozole inhibits 1A2, 2C9, and 3A4. Manufacturing data also state that anastrozole can inhibit 2C8. No clinical or *in vivo* studies have yet been performed. On the basis of current understanding, however, if the usual dose of anastrozole were administered *in vivo*, inhibition of P450 enzymes *in vivo* would be minimal and likely not clinically relevant. There are currently no case reports of clinically relevant drug interactions due to P450 inhibition by anastrozole.

Bicalutamide

Bicalutamide (Casodex) is a nonsteroidal androgen receptor antagonist used for early prostate cancer treatment. Cockshott (2004) found that the *R*-enantiomer confers almost all of the antiandrogenic activity. The metabolism of the *R*- and *S*-enantiomers is different, with *R*-bicalutamide metabolized via 3A4 while *S*-bicalutamide is metabolized via glucuronidation. *R*-bicalutamide may be a weak inhibitor of 3A4, 2C9, 2C19, and 2D6, but *in vivo* data are lacking.

Flutamide

Flutamide (Eulexin) is an androgen receptor antagonist often used in the treatment of prostate cancer. Using human liver microsome preparations, Shet and colleagues (1997) demonstrated flutamide is primarily metabolized via 1A2. A minor metabolite was formed by 3A4. The primary metabolite of flutamide, 2-hydroxyflutamide, is a more potent androgen receptor antagonist and can inhibit 1A2. No *in vivo* studies were found that focused on this phenomenon, and the clinical implications of 1A2 inhibition by 2-hydroxyflutamide cannot be estimated adequately from these data. There have been reports of hepatotoxicity with flutamide, and Matsuzaki et al. (2006) studied the effect of a different metabolite, FLU-1, in 1A2-deficient knockout mice. They showed that in glutathione-depleted diets, mice lacking 1A2 activity experienced significant hepatoxicity. Studies are ongoing to further delineate the role of this alternative metabolite.

Tamoxifen

Tamoxifen (Nolvadex) is an older chemotherapy drug that works as an anti-estrogen. Tamoxifen is metabolized through both *N*-demethylation and 4-hydroxylation. Tamoxifen's 4-hydroxylation metabolite, predominantly created via 2D6 metabolism, is intrinsically 100 times more potent than tamoxifen (Crewe et al. 1997). 2D6 also creates the metabolite endoxifen. Human liver microsome studies showed that *N*-demethylation of tamoxifen by 3A4 produces a less potent antiestrogen (Dehal and Kupfer 1997). The relative contributions of 3A4 and 2D6 may dictate potency and influence drug interactions, so Scripture and Figg (2006) advocate the use of pharmacogenetics to guide therapy.

Tamoxifen is susceptible to inhibitors and inducers of the P450 system. Stearns et al. (2002) reviewed data from women who were taking tamoxifen and paroxetine (Paxil) concurrently. Paroxetine, an SSRI used for the treatment of hot flashes resulting from tamoxifen treatment, is a 2D6 inhibitor and affected the production of endoxifen. The magnitude of the effect differed when women with the wild-type 2D6 genotype (64% reduction) were compared with women expressing nonfunctional 2D6 (24% reduction). A study of the interaction between rifampin (Rifadin) and tamoxifen found a marked reduction in the serum levels of tamoxifen in patients taking rifampin, a potent inducer of 3A4 (Kivisto et al. 1998).

Christians et al. (1996) determined that several drugs, including tamoxifen, may inhibit P450 metabolism of tacrolimus, a known 3A4 substrate. Additional *in vitro* data from Zhao et al. (2002) further suggest that tamoxifen is a 3A4 inhibitor. Complicating the pharmacokinetic profile are data suggesting tamoxifen is also a 3A4 inducer (Desai et al. 2002). The overall effect of tamoxifen on 3A4, and therefore assessment of any clinical impact, remains uncertain.

Toremifene

In a study involving healthy male volunteers, plasma concentrations of toremifene (Fareston) were decreased when the drug was given in concert with rifampin, a known 3A4 and 1A2 inducer (Kivisto et al. 1998). This finding suggests that toremifene is metabolized via 3A4 and 1A2. *In vitro* studies using human liver microsome preparations with known inhibitors showed that the majority of the metabolites of toremifene are created by these two isoforms (Berthou et al. 1994).

Miscellaneous Agents

Retinoids

Retinoic acid, a vitamin A derivative, plays an important role in maintaining normal cell growth and structure but has been shown to induce cell differentiation and development in healthy cells and cancer cells, leading to early cell death. Retinoic acid was initially administered only in cases of acute promyelocytic leukemia, but the drug is now used in patients with other tumors. Han and Choi (1996) found that retinoic acid is metabolized to 4- and 18-hydroxy metabolites. Retinoic acid induces its own metabolism, but human liver microsome assays of known inhibitors and inducers revealed no P450-mediated interactions.

All-*trans*-retinoic acid (ATRA) is metabolized by a variety of enzymes, including 2C8, 2C9, 2B6, and CYP26. An *in vitro* study of several head and neck cancer cell lines showed that the oxidative catabolism of retinoids is inhibited by fluconazole (suggesting 3A4 or 2C9 involvement) and induced by 13-*cis*-retinoic acid, 9-*cis*-retinoic acid, and retinal but not retinol (Kim et al. 1998). These findings corroborated those of Schwartz and colleagues (1995) from a prospective study involving patients with acute promyelocytic leukemia. However, Lee et al. (1995) found no differences in the area under the curve of ATRA with or without ketoconazole coadministration. 3A4 may or may not be involved in ATRA metabolism, but extrapolation of the data from these studies suggests that ATRA is metabolized by 2C9 to a much greater extent than by 3A4. CYP26 is a P450 enzyme that has not been extensively characterized. Sonneveld et al. (1998) reviewed retinoid metabolism and its probable hydroxylation via the CYP26 family, which is induced by ATRA. The CYP26 family was specific for the hydroxylation of ATRA only, not other isomers of retinoic acid. Although the CYP26 family is not prominent in P450-mediated metabolism, its effect on retinoids is notable.

ATRA also induces its own metabolism. In a phase II study of ATRA in prostate cancer patients, subjects who underwent 14 days of ATRA therapy had an 83% increase in activity of 2E1 as well as of phase II *N*-acetyltransferase (Adedoyin et al. 1998). Subjects showed no appreciable differences in

1A2, 2C19, 2D6, or 3A4 activity. Krekels et al. (1997) performed *in vitro* assays with breast cancer cells and found that autoinduction of ATRA metabolism is dose dependent.

Steroids

Glucocorticoids are a common addition to many chemotherapeutic regimens. Many studies, including those by Christians et al. (1996), Liddle et al. (1998), and El-Sankary et al. (2002), show both 3A4 metabolism and induction. Given the frequency of use of glucocorticoids, care must be taken when these drugs are used with either pro-drugs requiring activation via 3A4 or drugs requiring 3A4 for deactivation.

Immunomodulators

Much of the communication between cells, both healthy and diseased, is accomplished chemically. Cytokines are intercellular mediators released by cells in response to antigens or disease states in order to communicate with other cells, the immune system, or the body in general. Cytokines include interleukins, interferons, and other nonantibody proteins.

There has been a rapid increase in the number of drugs and synthetic antibodies routinely used in oncology as well as other areas of medicine. Many of these compounds are not metabolized by P450 enzymes but can have significant effects on the regulation of the overall activity level of these enzymes. Additionally, P450 activity has been noted to fluctuate in many disease states, a concept supported by several *in vitro* human liver microsome studies showing changes in enzyme activity due to the effects of cytokines.

Interferon-α

Dorr (1993) reviewed available literature on the effects of interferon-α (Intron-A) in multiple disease states and cited evidence that P450 enzymes, along with general cellular protein synthesis, are inhibited by interferon-α. Both Leeder and Kearns (1997) and Dorr (1993) noted increased clearance of theophylline in inflammatory states, suggesting possible 1A2 induction, although no reference was made to specific enzymes.

Cytokines and Interleukins

Gorski et al. (2000) briefly reviewed the decreases in P450 function observed with IL-6, tumor necrosis factor-α, and IL-1β therapy. In a study of IL-10 in healthy human volunteers, these investigators demonstrated a decrease in 3A4 activity without any effect on 1A2, 2C9, or 2D6. Observations in more clinical settings suggest that theophylline clearance is decreased during serologically confirmed upper respiratory tract infections (Leeder and Kearns 1997).

Elkahwaji et al. (1999) showed that high doses of IL-2 in patients with metastatic disease in the liver resulted in a decrease of total P450 and monooxygenase activity—specifically, the activity of 1A2, 2C, 2E1, and 3A4.

Oncostatin M is a cytokine in the IL-6 receptor family. Guillen et al. (1998) compared the effects of oncostatin M with those of interferon-γ and IL-6. The activity of 1A2 was noted to be significantly reduced by all three cytokines but most strongly by oncostatin M. Similar reductions in activity of 2A6, 2B6, and 3A4 were noted with exposure to oncostatin M.

Monoclonal Antibodies

Since the 1990s, numerous monoclonal antibodies have been developed to bind and modify proteins relevant to specific malignancies in efforts to offer novel approaches to cancer treatment and bolster current regimens. Cetuximab (Erbitux) and panitumumab (Vectibix), both epidermal growth factor receptor binders, and bevacizumab (Avastin), a vascular endothelial growth factor binder, have all been used in metastatic colon cancer regimens, but to date little is known about their effects on metabolism. Ettlinger et al. (2006) performed one *in vivo* study analyzing the pharmacokinetics of a combination of irinotecan with cetuximab and noted no impact on irinotecan metabolism.

Alemtuzumab (Campath), a CD52 binder used in chronic lymphocytic leukemia, gemtuzumab ozogamicin (Mylotarg), a CD33 binder used in acute myelogenous leukemia, trastuzumab (Herceptin), a Her2/neu binder used in breast cancer, and rituximab (Rituxan), a CD20 binder used in non-Hodgkin's lymphoma, all have unknown metabolic implications and require further investigation.

Tyrosine Kinase Inhibitors

Interest in tyrosine kinase inhibitors in the treatment of cancer has greatly increased since the development of the prototype imatinib (Gleevec) in the late 1990s for the treatment of chronic myelogenous leukemia (CML). The now well known Philadelphia Chromosome inherent to CML produces a perpetually active tyrosine kinase called bcr-abl, which is inhibited by imatinib with resultant dramatic remissions in disease. Multiple tyrosine kinase inhibitors have subsequently been developed, and early studies indicate that many of them are metabolized primarily via P450-related enzymes (see Table 16–5).

Imatinib, along with other studied tyrosine kinase inhibitors such as gefitinib (Iressa) and erlotinib (Tarceva), appears to be primarily metabolized via 3A4, with varying contributions from other P450 cytochromes. Peng et al. (2005) showed that imatinib is metabolized via 3A4 while also acting as a competitive inhibitor at 3A4. In a study of gefitinib metabolism by McKillop et al. (2005), gefitinib was found to be metabolized extensively via 3A4 with minor contribution via 2D6 and evidence of weak inhibition at 2D6. A two-part *in vivo* study by Swaisland et al. (2005, 2006) studied the effect of the 3A4 inducer rifampin and the 3A4 inhibitor itraconazole on gefitinib pharmacokinetics. Rifampin was shown to significantly decrease the maximum plasma concentration and AUC, whereas itraconazole increased these parameters. The same group noted a moderate increase in the AUC of the 2D6 substrate metoprolol (Lopressor) when coadministered with imatinib. These results were confirmed with another study by Li et al. (2007), which compared the pharmacokinetics of gefitinib and erlotinib. In this study, erlotinib was found to be metabolized by 3A4 with considerable contribution from 1A2.

Sorafenib (Nexavar), a tyrosine kinase inhibitor used in the treatment of renal cell carcinoma, is metabolized primarily via 1A9 with contribution from 3A4 (Lathia et al. 2006). Many other agents such as dasatinib (Sprycel), lapatinib (GW572016), nilotinib (AMN107), sunitinib (Sutent), and vandetanib (Zactima) are still under investigation, with early indications of metabolism via 3A4.

Summary

There are many classes and subclasses of chemotherapeutic agents. Many of these drugs are old enough or new enough that complete pharmacokinetic

TABLE 16–5. Tyrosine kinase inhibitors

Drug	Metabolism site(s)	Enzyme(s) inhibited	Enzyme(s) induced
Dasatinib (Sprycel)	3A4	3A4	None known
Erlotinib (Tarceva)	3A4, 1A2	3A4	None known
Gefitinib (Iressa)	3A4, 2D6, 2C19	2D6	None known
Imatinib (Gleevec)	3A4, 2D6, 2C19	3A4	None known
Lapatinib (Tykerb)	Unknown	None known	None known
Nilotinib (Tasigna)	Unknown	None known	None known
Sorafenib (Nexavar)	3A4, UGT1A9	None known	None known
Sunitinib (Sutent)	3A4	None known	None known
Vandetanib (Zactima)	Unknown	None known	None known

understanding does not exist. Oncology patients are routinely on polypharmacy to address the many side effects of these agents. Strategies to limit the potential for complicated interactions include avoiding herbal medications such as St. John's wort, a 3A4 inducer, and avoiding grapefruit juice, a potent inhibitor of intestinal 3A4. The interindividual variability of oxidative metabolism argues for the possible use of pharmacogenetic testing, particularly when chemotherapeutics with known potential for interaction (e.g., flutamide, cyclophosphamide, paclitaxel, or tamoxifen) are to be administered (van Schaik 2005).

References

Adedoyin A, Stiff DD, Smith DC, et al: All-*trans*-retinoic acid modulation of drug-metabolizing enzyme activities: investigation with selective metabolic drug probes. Cancer Chemother Pharmacol 41:133–139, 1998

Ando Y, Shimizu T, Nakamure K, et al: Potent and non-specific inhibition of cytochrome P450 by JM216: a new oral platinum agent. Br J Cancer 18:1170–1174, 1998

Arboix M, Paz OG, Colombo T, D'Incalci M: Multidrug resistance-reversing agents increase vinblastine distribution in normal tissues expressing the P-glycoprotein but do not enhance drug penetration in brain and testes. J Pharmacol Exp Ther 281:1226–1230, 1997

Bagniewski PG, Reid JM, Ames MM, et al: Increased etoposide clearance in patients with glioma may be associated with concurrent glucocorticoid or anticonvulsant treatment (abstract). Proceedings of the Annual Meeting of the American Association for Cancer Research 37:A1224, 1996

Balis FM: Pharmacokinetic drug interactions of commonly used anticancer drugs. Clin Pharmacokinet 11:223–235, 1986

Baumhakel M, Kasel D, Rao-Schymanski RA, et al: Screening for inhibitory effects of antineoplastic agents on CYP3A4 in human liver microsomes. Int J Clin Pharmacol Ther 39:517–528, 2001

Berkery R, Cleri LB, Skarin AT: Oncology: Pocket Guide to Chemotherapy. St. Louis, MO, Mosby Year–Book, 1997

Berthou F, Dreano Y, Belloc C, et al: Involvement of cytochrome P450 3A family in the major metabolic pathways of toremifene in human liver microsomes. Biochem Pharmacol 47:1883–1895, 1994

Brain EG, Yu LJ, Gustafsson K, et al: Modulation of P450 dependent ifosfamide pharmacokinetics: a better understanding of drug activation in vivo. Br J Cancer 77:1768–1776, 1998

Brinker A, Kisro J, Letsch C, et al: New insights into the clinical pharmacokinetics of trofosfamide. Int J Clin Pharmacol Ther 40:376–381, 2002

Britten CD, Baker SD, Denis LJ, et al: Oral paclitaxel and concurrent cyclosporin A: targeting clinically relevant systemic exposure to paclitaxel. Clin Cancer Res 6:3459–3468, 2000

Buggia I, Zecca M, Alessandrino EP, et al: Itraconazole can increase systemic exposure to busulfan in patients given bone marrow transplantation. GITMO (Gruppo Italiano Trapianto di Midollo Osseo). Anticancer Res 16:2083–2088, 1996

Cagnoni PJ, Matthes S, Day TC, et al: Modification of the pharmacokinetics of high-dose cyclophosphamide and cisplatin by antiemetics. Bone Marrow Transplant 24:1-4, 1999

Chan JD: Pharmacokinetic drug interactions of vinca alkaloids: summary of case reports. Pharmacotherapy 18:1304–1307, 1998

Chang P, Riggs CE Jr, Scheerer MT, et al: Combination chemotherapy with Adriamycin and streptozotocin, II: clinicopharmacologic correlation of augmented Adriamycin toxicity caused by streptozotocin. Clin Pharmacol Ther 20:611–616, 1976

Chang TK, Yu L, Maurel P, et al: Enhanced cyclophosphamide and ifosfamide activation in primary human hepatocyte cultures: response to cytochrome P-450 inducers and autoinduction by oxazaphosphorines. Cancer Res 57:1946–1954, 1997

Chladek J, Martinkova J, Sispera L: An in vitro study on methotrexate hydroxylation in rat and human liver. Physiol Res 46:371–379, 1997

Christians U, Schmidt G, Bader A, et al: Identification of drugs inhibiting the in-vitro metabolism of tacrolimus by human liver microsomes. Br J Clin Pharmacol 41:187–190, 1996

Clarke SJ, Rivory LP: Clinical pharmacokinetics of docetaxel. Clin Pharmacokinet 36:99–114, 1999

Cockshott ID: Bicalutamide: clinical pharmacokinetics and metabolism. Clin Pharmacokinet 43:855–878, 2004

Colburn DE, Giles FJ, Oladovich D, et al: In vitro evaluation of cytochrome P450-mediated drug interactions between cytarabine, idarubicin, itraconazole and caspofungin. Hematology 9:217–221, 2004

Cresteil T, Monsarrat B, Dubois J, et al: Regioselective metabolism of taxoids by human CYP3A4 and 2C8: structure-activity relationship. Drug Metab Dispos 30:438–445, 2002

Crewe HK, Ellis SW, Lennard MS, et al: Variable contribution of cytochromes P450 2D6, 2C8 and 3A4 to the 4-hydroxylation of tamoxifen by human liver microsomes. Biochem Pharmacol 53:171–178, 1997

Dai D, Zeldin D, Blaisdell JA, et al: Polymorphisms in human CYP2C8 decrease metabolism of the anticancer drug paclitaxel and arachidonic acid. Pharmacogenetics 11:597–607, 2001

Dean R, Nachman J, Lorenzana AN: Possible methotrexate-mezlocillin interaction. Am J Pediatr Hematol Oncol 14:88–89, 1992

Dehal SS, Kupfer D: CYP2D6 catalyzes tamoxifen 4-hydroxylation in human liver. Cancer Res 57:3402–3406, 1997

Desai PB, Duan JZ, Zhu YW, et al: Human liver microsomal metabolism of paclitaxel and drug interactions. Eur J Drug Metab Pharmacokinet 23:417–424, 1998

Desai PB, Nallani SC, Sane RS, et al: Induction of cytochrome P450 3A4 in primary human hepatocytes and activation of the human pregnane X receptor by tamoxifen and 4-hydroxytamoxifen. Drug Metab Dispos 30:608–612, 2002

Dorr RT: Interferon-alpha in malignant and viral diseases: a review. Drugs 45:177–211, 1993

Elkahwaji J, Robin MA, Berson A, et al: Decrease in hepatic cytochrome P450 after interleukin-2 immunotherapy. Biochem Pharmacol 57:951–954, 1999

El-Sankary W, Bombail V, Gibson GG, et al: Glucocorticoid mediated induction of CYP3A4 is decreased by disruption of a protein: DNA interaction distinct from the pregnane X receptor response element. Drug Metab Dispos 30:1029–1034, 2002

Ettlinger DE, Mitterhauser M, Wadsak W, et al: In vivo disposition of irinotecan (CPT-11) and its metabolites in combination with the monoclonal antibody cetuximab. Anticancer Res 26:1337–1441, 2006

Frenia ML, Long KS: Methotrexate and nonsteroidal anti-inflammatory drug interactions. Ann Pharmacother 26:234–237, 1992

Gorski JC, Hall SD, Becker P, et al: In vivo effects of interleukin-10 on human cytochrome P450 activity. Clin Pharmacol Ther 67:32–43, 2000

Grimm SW, Dyroff MC: Inhibition of human drug metabolizing cytochromes P450 by anastrozole, a potent and selective inhibitor of aromatase. Drug Metab Dispos 25:598–601, 1997

Guillen MI, Donato MT, Jover R, et al: Oncostatin M down-regulates basal and induced cytochromes P450 in human hepatocytes. J Pharmacol Exp Ther 285:127–134, 1998

Gunes A, Coskun U, Boruban C, et al: Inhibitory effect of 5-fluorouracil on cytochrome P450 2C9 activity in cancer patients. Basic Clin Pharmacol Toxicol 98:197–200, 2006

Han IS, Choi JH: Highly specific cytochrome P450-like enzymes for all-trans-retinoic acid in T47D human breast cancer cells. J Clin Endocrinol Metab 81:2069–2075, 1996

Hassan M, Oberg G, Bjorkholm M, et al: Influence of prophylactic anticonvulsant therapy on high-dose busulfan kinetics. Cancer Chemother Pharmacol 33:181–186, 1993

Hempel G, Krumpelman S, May-Manke A, et al: Pharmacokinetics of trofosfamide and its dechloroethylated metabolites. Cancer Chemother Pharmacol 40:45–50, 1997

Hill DL, Kirk MC, Struck RF: Microsomal metabolism of nitrosoureas. Cancer Res 35:296–301, 1975

Huang Z, Raychowdhury MK, Waxman DJ: Impact of liver P450 reductase suppression on cyclophosphamide activation, pharmacokinetics and antitumoral activity in a cytochrome P450-based cancer gene therapy model. Cancer Gene Ther 7:1034–1042, 2000a

Huang Z, Roy P, Waxman DJ: Role of human liver microsomal CYP 3A4 and CYP 2B6 in catalyzing N-dechloroethylation of cyclophosphamide and ifosfamide. Biochem Pharmacol 59:961–972, 2000b

Huitema AD, Mathot RA, Tibben MM, et al: A mechanism based model for the cytochrome P450 drug-drug interaction between cyclophosphamide and thioTEPA and the autoinduction of cyclophosphamide. J Pharmacokinet Pharmacodyn 28:211–230, 2001

Innocenti F, Undevia SD, Ramirez J, et al: A phase I trial of pharmacologic modulation of irinotecan with cyclosporine and phenobarbital. Clin Pharmacol Ther 76:490–502, 2004

Jacobson PA, Green K, Birnbaum A, et al: Cytochrome P450 isozymes 3A4 and 2B6 are involved in the in vitro human metabolism of thiotepa to TEPA. Cancer Chemother Pharmacol 49:461–467, 2002

Jamis-Dow CA, Pearl ML, Watkins PB, et al: Predicting drug interactions in-vivo from experiments in-vitro: human studies with paclitaxel and ketoconazole. Am J Clin Oncol 20:592–599, 1997

Janney LM, Waterbury NV: Capecitabine-warfarin interaction. Ann Pharmacother 39:1546–1551, 2005

Kajita J, Kuwabara T, Kobayashi H, et al: CYP3A4 is mainly responsible for the metabolism of a new vinca alkaloid, vinorelbine, in human liver microsomes. Drug Metab Dispos 28:1121–1127, 2000

Kawashiro T, Yamashita K, Zhao XJ, et al: A study on the metabolism of etoposide and possible interactions with antitumor or supporting agents by human liver microsomes. J Pharmacol Exp Ther 286:1294–1300, 1998

Kehrer DF, Mathijssen RH, Verweij J, et al: Modulation of irinotecan metabolism by ketoconazole. J Clin Oncol 20:3122–3129, 2002

Kim SY, Han IS, Yu HK, et al: The induction of P450-mediated oxidation of all-trans retinoic acid by retinoids in head and neck squamous cell carcinoma cell lines. Metabolism 47:955–958, 1998

Kivisto KT, Villikka K, Nyman L, et al: Tamoxifen and toremifene concentrations in plasma are greatly decreased by rifampin. Clin Pharmacol Ther 64:648–654, 1998

Krekels MD, Verhoeven A, van Dun J, et al: Induction of the oxidative catabolism of retinoid acid in MCF-7 cells. Br J Cancer 75:1098–1104, 1997

Lathia C, Lettieri J, Cihon F, et al: Lack of effect of ketoconazole-mediated CYP3A inhibition on sorafenib clinical pharmacokinetics. Cancer Chemother Pharmacol 57:685-692, 2006

Lee JS, Newman RA, Lippman SM, et al: Phase I evaluation of all-trans retinoic acid with and without ketoconazole in adults with solid tumors. J Clin Oncol 13:1501–1508, 1995

Leeder JS, Kearns GL: Pharmacogenetics in pediatrics. Implications for practice. Pediatr Clin North Am 44:55–77, 1997

Le Guellec C, Lacarelle B, Catalin J, et al: Inhibitory effects of anticancer drugs on dextromethorphan-O-demethylase activity in human liver microsomes. Cancer Chemother Pharmacol 32:491–495, 1993

Levin VA, Stearns J, Byrd A, et al: The effect of phenobarbital pretreatment on the antitumor activity of 1,3-bis(2-chloroethyl)-1-nitrosourea (BCNU), 1-(2-chloroethyl)-3-cyclohexyl-1-nitrosourea (CCNU) and 1(2-chloroethyl)-3-(2,6-dioxo-3-piperidyl)-1-nitrosourea (PCNU), and on the plasma pharmacokinetics and biotransformation of BCNU. J Pharmacol Exp Ther 208:1–6, 1979

Li J, Zhao M, He P, et al: Differential metabolism of gefitinib and erlotinib by human cytochrome P450 enzymes. Clin Cancer Res 13:3731–3737, 2007

Liddle C, Goodwin BJ, George J, et al: Separate and interactive regulation of cytochrome P450 3A4 by triiodothyronine, dexamethasone, and growth hormone in cultured hepatocytes. J Clin Endocrinol Metab 83:2411–2416, 1998

Malingre MM, Ten Bokkel Huinink WW, Mackay M, et al: Pharmacokinetics of oral cyclosporin A when co-administered to enhance the absorption of orally administered docetaxel. Eur J Clin Pharmacol 57:305–307, 2001

Masuyama H, Suwaki N, Tateishi Y, et al: The pregane X receptor regulates gene expression in a ligand- and promoter- selective fashion. Mol Endocrinol 19:1170–1180, 2005

Mathijssen RH, Verweij J, de Bruijn P, et al: Effects of St. John's wort on irinotecan metabolism. J Natl Cancer Inst 16:1247–1249, 2002

Matsuzaki Y, Nagai D, Ichimura E, et al: Metabolism and hepatic toxicity of flutamide in cytochrome P450 1A2 knockout SV129 mice. J Gastroenterol 41:231–239, 2006

May-Manke A, Kroemer H, Hempel G, et al: Investigation of the major human hepatic cytochrome P450 involved in 4-hydroxylation and N-dechloroethylation of trofosfamide. Cancer Chemother Pharmacol 44:327–334, 1999

McKillop D, McCormick AD, Millar A, et al: Cytochrome P450-dependent metabolism of gefitinib. Xenobiotica 35:39–50, 2005

McLeod HL: Clinically significant drug-drug interactions in oncology. Br J Clin Pharmacol 45:539–544, 1998

Neuman MG, Cameron RG, Haber JA, et al: Inducers of cytochrome P450 2E1 enhance methotrexate-induced hepatotoxicity. Clin Biochem 32:519–536, 1999

Patterson LH, Murray GI: Tumour cytochrome P450 and drug activation. Curr Pharm Des 8:1335–1347, 2002

Peng B, Lloyd P, Schran H: Clinical pharmacokinetics of imatinib. Clin Pharmacokinet 44:879–894, 2005

Primavera A, Audenino D, Cocito L: Ifosfamide encephalopathy and nonconvulsive status epilepticus. Can J Neurol Sci 29:180–183, 2002

Rahman A, Korzekwa KR, Grogan J, et al: Selective biotransformation of taxol to 6-alpha-hydroxytaxol by human cytochrome P450 2C8. Cancer Res 54:5543–5546, 1994

Rae JM, Soukhova NV, Flockhart DA: Triethylenethiophosphamide is a specific inhibitor of cytochrome P450 2B6: implications for cyclophosphamide metabolism. Drug Metab Dispos 30:525–530, 2002

Reid JM, Kuffel MJ, Miller JK, et al: Metabolic activation of dacarbazine by human cytochromes P450: the role of CYP1A1, CYP1A2, and CYP2E1. Clin Cancer Res 5:2192–2197, 1999

Relling MV, Nemec J, Schuetz EG, et al: O-Demethylation of epipodophyllotoxins is catalyzed by human cytochrome P450 3A4. Mol Pharmacol 45:352–358, 1994

Relling MV, Pui CH, Sandlund JT, et al: Adverse effect of anticonvulsants on efficacy of chemotherapy for acute lymphoblastic leukaemia. Lancet 356:285–290, 2000

Royer I, Monsarrat B, Sonnier M, et al: Metabolism of docetaxel by human cytochromes P450: interactions with paclitaxel and other antineoplastic drugs. Cancer Res 56:58–65, 1996

Rozman B: Clinical pharmacokinetics of leflunomide. Clin Pharmacokinet 41:421–430, 2002

Santner SJ, Rosen H, Osawa Y, et al: Additive effects of aminoglutethimide, testololactone, and 4-hydroxyandrostenedione as inhibitors of aromatase. J Steroid Biochem 20:1239–1242, 1984

Santos A, Zanetta S, Cresteil T, et al: Metabolism of irinotecan (CPT-11) by CYP3A4 and CYP3A5 in humans. Clin Cancer Res 6:2012–2020, 2000

Schmidt R, Baumann F, Hanschmann H, et al: Gender differences in ifosfamide metabolism by human liver microsomes. Eur J Drug Metab Pharmacokinet 26:193–200, 2001

Schwartz EL, Hallam S, Gallagher RE, et al: Inhibition of all-trans-retinoic acid metabolism by fluconazole in-vitro and in patients with acute promyelocytic leukemia. Biochem Pharmacol 50:923–928, 1995

Schwartz JD, Howard W, Scadden DT: Potential interaction of antiretroviral therapy with paclitaxel in patients with AIDS-related Kaposi's sarcoma. AIDS 13:283–284, 1999

Scripture CD, Figg WD: Drug interactions in cancer therapy. Nat Rev Cancer 6:546-558, 2006

Shet MS, McPhaul M, Fisher CW, et al: Metabolism of the antiandrogenic drug (flutamide) by human CYP1A2. Drug Metab Dispos 25:1298–1303, 1997

Shou M, Martinet M, Korzekwa KR, et al: Role of human cytochrome P450 3A4 and 3A5 in the metabolism of Taxotere and its derivatives: enzyme specificity, interindividual distribution and metabolic contribution in human liver. Pharmacogenetics 8:391–401, 1998

Sonneveld E, van den Brink CE, van der Leede BM, et al: Human retinoic acid (RA) 4-hydroxylase (CYP26) is highly specific for all-trans-RA and can be induced through RA receptors in human breast and colon carcinoma cells. Cell Growth Differ 9:629–637, 1998

Spratlin J, Sawyer MB: Pharmacogenetics of paclitaxel metabolism. Crit Rev Oncol Hematol 61:222–229, 2007

Stearns V, Ullmer L, Lopez JF, et al: Hot flushes. Lancet 360:1851–1861, 2002

Swaisland HC, Ranson M, Smith RP, et al: Pharmacokinetic drug interactions of gefitinib with rifampicin, itraconazole, and metoprolol. Clin Pharmacokinet 44:1067–1081, 2005

Swaisland HC, Cantarini MV, Fuhr R, et al: Exploring the relationship between expression of cytochrome P450 enzymes and gefitinib pharmacokinetics. Clin Pharmacokinet 45:633–644, 2006

van Meerten E, Verweij J, Schellens JH: Antineoplastic agents: drug interactions of clinical significance. Drug Saf 12:168–182, 1995

van Schaik RH: Cancer treatment and pharmacogenetics of cytochrome P450 enzymes. Invest New Drugs 23:513–522, 2005

Villikka K, Kivisto KT, Maenpaa H, et al: Cytochrome P450-inducing antiepileptics increase the clearance of vincristine in patients with brain tumors. Clin Pharmacol Ther 66:589–593, 1999

Weber GF, Waxman DJ: Denitrosation of the anti-cancer drug 1,3-bis(2-chloroethyl)-1-nitrosourea catalyzed by microsomal glutathione S-transferase and cytochrome P450 monooxygenases. Arch Biochem Biophys 307:369–378, 1993

Yu DK: The contribution of P-glycoprotein to pharmacokinetic drug-drug interactions. J Clin Pharmacol 39:1203–1211, 1999

Zhao XJ, Jones DR, Wang YH, et al: Reversible and irreversible inhibition of CYP3A enzymes by tamoxifen and metabolites. Xenobiotica 32:863–878, 2002

17

Pain Management I: Nonnarcotic Analgesics

Gary H. Wynn, M.D.

Scott C. Armstrong, M.D., D.F.A.P.A., F.A.P.M.

Reminder: This chapter is dedicated primarily to metabolic inter-actions and effects on transporters. Interactions due to displaced protein-binding, alterations in absorption or excretion, and pharma-codynamics are not covered. In addition, the two most common and potentially serious side effects secondary to most nonnarcotic anal-gesic use—gastrointestinal bleeding and liver toxicity—are not cov-ered in this chapter.

The nonnarcotic analgesics discussed in this chapter include a wide array of medications: "coxib" cyclooxygenase-2 (COX-2) inhibitors, noncoxib COX-2 inhibitors, aspirin, aspirin-like products, acetaminophen (Tylenol), and nonsteroidal anti-inflammatory drugs (NSAIDs). Generally, little has been written about pharmacokinetic drug interactions involving these medications, for two reasons:

1. Nonnarcotic analgesics have broad safety margins, and any drug interactions that do occur rarely have dire consequences. Indeed, many of these drugs are available over the counter in the United States.
2. What is known about the metabolism of nonnarcotic analgesics is sketchy. Many nonnarcotic analgesics are older medications that have gone off patent, and drug companies and researchers thus have been less interested in profiling their metabolisms. In addition, determining exactly how these drugs are metabolized can be difficult because many are metabolized via less well delineated non–cytochrome P450 pathways.

Five overriding themes are important in regard to potential pharmacokinetic drug interactions of nonnarcotic analgesics:

1. Some nonnarcotic analgesics (e.g., aspirin, salsalate [Disalcid], and nabumetone [Relafen]) are pro-drugs. Such drugs require an oxidative reaction to become "active" and exert their pharmacologic effect. If that process is inhibited, the drugs' effectiveness may be diminished.
2. Many nonnarcotic analgesics are oxidatively metabolized by 2C9, and so, theoretically, levels and side effects of the drugs could be increased in poor metabolizers (PMs) at 2C9 or with the addition of a 2C9 inhibitor (such as fluvoxamine [Luvox] or fluconazole [Diflucan]). This result is rarely, if ever, reported, since many nonnarcotic analgesics that are 2C9 substrates also progress directly to UGT conjugation even if they are not oxidatively changed by 2C9. In addition, nonnarcotic analgesics have wide safety margins, so inhibition of metabolism may have only modest consequences, resulting in mild side effects (or ineffective therapy in the case of pro-drugs) rather than deleterious effects that result in case reports.
3. Some nonnarcotic analgesics are inhibitors of UGT enzymes. Many traditional NSAIDs inhibit UGT2B7. In particular, some have been shown to inhibit conjugation of zidovudine (AZT) and oxazepam by this mechanism. However, there is little evidence that these interactions are clinically relevant.

4. Most nonnarcotic analgesics that are conjugated go through a process of recirculation in the hepatic system and are deconjugated and become active again. This could be important if an inhibitor of conjugation is added, but, again, little to no information exists on this potential drug interaction.
5. Nonnarcotic analgesics are not known to induce any metabolic enzymes.

Cyclooxygenase-2 Inhibitors

Celecoxib

Celecoxib (Celebrex), commonly used for chronic pain due to osteoarthritis, is metabolized chiefly through hydroxylation by 2C9 (Celebrex 2008) and by one or more unspecified UGT enzymes to form a 1-*O*-glucuronide metabolite (Paulson et al. 2000). Because 2C9 is considered the major enzyme for clearance of celecoxib, there has been some concern that celecoxib's effectiveness may be altered in PMs at 2C9 or by inhibitors or inducers of 2C9 (see Table 17–1). One study by Tang et al. (2001) demonstrated that celecoxib's area under the plasma concentration-time curve (AUC) was increased 2.2-fold in three volunteers who were PMs at 2C9 and received a single 200-mg dose. Werner et al. (2002) obtained a similar finding in one PM: AUC was doubled compared with the AUCs in extensive metabolizers at 2C9, although the half-life of the drug was unchanged. The manufacturer has recommended caution when the drug is used in known or suspected PMs at 2C9 (Celebrex 2008)—and, we would add, with potent 2C9 inhibitors such as fluconazole. Finally, in a small study of healthy volunteers, celecoxib was shown to inhibit 2D6 (Werner et al. 2003), but no further evidence has addressed this possibility.

Rofecoxib

Rofecoxib (Vioxx) was voluntarily removed from the U.S. market in September 2004 because of concerns about increased rates of stroke and myocardial infarctions associated with rofecoxib administration. Rofecoxib is not appreciably metabolized via P450 enzymes (Garnett 2001; Vioxx 2004), since it undergoes reduction rather than oxidation and conjugation. The enzymes responsible for this reduction are non-P450 cytosolic enzymes; however, "pan-inducers" of hepatic enzymes, such as rifampin and phenobarbital, may decrease serum rofecoxib levels.

TABLE 17–1.   Cyclooxygenase-2 (COX-2) inhibitors

Drug	P450 enzyme(s) that metabolize drug	UGT(s) that metabolize drug	Enzyme(s) inhibited	Pro-drug?[1]
"Coxibs"				
Celecoxib (Celebrex)	2C9	Unspecified UGT(s)	2D6[4]	No
Rofecoxib (Vioxx)	None	None	None	No
Valdecoxib (Bextra)	3A4, 2C9	Unspecified UGTs	2C9, 2C19, 3A4	No[2]
Other COX-2 inhibitors				
Diclofenac (Voltaren)	2C9, 3A4	UGT1A9, UGT2B7	UGT2B7	No
Etodolac (Lodine)	2C9	1A9	2C9[4]	No
Meloxicam (Mobic)	2C9, 3A4	None	2C9, 3A4[4]	No
Nabumetone (Relafen)	Unspecified phase I enzyme	Unspecified UGTs	None	Yes[3]

Note.   SULT=sulfotransferase; UGT=uridine 5'-diphosphate glucuronosyltransferase.
[1]Notes whether drug is or is not metabolized to an active analgesic metabolite.
[2]Parenteral parecoxib is rapidly metabolized to valdecoxib.
[3]Metabolized by phase I enzymes to the active analgesic 6-methoxy-2-naphthylacetic acid (6-MNA).
[4]Weak inhibition or limited evidence of inhibition.

Valdecoxib

In April 2005 Pfizer voluntarily withdrew valdecoxib (Bextra) from the U.S. market due to concerns regarding increased stroke and myocardial infarctions similar to concerns for other NSAIDs, as well as the risk of severe skin reactions (U.S. Food and Drug Administration 2005). Oral formulations of valdecoxib are no longer available; however, current studies are evaluating topical formulations of valdecoxib for analgesic use. Topical administration does not appear to result in any significant systemic absorption, negating concerns regarding drug interactions. Oral preparations (no longer available) and any systemically absorbed valdecoxib undergo metabolism via 3A4 and 2C9 (Bextra 2002). Further metabolism of the P450 products occurs with conjugation by UGTs (Bextra 2002; Yuan et al. 2002). In addition, valdecoxib is a moderate inhibitor of 2C19, 3A4, and 2C9.

Other Cyclooxygenase-2 Inhibitors

Diclofenac

Diclofenac (Voltaren) is metabolized by both oxidative and conjugation processes. Oxidative clearance is considered a minor role. 2C9 appears to be the main enzyme involved in 4-hydroxylation of diclofenac, with minor contribution by 3A4 (Crespi et al. 2006). The bulk of diclofenac, including its 4-hydroxylated metabolite, is metabolized via UGT1A9 and 2B7 (Kaji et al. 2005; King et al. 2001). Metabolism via UGT2B7 produces a potentially hepatotoxic acylglucuronide metabolite (Tang 2003). UGT2B7 production of acylglucuronide, causing hepatotoxicity, is complicated by multiple factors, including underlying enzymatic activity (Aithal et al. 2000). PMs at 2C9 may be at increased risk of hepatotoxicity because of the increased "shunting" of diclofenac metabolism via UGT2B7 and an accumulation of the potentially hepatotoxic metabolite. Coadministration of 2C9 or pan-inhibitors of oxidative clearance (e.g., fluoxetine) may also increase an individual's risk for hepatotoxicity from diclofenac.

Diclofenac inhibits UGT2B7 (Mano et al. 2007). *In vitro* studies have shown significant inhibition when diclofenac is coadministered with UGT2B7 substrates, including morphine (Hara et al. 2007). This finding has potentially important clinical application, given the overlap between morphine and diclofenac administration for the treatment of a variety of different types of pain.

Etodolac

Etodolac (Lodine) undergoes oxidation and glucuronidation. Oxidation of etodolac occurs via 2C9, resulting in multiple metabolites, whereas glucuronidation occurs primarily via UGT1A9 (Tougou et al. 2004). It appears that etodolac's metabolism is subject to stereoselective preference, with 2C9 preferentially metabolizing R-etodolac and UGT1A9 preferential to S-etodolac. In addition, *in vitro* data suggest that etodolac is a weak inhibitor of 2C9, although the clinical impact of this inhibition is unknown (Nakamura et al. 2005).

Meloxicam

The major metabolites of meloxicam (Mobic) are produced by 2C9 and 3A4 (Gates et al. 2005). 2C9 appears to be the major enzyme involved, with 3A4 playing a lesser role. The two metabolites, 5-hydroxymethylmeloxicam and 5-carboxymeloxicam, are the primary forms found in urine and feces (Mobic 2005), indicating that phase II has little involvement in the metabolism of meloxicam. There appears to be some evidence that meloxicam inhibits 2C9 and 3A4, although the extent of inhibition and clinical impact are not yet understood.

Nabumetone

Nabumetone (Relafen) is a pro-drug that is metabolized to 6-methoxy-2-naphthylacetic acid (6-MNA), the active compound that has inhibitory effects on COX-2 (Davies and McLachlan 2000). This transformation is very rapid, and little of the parent compound is found in the serum after ingestion of the drug (Relafen 2006). The advantage of this rapid transformation is that nabumetone is inactive as it enters the gastrointestinal tract, and so, unlike other NSAIDs, it is less likely to cause gastrointestinal distress and bleeding. It has other metabolites, including inactive oxidative metabolites and glucuronides, although the specific enzymes and resultant metabolites have not yet been fully elucidated. Liver failure may decrease the rate of 6-MNA formation, affecting efficacy. Nabumetone is highly protein bound, and most of the drug interactions that have been described have been presumed to be connected with this mechanism.

Aspirin and Acetaminophen

Aspirin

Aspirin (acetylsalicylic acid) is rapidly deacetylated upon ingestion to salicylic acid, then metabolized via glucuronidation by multiple UGTs to acyl- and phenolic glucuronide products (see Table 17–2). Current evidence suggests that UGT2B7 catalyzes the production of the acyl product and that numerous UGTs are involved in phenolic glucuronidation (Kuehl et al. 2006). The complex pharmacodynamics and pharmacokinetics of aspirin complicate the understanding of potential interactions. Salicylic acid interacts with other drugs because of factors including extensive protein binding and enzymatic inhibition. For instance, coadministration of aspirin and valproate results in an increase in free valproate levels due to both protein-binding displacement and inhibition of β-oxidation by aspirin (Sandson et al. 2006). Zidovudine levels have been increased with acetylsalicylic acid use—evidence that UGT2B7 may be inhibited (Sim et al. 1991). There is also evidence that salicylic acid may inhibit sulfation conjugation (Wang et al. 2006), which could increase acetaminophen levels. Finally, levels of salicylic acid may be reduced when it is used with phase II inducers, such as some oral contraceptives, corticosteroids, or rifampin (Miners 1989). There is no evidence that salicylic acid can induce any metabolic enzymes.

Salsalate (Disalcid), a related compound, is inactive as an analgesic. However, it is metabolized by esterases in the liver and elsewhere to two moieties of salicylic acid.

Acetaminophen

Acetaminophen (Tylenol) is metabolized through glucuronidation (50%; via UGT1A1, UGT1A6, UGT1A9, and UGT2B15 [Mutlib et al. 2006]), sulfation conjugation (30%–40%; via sulfotransferase [SULT] 1A1 [Dooley 1998; Nagar et al. 2006]), and oxidative metabolism (via 2E1 and, to a lesser extent, 3A4 and 1A2 [Manyike et al. 2000]) (see Table 17–2). 2E1 is typically a minor pathway, but creates a metabolite toxic to the liver. If acetaminophen is taken in overdose or if 2E1 is induced, there is risk of hepatic injury.

TABLE 17–2. Aspirin and acetaminophen

Drug	P450 enzyme(s) that metabolize drug	UGT(s) that metabolize drug	Other metabolism	Enzyme(s) inhibited	Pro-drug?[1]
Acetaminophen (Tylenol)	2E1, 3A1[2], 1A2[2]	UGT1A1, UGT1A6, UGT1A9, UGT2B15	Sulfation by SULT1A1	None	No
Acetylsalicylic acid (aspirin)	None known	None known	Deacetylated to salicylic acid	None	Yes
Salicylic acid (Salacid)	None	UGT2B7, others	Glycine conjugation	UGT2B7, beta-oxidation, ?other UGTs, SULTs	No
Salsalate (Disalcid)	None	Unspecified	Split by esterases into two salicylic acids	UGT2B7, ?other UGTs, SULTs	Yes

Note. SULT=sulfotransferase; UGT=uridine 5′-diphosphate glucuronosyltransferase.
[1]Notes whether drug is or is not metabolized to an active analgesic metabolite.
[2]Minor contribution to metabolism.

Nonsteroidal Anti-Inflammatory Drugs

Diflunisal

Diflunisal (Dolobid) is metabolized chiefly through glucuronidation, with two soluble glucuronides in the urine accounting for 90% of the administered dose (Dolobid 1998). UGT1A3 and UGT1A9 seem to be the predominant enzymes in this process (Sallustio et al. 2000). There appears to be no metabolism by P450 enzymes. Diflunisal inhibits UGT1A9 (Mano et al. 2006), UGT2B7 (Mano et al. 2007), and sulfation enzymes (Vietri et al. 2000), although the clinical impact of this inhibition is not known (see Table 17–3).

Fenoprofen

Fenoprofen (Nalfon) is metabolized by UGT1A3 and UGT2B7 (Patel et al. 1995). There is some evidence that fenoprofen inhibits UGT2B7; when administered with oxazepam (Serax), fenoprofen decreased glucuronidation of oxazepam (which is conjugated by UGT2B7) (Patel et al. 1995). However, other NSAIDs decreased glucuronidation of that drug to a greater extent.

Flurbiprofen

Flurbiprofen (Ansaid) is 4-hydroxylated by 2C9 (Greenblatt et al. 2006). The specificity of this reaction is such that flurbiprofen is used as a probe substrate for 2C9. Flurbiprofen also undergoes glucuronidation via UGT1A3, 1A9, 2B4, and 2B7 (Kuehl et al. 2005). There is no evidence that flurbiprofen causes any inhibition or induction.

Ibuprofen

Ibuprofen (Advil, Motrin) is a chiral NSAID hydroxylated from its racemic isomers R- and S-ibuprofen by 2C8 and 2C9, respectively (Hynninen et al. 2006). 2C9 PMs receiving ibuprofen may experience more pharmacological activity from thromboxane B_2 formation (Kirchheiner et al. 2002), reflecting COX-1 inhibition. This increased activity comes with the potential for increased side effects, such as gastroduodenal bleeding (Pilotto et al. 2007). In addition, ibuprofen is glucuronidated via UGT1A3, UGT1A9, UGT2B4, and UGT2B7 (Kuehl et al. 2005). Ibuprofen also appears to inhibit UGT2B7, as demonstrated by its inhibition of oxazepam glucuronidation (Patel et al. 1995).

TABLE 17–3.   Nonsteroidal anti-inflammatory drugs

Drug	P450 enzyme(s) that metabolize drug	UGT(s) that metabolize drug	Other metabolism	Enzyme(s) inhibited
Diflunisal (Dolobid)	None	UGT1A3, UGT1A9	None	UGT1A9, UGT2B7, SULTs
Fenoprofen (Nalfon)	None	UGT2B7	None	UGT2B7
Flurbiprofen (Ansaid)	2C9	UGT1A3, UGT1A9, UGT2B4, UGT2B7	None	None
Ibuprofen (Advil)	2C8 (R-ibuprofen), 2C9 (S-ibuprofen)	UGT1A1, UGT1A9, UGT2B4, UGT2B7	None	UGT2B7
Indomethacin (Indocin)	2C9, 2C19	UGT1A1, UGT1A3, UGT1A9, UGT2B7	None	UGT2B7
Ketoprofen (Actron, Orudis)	?2C9	UGT1A3, UGT1A9, UGT2B4, UGT2B7	None	UGT2B7
Ketorolac (Toradol)	Unspecified P450 enzyme	Unspecified UGT	60% excreted in urine unchanged	None
Meclofenamate (Meclomen)	2C9	UGT1A9, UGT2B7	None	UGT1A9, UGT2B7, SULTs
Mefenamic acid (Ponstel)	2C9	UGT1A9, UGT2B7	None	UGT1A9, UGT2B7, SULTs
Naproxen (Aleve, Naprosyn)	2C9, 2C8, 1A2	UGT1A3, UGT1A9, UGT2B4, UGT2B7	None	UGT2B7

TABLE 17–3. Nonsteroidal anti-inflammatory drugs *(continued)*

Drug	P450 enzyme(s) that metabolize drug	UGT(s) that metabolize drug	Other metabolism	Enzyme(s) inhibited
Oxaprozin (Daypro)	Unspecified P450 enzymes	Unspecified UGTs	None	None
Piroxicam (Feldene)	2C9, ?others	None	None	None
Sulindac (Clinoril)[1]	None known	UGT1A1, UGT1A3, UGT1A9, UGT2B7	Activated by FMO$_3$	SULT1E1
Tolmetin (Tolectin)	3A4	Unspecified UGTs	None	None

Note. FMO$_3$ = flavin-containing monooxygenase 3; SULT = sulfotransferase; UGT = uridine 5'-diphosphate glucuronosyltransferase.
[1] Sulindac is a pro-drug activated by enzymatic activity of FMO$_3$.

Indomethacin

Indomethacin (Indocin) is metabolized through O-demethylation by 2C9 (Nakajima et al. 1998), although 2C19 also plays a minor role in production of the metabolite. O-Demethylation by 2C9 appears to be the major metabolic pathway for indomethacin. Glucuronidation occurs via UGT1A1, UGT1A3, UGT1A9, and UGT2B7 (Kuehl et al. 2005). Urine recovery after drug ingestion indicates that most of the drug is eliminated as the parent compound or as the oxidative metabolite, with a small portion as glucuronide metabolites (Indocin 2006). Additionally, indomethacin inhibits UGT2B7 *in vitro* (Mano et al. 2007), though *in vivo* evidence or clinical impact has not been reported.

Ketoprofen

Ketoprofen (e.g., Actron, Orudis) is glucuronidated via UGT1A3, 1A9, 2B4, and 2B7 (Kuehl et al. 2005; Sabolovic et al. 2000). According to the package insert, the acyl glucuronide formed by UGTs is unstable (Ketoprofen 2006). This implies that the glucuronide creates a reservoir for the parent compound, with metabolite changing back to the parent compound. Indeed, much of the parent compound is found unchanged in the urine along with the acyl glucuronide. There is some suggestion of metabolism via 2C9, but this remains unsubstantiated. Ketoprofen inhibits UGT2B7 *in vitro* (Grancharov et al. 2001; Mano et al. 2007; Patel et al. 1995), although neither *in vivo* evidence nor clinical impact has been reported.

Ketorolac

Ketorolac (e.g., Toradol) is hydroxylated and conjugated in the liver, although the specific cytochromes and UGTs responsible have not yet been elucidated. After ingestion, 60% is found unchanged in the urine, with 28% as glucuronides and 12% as p-hydroxyketorolac (Mroszczak et al. 1990). Pharmacokinetic drug interactions have not been well established, but the agent is highly protein bound, so displacement of other compounds is the most likely interaction (Ketorolac 2006).

Mefenamic Acid

Mefenamic acid (Ponstel) (as well as meclofenamate [Meclomen]) has long been known to be primarily metabolized via oxidation at 2C9 (Leemann et al. 1993). Mefenamic acid is metabolized by 2C9 to two metabolites: 3-hydroxymethyl mefenamic acid and 3-carboxymefenamic acid (Ponstel 2000). Additionally, mefenamic acid and its 2C9 metabolites undergo glucuronidation via UGT1A9 and UGT2B7 (Gaganis et al. 2007). Mefenamic acid inhibits UGT1A9 and UGT2B7 *in vitro* (Mano et al. 2007; Tachibana et al. 2005). Mefenamic acid also appears to inhibit the activity of SULTs (Wang et al. 2006), although the specifics of this have not yet been determined.

Naproxen

Naproxen (Naprosyn, Aleve) is oxidatively metabolized through 6-O-demethylation, primarily by 2C9, but 2C8 and 1A2 also contribute (Tracy et al. 1997). Naproxen and its oxidative metabolite are further metabolized by several UGTs, including UGT1A3, UGT1A9, UGT2B4, and UGT2B7 (Kuehl et al. 2005). Naproxen inhibits UGT2B7 *in vitro* (Mano et al. 2007) to include specific inhibition of the metabolism of AZT (Grancharov et al. 2001; Veal and Back 1995) and oxazepam (Patel et al. 1995).

Oxaprozin

Oxaprozin (Daypro) is metabolized by unspecified oxidative reactions (65%) and glucuronidation (35%) and has a long half-life (Daypro 2002). It is also highly protein bound. One unique aspect of oxaprozin's metabolism is that several of the oxidative metabolites can produce false-positive results on urine benzodiazepine assays (Fraser and Howell 1998).

Piroxicam

Piroxicam (Feldene) is oxidatively metabolized through 5-hydroxylation by 2C9 (Leemann et al. 1993). However, it also appears to go through many other oxidative reactions (Feldene 1999); these may occur via other P450 enzymes and non-P450 enzymes that have not been clearly identified. Piroxicam does not appear to cause any inhibition or induction.

Sulindac

Sulindac (Clinoril) is structurally similar to indomethacin and has similar analgesic and antipyretic properties. It is actually a pro-drug and requires sulfation for activation (Clinoril 1998). This reaction is carried out by flavin-containing monooxygenase 3 (Hamman et al. 2000) and creates sulindac sulfone. Both the parent compound and the active metabolite are conjugated by UGT1A1, UGT1A3, UGT1A9, and UGT2B7 (Kuehl et al. 2005). Sulindac inhibits SULT1E1 *in vitro* (King et al. 2006).

Tolmetin

Tolmetin (Tolectin) is oxidatively metabolized by 3A4 (Chen et al. 2006), and both the parent compound and the metabolite are conjugated to an acyl glucuronide. Sixty percent of the oxidative metabolite, 20% of the glucuronides, and 20% of unchanged tolmetin are found in the urine 24 hours after ingestion. Drug interactions have not been well established.

References

Aithal GP, Day CP, Leathart JB, et al: Relationship of polymorphism in CYP2C9 to genetic susceptibility to diclofenac-induced hepatitis. Pharmacogenetics 10:511–518, 2000

Bextra (package insert). New York, Pfizer Inc., 2002

Celebrex (package insert). New York, Pfizer Inc., 2008

Chen Q, Doss G, Tung E, et al: Evidence for the bioactivation of zomepirac and tolmetin by an oxidative pathway: identification of glutathione adducts in vitro in human liver microsomes and in vivo in rats. Drug Metab Dispos 34:145–151, 2006

Clinoril (package insert). West Point, PA, Merck & Co., 1998

Crespi CL, Chang TK, Waxman DJ: Determination of CYP2C9-catalyzed diclofenac 4′-hydroxylation by high-performance liquid chromatography. Methods Mol Biol 320:109–113, 2006

Davies NM, McLachlan AJ: Properties and features of nabumetone (in French). Drugs 59:25–33, 2000

Daypro (package insert). Chicago, IL, GD Searle LLC, January 2002

Dolobid (package insert). West Point, PA, Merck & Co., 1998

Dooley TP: Molecular biology of the human phenol sulfotransferase gene family. J Exp Zool 282:223–230, 1998

Feldene (package insert). New York, Pfizer Inc., 1999

Fraser AD, Howell P: Oxaprozin cross-reactivity in three commercial immunoassays for benzodiazepines in urine. J Anal Toxicol 22:50–54, 1998

Gaganis P, Miners JO, Knights KM: Glucuronidation of fenamates: kinetic studies using human kidney cortical microsomes and recombinant UDP-glucuronosyltransferase (UGT) 1A9 and 2B7. Biochem Pharmacol 73:1683–1691, 2007

Garnett WR: Clinical implications of drug interactions with coxibs. Pharmacotherapy 21:1223–1232, 2001

Gates BJ, Nguyen TT, Setter SM, et al: Meloxicam: a reappraisal of pharmacokinetics, efficacy and safety. Expert Opin Pharmacother 6:2117–2140, 2005

Grancharov K, Naydenova Z, Lozeva S, et al: Natural and synthetic inhibitors of UDP-glucuronosyltransferases. Pharmacol Ther 89:171–186, 2001

Greenblatt DJ, von Moltke LL, Luo Y, et al: Gingko biloba does not alter clearance of flurbiprofen, a cytochrome P450–2C9 substrate. J Clin Pharmacol 46:214–221, 2006

Hamman MA, Haehner-Daniels BD, Wrighton SA, et al: Stereoselective sulfoxidation of sulindac sulfide by flavin-containing monooxygenases: comparison of human liver and kidney microsomes and mammalian enzymes. Biochem Pharmacol 60:7–17, 2000

Hara Y, Nakajima M, Miyamoto K, et al: Morphine glucuronosyltransferase activity in human liver microsomes is inhibited by a variety of drugs that are co-administered with morphine. Drug Metab Pharmacokinet 22:103–112, 2007

Hynninen V, Olkkola K, Leino K, et al: Effects of the antifungals voriconazole and fluconazole on the pharmacokinetics of S-(+)- and R(-)-Ibuprofen. Antimicrob Agents Chemother 50:1967–1972, 2006

Ibrahim A, Karim A, Feldman J, et al: The influence of parecoxib, a parenteral cyclooxygenase-2 specific inhibitor, on the pharmacokinetics and clinical effects of midazolam. Anesth Analg 95:667–673, 2002a

Indocin (package insert). Whitehouse Station, NJ, Merck & Co., 2006

Kaji H, Kume T: Identification of human UDP-glucuronosyltransferase isoform(s) responsible for the glucuronidation of 2-(4-chlorophenyl)-5-(2-furyl)(-4-oxazoleacetic acid (TA-1801A). Drug Metab Pharmacokinet 20:212–218, 2005

Ketoprofen (package insert). Sellersville, PA, Teva Pharmaceuticals USA, 2006

Ketorolac (package insert). Bedford, OH, Ben Venue Laboratories, Inc., 2006

King C, Tang W, Ngui J, et al: Characterization of rat and human UDP-glucuronosyltransferases responsible for the in vitro glucuronidation of diclofenac. Toxicol Sci 61:49–53, 2001

King RS, Ghosh AA, Wu J: Inhibition of human phenol and estrogen sulfotransferase by certain non-steroidal anti-inflammatory agents. Curr Drug Metab 7:745–753, 2006

Kirchheiner J, Meineke I, Freytag G, et al: Enantiospecific effects of cytochrome P450 2C9 amino acid variants on ibuprofen pharmacokinetics and on the inhibition of cyclooxygenases 1 and 2. Clin Pharmacol Ther 72:62–75, 2002

Kuehl GE, Lampe JW, Potter JD, et al: Glucuronidation of nonsteroidal anti-inflammatory drugs: identifying the enzymes responsible in human liver microsomes. Drug Metab Dispos 33:1027–1035, 2005

Kuehl GE, Bigler J, Potter JD, et al: Glucuronidation of the aspirin metabolite salicylic acid by expressed UDP-glucuronosyltransferases and human liver microsomes. Drug Metab Dispos 34:199–202, 2006

Leemann TD, Transon C, Bonnabry P, et al: A major role for cytochrome P450TB (CYP2C subfamily) in the actions of non-steroidal antiinflammatory drugs. Drugs Exp Clin Res 19:189–195, 1993

Mano Y, Usui T, Kamimura H: In vitro inhibitory effects of non-steroidal anti-inflammatory drugs on 4-methylumbelliferone glucuronidation in recombinant human UDP-glucuronosyltransferase 1A9: potent inhibition by niflumic acid. Biopharm Drug Dispos 27:1–6, 2006

Mano Y, Usui T, Kamimura H: Inhibitory potential of nonsteroidal anti-inflammatory drugs on UDP-glucuronosyltransferase 2B7 in human liver microsomes. Eur J Clin Pharmacol 63:211–216, 2007

Manyike PT, Kharasch ED, Kalhorn TF, et al: Contribution of CYP2E1 and CYP3A to acetaminophen reactive metabolite formation. Clin Pharmacol Ther 67:275–282, 2000

Miners JO: Drug interactions involving aspirin (acetylsalicylic acid) and salicylic acid. Clin Pharmacokinet 17:327–344, 1989

Mobic (package insert). Ridgefield, CT, Boehinger Ingelheim Pharma KG, 2005

Mroszczak EJ, Jung D, Yee J, et al: Ketorolac tromethamine pharmacokinetics and metabolism after intravenous, intramuscular, and oral administration in humans and animals. Pharmacotherapy 10:33S–39S, 1990

Mutlib AE, Goosen TC, Bauman JN, et al: Kinetics of acetaminophen glucuronidation by UDP-glucuronosyltransferases 1A1, 1A6, 1A9, and 2B15: potential implications in acetaminophen-induced hepatotoxicity. Chem Res Toxicol 19:701–709, 2006

Nagar S, Walther S, Blanchard RL: Sulfotransferase (SULT) 1A1 polymorphic variants *1, *2, and *3 are associated with altered enzymatic activity, cellular phenotype, and protein degradation. Mol Pharmacol 69:2084–2092, 2006

Nakajima M, Inoue T, Shimada N, et al: Cytochrome P450 2C9 catalyzes indomethacin O-demethylation in human liver microsomes. Drug Metab Dispos 26:261–266, 1998

Nakamura A, Tougou K, Kitazumi H, et al: Effects of etodolac on P450 isoform-specific activities in human hepatic microsomes. Arzneimittelforschung 55:744–748, 2005

Naprosyn (package insert). Nutley, NJ, Roche Laboratories Inc., 2001

Patel M, Tang BK, Kalow W: (S)oxazepam glucuronidation is inhibited by ketoprofen and other substrates of UGT2B7. Pharmacogenetics 5:43–49, 1995

Paulson SK, Hribar JD, Liu NW, et al: Metabolism and excretion of [(14)C]celecoxib in healthy male volunteers. Drug Metab Dispos 28:308–314, 2000

Pilotto A, Seripa D, Franceschi M, et al: Genetic susceptibility to nonsteroidal anti-inflammatory drug-related gastroduodenal bleeding: role of cytochrome P450 2C9 polymorphisms. Gastroenterology 133:465–471, 2007

Ponstel (package insert). Morris Plains, NJ, Parke-Davis, 2000

Relafen (package insert). Research Triangle Park, NC, GlaxoSmithKline, 2006

Sabolovic N, Magdalou J, Netter P, et al: Nonsteroidal anti-inflammatory drugs and phenols glucuronidation in Caco-2 cells: identification of the UDP-glucuronosyltransferases UGT1A6, 1A3 and 2B7. Life Sci 67:185–196, 2000

Sallustio BC, Sabordo L, Evans AM, et al: Hepatic disposition of electrophilic acyl glucuronide conjugates. Curr Drug Metab 1:163–180, 2000

Sandson N, Marcucci C, Bourke D, et al: An interaction between aspirin and valproate: the relevance of plasma protein displacement drug-drug interactions. Am J Psychiatry 163:1891–1896, 2006

Sim SM, Back DJ, Breckenridge AM: The effect of various drugs on the glucuronidation of zidovudine (azidothymidine; AZT) by human liver microsomes. Br J Clin Pharmacol 32:17–21, 1991

Tachibana M, Tanaka M, Masubuchi Y, et al: Acyl glucuronidation of fluoroquinolone antibiotics by the UDP-glucuronosyltransferase 1A subfamily in human liver microsomes. Drug Metab Dispos 33:803–811, 2005

Tang C, Shou M, Rushmore TH, et al: In-vitro metabolism of celecoxib, a cyclooxygenase-2 inhibitor, by allelic variant forms of human liver microsomal cytochrome P450 2C9: correlation with CYP2C9 genotype and in-vivo pharmacokinetics. Pharmacogenetics 11:223–235, 2001

Tang W: The metabolism of diclofenac: enzymology and toxicology perspectives. Curr Drug Metab 4:319–329, 2003

Tougou K, Gotou H, Ohno Y, et al: Stereoselective glucuronidation and hydroxylation of etodolac by UGT1A9 and CYP2C9 in man. Xenobiotica 34:449–461, 2004

Tracy TS, Marra C, Wrighton SA, et al: Involvement of multiple cytochrome P450 isoforms in naproxen O-demethylation. Eur J Clin Pharmacol 52:293–298, 1997

U.S. Food and Drug Administration: FDA Public Health Advisory. April 7, 2005. Available at http://www.fda.gov/CDER/Drug/advisory/COX2.htm.

Veal GJ, Back DJ: Metabolism of zidovudine. Gen Pharmacol 26:1469–1475, 1995

Vietri M, Pietrabissa A, Mosca F, et al: Mycophenolic acid glucuronidation and its inhibition by non-steroidal anti-inflammatory drugs in human liver and kidney. Eur J Clin Pharmacol 56:659–664, 2000

Vioxx (package insert). West Point, PA, Merck & Co., 2004

Wang LQ, James MO: Inhibition of sulfotransferases by xenobiotics. Curr Drug Metab 7:83–104, 2006

Werner U, Werner D, Pahl A, et al: Investigation of the pharmacokinetics of celecoxib by liquid chromatography-mass spectrometry. Biomed Chromatogr 16:56–60, 2002

Werner U, Werner D, Rau T, et al: Celecoxib inhibits metabolism of cytochrome P450 2D6 substrate metoprolol in humans. Clin Pharmacol Ther 74:130–137, 2003

Yuan JJ, Yang DC, Zhang JY, et al: Disposition of a specific cyclooxygenase-2 inhibitor, valdecoxib, in human. Drug Metab Dispos 30:1013–1021, 2002

18

Pain Management II: Narcotic Analgesics

Gary H. Wynn, M.D.

Scott C. Armstrong, M.D., D.F.A.P.A., F.A.P.M.

Reminder: This chapter is dedicated primarily to metabolic and transporter interactions. Interactions due to displaced protein-binding, alterations in absorption or excretion, and pharmacodynamics are not covered.

Synthetic Opiates: Phenylpiperidines

The phenylpiperidines are a class of chemicals that have been developed as opiate narcotic analgesics. They are synthetic in that they are not chemically related to naturally occurring morphine and are strictly created in the laboratory. Meperidine (Demerol) was the first phenylpiperidine analgesic to come on the market in the United States. The basic structure of phenylpiperidines is shown in Figure 18–1.

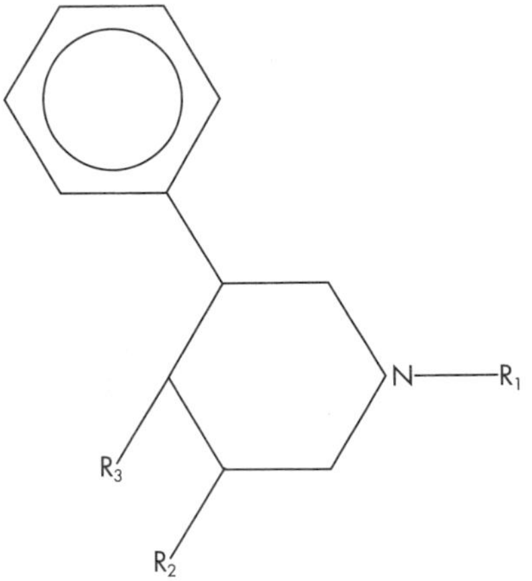

FIGURE 18–1.   Basic structure of phenylpiperidines.

The very short-acting phenylpiperidines, such as alfentanil, also change or add complex moieties at the site with the benzyl group attached. Meperidine has the simplest structure, with a $-CH_3$ (methyl) group at R_1, no moiety at R_2, and a 3-carbon ester group at R_3.

Because phenylpiperidines differ from morphine and morphine-related compounds, they are metabolized differently. 3A4 appears to metabolize most of the phenylpiperidines, but there are some notable exceptions (see Table 18–1). Two phenylpiperidines, diphenoxylate and loperamide, are also known as ABCB1 (P-gp) substrates. Therefore, inhibitors and inducers of both 3A4 and ABCB1 are likely candidates for causing drug interactions with these compounds. The phenylpiperidines do not significantly induce any known metabolic enzymes, and none of them are known to be pro-drugs (i.e., drugs that require enzymatic metabolism to enhance analgesic efficacy).

Alfentanil

Alfentanil (Alfenta) is a parenteral or intrathecal synthetic opiate whose clearance has been well established as depending on 3A4 (Kharasch et al. 2007; Tateishi et al. 1996). 3A4 metabolizes alfentanil to two separate metabolites through dealkylation (Labroo et al. 1995). Alfentanil is extremely short acting

TABLE 18–1.   Phenylpiperidine synthetic opiates

Drug	Metabolic site(s)	Enzyme(s) inhibited	Transporter substrate
Alfentanil (Alfenta)	3A4	No	?
Diphenoxylate (Lomotil[1])	?	No	?
Fentanyl (Duragesic)	3A4	3A4?	ABCB1
Loperamide (Imodium)	2C8, 3A4	3A4?	ABCB1
Meperidine (Demerol)	2B6, 3A4, 2C19, hydrolysis (CYP?), ?UGT1A4, ?other UGTs	No	?
Remifentanil (Ultiva)	Esterases	No	No
Sufentanil (Sufenta)	3A4, 2D6[2]	No	No

Note.   ?=unknown; UGT=uridine 5′-diphosphate glucuronosyltransferase.
[1]With atropine.
[2]Minor contribution to metabolism.

in normal circumstances, having a half-life of 1–2 hours. Because it relies on 3A4 for clearance, studies have been done to determine whether 3A4 inducers and inhibitors can significantly alter alfentanil's pharmacokinetics and effectiveness. Kharasch et al. (1997) found that adding rifampin (a 3A4 inducer) or troleandomycin (a 3A4 inhibitor) significantly alters alfentanil's elimination half-life. Adding rifampin decreased the half-life from 58 to 35 minutes, but more significantly, the inhibitor troleandomycin increased the half-life to an average of 630 minutes. In addition, Palkama et al. (1998a) found that adding fluconazole (a moderate inhibitor of 3A4) decreased clearance of alfentanil by 55%.

Given these findings, caution is advised when alfentanil is administered with *any* 3A4 inhibitor, of which there are many (see Chapter 6, "3A4"). A number of confirmed cases and studies of this interaction exist, including the commonly used drugs diltiazem (Ahonen et al. 1996), erythromycin (Bartkowski et al. 1989), and voriconazole (Saari et al. 2006).

Diphenoxylate and Loperamide

Diphenoxylate and loperamide (Imodium A-D) are phenylpiperidine derivatives, but they are not normally CNS-acting analgesics. Both compounds are used to relieve diarrhea and are taken orally. Diphenoxylate is commonly combined with atropine in a formulation called Lomotil.

Diphenoxylate and loperamide have opiate activity outside the CNS. Loperamide is a substrate of 2C8, 3A4, and the transporter ABCB1 (Niemi et al 2006). ABCB1 is a major component of the blood-brain barrier, and therefore ABCB1 regulates substrates entering the CNS. Diphenoxylate does not appear to be a substrate of ABCB1 (Crowe et al. 2003). Other phenylpiperidine analogs are *not* ABCB1 substrates, which may be why they are effective in the CNS (Wandel et al. 2002). Neither loperamide nor diphenoxylate is known to inhibit or induce enzymatic activity.

In eight healthy control subjects, use of loperamide with a known ABCB1 inhibitor, quinidine, resulted in respiratory depression, despite the fact that serum loperamide levels were unchanged (Sadeque et al. 2000). This effect did not occur with administration of loperamide alone. The implications are clear: use of loperamide (and perhaps diphenoxylate) with an ABCB1 inhibitor can lead to CNS opiate effects by inhibiting or disabling the blood-brain barrier. Some ABCB1 inhibitors (such as cyclosporine, quinidine, ritonavir, and verapamil [Bendayan et al. 2002]) are widely available by prescription. Thus, there is the potential for an untoward drug interaction or for abuse, with use of an ABCB1 inhibitor and either of these two over-the-counter antidiarrheals.

Fentanyl

Fentanyl (Duragesic) is a synthetic opiate whose metabolism is similar to alfentanil's in that the predominant route of clearance is through dealkylation by 3A4 (Jin et al. 2005; Labroo et al. 1997). Fentanyl also undergoes transport via ABCB1 (Park et al. 2007). Unlike alfentanil, however, fentanyl is available by injection or transdermally for long-term delivery (up to 72 hours). The metabolites from 3A4 are not active. In 12 healthy volunteers, pretreatment with ritonavir, a potent 3A4 inhibitor, reduced the clearance of a single intravenous dose of fentanyl (5 mg/kg) by 67% and increased the area under the curve from 4.8 to 8.8 hours (Olkkola et al. 1999). Other findings involving 3A4 inhibitors have been inconsistent, however. For example, fentanyl pharmacokinetics did not change when the drug was administered with itraconazole, a potent 3A4 inhibitor (Palkama et al. 1998b) while fluconazole reportedly does inhibit fentanyl metabolism via 3A4 (Hallberg et al. 2006). Despite the inconsistencies, caution is advised when fentanyl is used with any 3A4 inhibitor or inducer.

Fentanyl may be a modest competitive inhibitor of 3A4. In an *in vitro* study, Oda et al. (1999) demonstrated that use with midazolam slowed midazolam clearance. Further study is needed to understand this potential interaction.

Meperidine

Meperidine (Demerol), available both orally and parenterally, is the oldest of the synthetic opioids discussed in this section. Therefore, knowledge of its metabolism is both extensive and limited: the metabolites created by meperidine, its pharmacokinetics in terms of half-life, and its routes of elimination are all well known, but the actual enzymes responsible for these reactions are not completely understood. It is thus difficult to draw any definitive conclusions regarding meperidine's metabolism and the potential for other drugs to inhibit or induce that metabolism. The route of meperidine metabolism is shown in Figure 18–2.

In vitro microsomal studies indicate that meperidine's primary metabolite normeperidine results via *N*-demethylation by 2B6 and 3A4 with minor contribution via 2C19 (Ramirez et al. 2004). Meperidine's hydrolysis reaction may also occur via P450 enzymes, although non-P450 enzymes catalyze this type of reaction for other drugs.

Meperidine also undergoes phase II glucuronidation and sulfation, although the specific enzymes involved in this process have not been elucidated. There is some evidence that UGT1A4 plays a role in the glucuronidation of meperidine, but this has not been confirmed (Hawes 1998).

Meperidine is not indicated for long-term use because normeperidine has a long half-life, stimulates the central nervous system (CNS), and can lead to seizures. Theoretically, if hydrolysis (or perhaps *N*-demethylation) were inhibited, both meperidine and normeperidine levels would increase, causing CNS toxicity or excessive opioid activity. An inducer of the *N*-demethylation enzyme would perhaps have an even more important effect—that of directly increasing the amount of toxic normeperidine. Both chronically administered ritonavir (Piscitelli et al. 2000)—which eventually induces 3A4 and perhaps other P450 enzymes—and phenobarbital (Stambaugh et al. 1978) have been shown to decrease meperidine levels. Normeperidine's area under the curve was increased by 47% in the ritonavir study by Piscitelli et al. (2000). Stambaugh et al. (1978) showed that more *N*-demethylation occurs and more normeperidine is made when meperidine is used with phenobarbital. Caution is advised when meperidine is used with 3A4 inducers or "pan-inducers." The package insert also warns that these inducers could decrease meperidine levels and cause opiate withdrawal (Demerol 2000).

Finally, there are no case reports of the use of inhibitors of 3A4 or other P450 enzymes with meperidine. One reason for this lack of data may be that the use of selective serotonin reuptake inhibitors, which are notorious for P450 inhibition, with meperidine is contraindicated because of the risk of serotonin syndrome. Meperidine blocks serotonin reuptake and therefore should not be used with selective serotonin reuptake inhibitors, monoamine oxidase inhibitors, or St. John's wort.

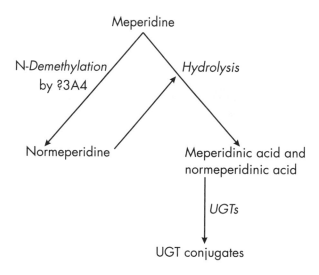

FIGURE 18–2. Metabolism of meperidine.

UGT=uridine 5′-diphosphate glucuronosyltransferase.

Source. Adapted from *Goodman and Gilman's The Pharmacological Basis of Therapeutics,* 6th Edition, New York, Macmillan, 1980, and from Demerol package insert (Demerol 2000). Used with permission.

Remifentanil

Remifentanil (Ultiva) is available as a very-short-acting anesthetic opiate with a half-life of 5–10 minutes. It is rapidly metabolized throughout the body by nonspecific esterases (Kato et al. 2007). There are no reports of pharmacokinetic drug interactions involving this drug.

Sufentanil

Sufentanil (Sufenta), like most of the "fentanils," is a short-acting anesthetic synthetic opiate used only parenterally or intrathecally. Its elimination half-life is 2–3 hours. 3A4 has been shown to *N*-dealkylate sufentanil (Tateishi et al. 1996), but a small amount of sufentanil may be *O*-demethylated by 2D6.

As with other fentanils that rely on 3A4 for clearance, caution is advised when sufentanil is administered with 3A4 inhibitors or inducers. To date, there are very few reports of interactions, and a potent 3A4 inhibitor had no effect on the terminal half-life of sufentanil in one small series (Bartkowski et al. 1993).

Morphine and Related Semisynthetic or Synthetic Opiates

Opium, derived from seeds of the poppy plant, is a combination of several distinct alkaloids. Many of these alkaloids—including morphine, heroin, codeine, and papaverine—have pharmacological activity. Morphine and nearly all of its related chemicals have some opiate receptor activity. The basic structure of morphine and all related opiates is found in Figure 18–3.

The 3-, 6-, and 17- positions are the three key sites where various substitutions by methyl, ester, and hydroxyl groups occur to form different drugs. Morphine has hydroxyl groups at carbon positions 3 and 6 and a methyl group at the 17-N position. Codeine is created by adding a methyl group to the 3- position hydroxyl group. (Other synthetic narcotics are made by substituting at the 3-, 6-, and 17- positions and with a single bond between carbons 7 and 8.) The 3-, 6-, and 17- positions are very important in metabolism because they are the sites oxidized by P450 enzymes. These oxidative reactions ease glucuronidation in the liver, by UGTs, at the 3- and 6- positions. The 3- and 6- positions are often referred to as the O and N sites, respectively, in regard to oxidative P450 metabolism.

Some morphine-related analgesics inhibit metabolic enzymes, but none are known to be potent inhibitors (see Table 18–2). In addition, none of the morphine-related analgesics induce any metabolic enzymes. Several of these agents are pro-drugs or are theorized to be pro-drugs. Details regarding the morphine-related analgesics are presented here.

Morphine

Morphine is the principal alkaloid obtained from unripened seed capsules of the opium poppy. It can also be made in the laboratory. Morphine is metabolized chiefly through glucuronidation via UGT2B7 and UGT1A3 (Stone et al. 2003). The glucuronides are 3-glucuronide (50%), 6-glucuronide, and

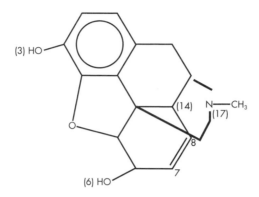

FIGURE 18–3. Basic structure of morphine and all related opiates.

3/6-glucuronide. Morphine-6-glucuronide (M6G) is more potent as an analgesic than is morphine itself (Lotsch and Geisslinger 2001)—possibly 50 times more potent (Christrup 1997). M6G is metabolized chiefly by UGT2B7 (Stone et al. 2003) and has been proposed as a possible analgesic. The 3-glucuronide has few analgesic properties but may cause CNS neuroexcitatory effects (Smith 2000). Inhibition of UGT2B7 could decrease morphine's analgesia as a result of a decrease in M6G production, but there are no reports of this interaction to date.

Morphine has been shown to competitively inhibit UGTs—probably UGT2B7 (Grancharov et al. 2001)—and morphine has been shown to increase zidovudine (AZT) levels via this mechanism.

Hydromorphone and Oxymorphone

Hydromorphone (Dilaudid) and oxymorphone (Numorphan) are chemically almost identical to morphine; hydromorphone differs only in that it has an =O (keto) group at the 6- position (morphine has an −OH, hydroxyl), and oxymorphone is similar to hydromorphone except that an −OH group is added to carbon position 14 in oxymorphone. In addition, unlike morphine, both hydromorphone and oxymorphone have a single bond between carbons 7 and 8.

Hydromorphone is metabolized at the 3- and 6- positions to form glucuronides via UGTs, including UGT1A3 and UGT2B7 (Radominska-Pandya

TABLE 18–2.   Morphine-like opiates, methadone-like opiates, and tramadol

Drug	P450 enzymes that metabolize drug	Other enzymes that metabolize drug	Enzyme(s) inhibited	Pro-drug?[1]
Buprenorphine (Buprenex)	3A4	UGT2B7, UGT1A3	None	No
Codeine	3A4 (10%), 2D6 (5%)	UGT2B7 (80%)	UGT2B7	No[2]
Hydrocodone (Vicodin [with acetaminophen])	2D6, 3A4	Reduction by unknown enzymes	None	Yes[3]
Hydromorphone (Dilaudid)	2D6, 3A4	UGT1A3, UGT2B7, others	None	No
Methadone	3A4, 2D6 (minor), 2B6	None known	3A4, UGT2B7	No
Morphine	Small amounts by 2D6 and 3A4	UGT2B7, UGT1A3	UGT2B7	Yes[4]
Nalbuphine (Nubain)	?Unspecified P450 enzymes	?	None	No
Naloxone (Narcan)	None	UGT2B7	None	No
Naltrexone (ReVia, Vivitrol)	?Unspecified P450 enzymes	Reduction by unknown enzymes; conjugated by UGT2B7	None	No

TABLE 18–2. Morphine-like opiates, methadone-like opiates, and tramadol *(continued)*

Drug	P450 enzymes that metabolize drug	Other enzymes that metabolize drug	Enzyme(s) inhibited	Pro-drug?[1]
Oxycodone (OxyContin)	2D6, 3A4	2D6 metabolite, oxymorphone, glucuronidated with UGT2B7	None	No[5]
Oxymorphone (Numorphan)	None	UGT2B7, others	None	No
Propoxyphene (Darvon)	3A4	None known	None known	Yes[6]
Tramadol (Ultram)	2D6, 3A4, 2B6	Unspecified UGTs	None	Yes[7]

Note.   UGT=uridine 5′-diphosphate glucuronosyltransferase.
[1] Drug is metabolized to an active analgesic metabolite.
[2] It is doubtful that codeine is a pro-drug from 2D6 activity. If codeine is a pro-drug, it may be from UGT2B7 conversion to codeine-6-glucuronide.
[3] Hydrocodone is metabolized by 2D6 to hydromorphone, which may be a significant component of hydrocodone's analgesia efficacy.
[4] Morphine is metabolized by UGT2B7 to morphine-6-glucuronide, which is more potent than morphine.
[5] Oxycodone is metabolized by 2D6 to hydrocodone; however, oxycodone is a very good analgesic independently.
[6] Propoxyphene is metabolized by 3A4 to norpropoxyphene—a potent analgesic along with the parent compound, but with central nervous system toxicity.
[7] Tramadol is metabolized by 2D6 to an active analgesic metabolite. The metabolite, M1, has been confirmed in controlled trials to be correlated with analgesic efficacy.

et al. 1999). Hydromorphone undergoes O-demethylation via 2D6 and N-demethylation via 3A4 (Hutchinson et al. 2004). Hydromorphone has not been shown to inhibit any enzymes. Similar to morphine, hydromorphone's 3-glucuronide may be CNS toxic (Smith 2000) and has been shown to cause seizures in rats (Wright et al. 1998).

Like hydromorphone, oxymorphone is metabolized by several UGTs, with the 6-glucuronide being created primarily by UGT2B7 (Numorphan 2002; Radominska-Pandya et al. 1999). Oxymorphone is only available parenterally and is a potent analgesic.

Codeine

Codeine is very similar in structure to morphine, with the only difference being a methyl group attached to the 3-hydroxyl site of morphine. They have a methyl group at the N-17 site of the molecule and a hydroxyl group at the 6-position. These three sites are the main sites of metabolism of all morphine-like derivatives.

Eighty percent of codeine is glucuronidated to codeine-6-glucuronide by UGT2B7, 5% is oxidatively metabolized by 2D6 via O-3-demethylation, and 10% is N-6-demethylated by 3A4 (Armstrong and Cozza 2003). The 2D6 O-demethylation creates morphine, and the 3A4 N-demethylation creates nor-codeine (Caraco et al. 1996). Norcodeine has few analgesic properties. Codeine itself is a poor analgesic, and therefore its metabolites are believed to be the key to the drug's analgesic efficacy.

Codeine may inhibit UGT2B7, perhaps through its metabolite C6G (Grancharov et al. 2001).

Codeine has an isomer. In the S form, it is codeine. In the R form, it is the over-the-counter antitussive dextromethorphan. Dextromethorphan lacks CNS opiate activity but does have antitussive activity. Its metabolism appears to be similar to that of codeine in that dextromethorphan is O-demethylated by 2D6 and N-demethylated by 3A4. The 2D6 reaction gives rise to dextrophan; the secretion of this product can be used to determine 2D6 activity because no other enzyme creates dextrophan from dextromethorphan, making the latter useful as a probe drug (Frank et al. 2007).

Hydrocodone and Oxycodone

Hydrocodone (Vicodin [with acetaminophen]) and oxycodone (OxyContin) are chemically similar to codeine. They both have a single bond between carbons 7 and 8 (codeine has a double bond there). Hydrocodone has an =O (keto) instead of a hydroxyl moiety at the 6-carbon site. Oxycodone is similar to hydrocodone, with the simple addition of a hydroxyl group at the 14-carbon site.

Hydrocodone is *O*-3-demethylated by 2D6 to hydromorphone (Susce et al. 2006). *N*-6-Demethylation occurs as well, probably via 3A4. Hydrocodone is also reduced at the 6- position to a keto group by an unknown enzyme or enzymes that create active metabolites.

Hydromorphone has more affinity at the μ receptor than does hydrocodone. Therefore, it is theorized that poorer analgesic responses to hydrocodone will occur in PMs at 2D6 or with coadministration of 2D6 inhibitors (a theory similar to one regarding codeine). Results of studies conducted to test this theory are conflicting. One small study showed that being an extensive metabolizer or PM at 2D6 does not predict hydrocodone abuse liability (Kaplan et al. 1997). In contrast, Tyndale et al. (1997) showed that PMs are underrepresented as opiate abusers. Unfortunately, the study did not distinguish among the three codeine-like drugs. The only published study of inhibition of 2D6 in analgesia was by Lelas et al. (1999). In that study, use of 2D6 inhibitors did change the analgesic response to hydrocodone, but the study was performed in rhesus monkeys, and the findings are not easily applied to humans. Nevertheless, it is plausible that hydrocodone's transformation in humans by 2D6 is more important for analgesic efficacy than codeine's similar transformation, because hydrocodone's clearance is more dependent on oxidative reactions than is codeine's and because little glucuronidation of hydrocodone occurs (Cone et al. 1978). Thus, we believe that there is some evidence that hydrocodone's effectiveness could be modified by 2D6 inhibitors or in PMs at 2D6.

Oxycodone is a potent analgesic itself, so inhibition of its metabolism should not decrease its effectiveness, but inhibition could increase its effectiveness or cause excessive CNS opioid effects. Oxycodone is *O*-3-demethylated by 2D6 to oxymorphone and probably *N*-6-demethylated by 3A4 to noroxycodone (Armstrong and Cozza 2003). In a study involving concomitant use of quinidine, a potent 2D6 inhibitor, oxycodone's metabolism to oxymorphone was completely stopped (Heiskanen et al. 1998). However, the effects of oxycodone (by subjective measures and in terms of psychomotor functions) did not differ in subjects who received quinidine and those who did not. We would argue that 2D6 inhibitors or PMs taking oxycodone could, however, have *increased* opiate effects over time, because oxycodone is often prescribed for long periods in clinical practice, unlike in this study.

Ultimately, oxycodone's oxidative metabolites are glucuronidated. The 2D6 metabolite of oxycodone, oxymorphone, is known to be glucuronidated with UGT2B7 (Lugo and Kern 2004).

Buprenorphine

Buprenorphine (Buprenex) is a powerful semisynthetic morphine-like drug with mixed agonist/antagonist properties. It is extensively metabolized by 3A4 via *N*-6-dealkylation (Picard et al. 2005). These metabolites are further conjugated by UGTs—most likely UGT2B7 and UGT1A3 (Green et al. 1998). Because of buprenorphine's reliance on 3A4, caution is advised when the drug is used with 3A4 inhibitors and 3A4 inducers (Buprenorphine 2004).

Nalbuphine

Nalbuphine (Nubain) is a semisynthetic morphine-like mixed opiate agonist/antagonist. Because of extensive first-pass elimination, it is used only parenterally. The enzyme or enzymes responsible for clearance have not been well characterized, but clearance appears to involve both oxidative enzymes and glucuronidation (Yoo et al. 1995). Pharmacokinetic drug interactions involving nalbuphine have been little studied.

Naloxone

Naloxone (Narcan) is a semisynthetic opioid compound with a structure similar to that of morphine but with opiate antagonist properties. It is metabolized chiefly by UGT2B7 (Coffman et al. 2001; de Wildt et al. 1999). Because naloxone does not rely on P450 enzymes for metabolism, untoward pharmacokinetic drug interactions have *not* been reported. The main concern is that a pharmacodynamic event might occur with inadvertent use in a patient who is dependent on opiates.

Naltrexone

Naltrexone (ReVia, Vivitrol) is a semisynthetic opiate antagonist available in both oral and long-acting injectable forms. It is very similar in structure to naloxone, but naltrexone has a much longer half-life and is effective orally, whereas naloxone is not effective orally because of extensive first-pass metabolism. Naltrexone is oxidatively metabolized to several metabolites by unknown enzymes. These products are then glucuronidated, presumably by UGT2B7 (Radominska-Pandya et al. 1999). Very little study has been done of pharmacokinetic drug interactions involving naltrexone.

Methadone and Propoxyphene

Methadone

Methadone's structure does not resemble that of morphine or morphine-like compounds. Steric factors appear to force the molecule into a pseudopiperidine ring, thought to be essential to its opioid activity.

Methadone is a powerful opiate agonist. It is N-demethylated by 2B6 and 3A4 (Totah et al. 2008). There appears to be a small contribution by 2D6 as well, particularly the (R) isomer (Begre et al. 2002). When this agent is administered with an inducer of 3A4, opiate withdrawal may occur. Altice et al. (1999) reported on a series of seven HIV-positive patients in a methadone maintenance program who were administered the antiviral nevirapine (Viramune), an inducer of 3A4. The patients developed opiate withdrawal symptoms. There are also reports of methadone withdrawal with coadministration of rifampin, another potent 3A4 inducer (Holmes 1990; Kreek et al. 1976). Altice and colleagues (1999) suggested that methadone and rifampin can be coadministered if higher doses of methadone are prescribed. Caution is needed if patients discontinue taking the inducing medication (nevirapine, rifampin, carbamazepine, phenobarbital, phenytoin), because it may take several weeks for the effects of induction to diminish, and opiate toxicity may gradually develop if patients are not simultaneously weaned from opiate therapy.

Methadone is also subject to inhibition by potent 3A4 inhibitors, but these findings have been inconsistent. Iribarne et al. (1997) reported *in vitro* evidence that methadone is an inhibitor at 3A4. Finally, methadone may inhibit UGT2B7 (Trapnell et al. 1998). The evidence that methadone inhibits enzymatic activity or undergoes inhibition of its metabolism is scant and controversial. Further study is needed to understand the complexities of these potential interactions.

Propoxyphene

Propoxyphene (Darvon) is a synthetic opiate agonist with a structure similar to that of methadone. Propoxyphenone undergoes metabolism via 3A4 (Somogyi et al. 2004). Norpropoxyphene is an active metabolite with a long half-life (30 hours), whereas the parent compound's half-life is only 6–12 hours. Norpropoxyphene, like normeperidine, has CNS stimulatory effects. Although no definitive studies or cases have been reported at the time of this writing, caution is advised when propoxyphene is used with 3A4 inhibitors or inducers.

Tramadol

Tramadol (Ultram) is an analgesic with a low affinity for opioid receptors. It is chemically related to codeine but has unique features that make it a very weak opiate agonist. Its mechanism of action with regard to analgesia is unclear.

Tramadol may be an inactive pro-drug that requires phase I metabolism to become efficacious.

Tramadol is O-demethylated by 2D6 (Laugesen et al. 2005) from parent drug to active metabolite, called M1. The (+)–M1 isomer is considered to be the more pharmacologically active compound. 3A4 and 2B6 are involved in the metabolism of tramadol to other metabolites, and the drug is also conjugated by UGTs (Subrahmanyam et al. 2001; Ultram 2000).

PMs at 2D6 may have little or no analgesic response to tramadol. Poulsen et al. (1996) showed a distinct difference in analgesic response could be detected in PMs at 2D6 compared with extensive metabolizers. We believe evidence for the reliance of tramadol on 2D6 for efficacy is fairly strong. Tramadol may truly be a pro-drug, requiring 2D6 activity for adequate efficacy. However, tramadol and its 2D6 metabolite, M1, are very poor μ agonists. Serotonin and norepinephrine receptor effects with M1 and other metabolites probably contribute greatly to tramadol's analgesic effects. Despite this uncertainty, we urge clinicians to use caution when administering tramadol with potent 2D6 inhibitors. Patients who have a poor analgesic response to tramadol may be phenotypic PMs at 2D6.

References

Ahonen J, Olkkola KT, Salmenpera M, et al: Effect of diltiazem on midazolam and alfentanil disposition in patients undergoing coronary artery bypass grafting. Anesthesiology 85:1246–1252, 1996

Altice FL, Friedland GH, Cooney EL: Nevirapine induced opiate withdrawal among injection drug users with HIV infection receiving methadone. AIDS 13:957–962, 1999

Armstrong S, Cozza K: Pharmacokinetic drug interactions of morphine, codeine, and their derivatives: theory and clinical reality, part II. Psychosomatics 44:515–520, 2003

Bartkowski RR, Goldberg ME, Larijani GE, et al: Inhibition of alfentanil metabolism by erythromycin. Clin Pharmacol Ther 46:99–102, 1989

Bartkowski RR, Goldberg ME, Huffnagle S, et al: Sufentanil disposition: is it affected by erythromycin administration? Anesthesiology 78:260–265, 1993

Begre S, von Bardeleben U, Ludwig D, et al: Paroxetine increases steady-state concentrations of methadone in CYP2D6 extensive but not in poor metabolizers. J Clin Psychopharmacol 22:211–215, 2002

Bendayan R, Lee G, Bendayan M: Functional expression and localization of P-glycoprotein at the blood brain barrier. Microsc Res Tech 57:365–380, 2002

Buprenorphine (package insert). Bedford, OH, Ben Venue Laboratories, Inc., 2004

Caraco Y, Tateishi T, Guengerich FP, et al: Microsomal codeine N-demethylation: cosegregation with cytochrome P4503A4 activity. Drug Metab Dispos 24:761–764, 1996

Christrup LL: Morphine metabolites. Acta Anaesthesiol Scand 41:116–122, 1997

Coffman BL, Rios GR, King CD, et al: Human UGT2B7 catalyzes morphine glucuronidation. Drug Metab Dispos 25:1–4, 1997

Coffman BL, Kearney WR, Green MD, et al: Analysis of opioid binding to UDP-glucuronosyltransferase 2B7 fusion proteins using nuclear resonance spectroscopy. Mol Pharmacol 59:1464–1469, 2001

Cone EJ, Darwin WD, Gorodetzky CW, et al: Comparative metabolism of hydrocodone in man, rat, guinea pig, rabbit, and dog. Drug Metab Dispos 6:488–493, 1978

Crowe A, Wong P: Potential roles of P-gp and calcium channels in loperamide and diphenoxylate transport. Toxicol Appl Pharmacol 15:127–137, 2003

Demerol (package insert). New York, NY, Sanofi-Synthelabo, 2000

de Wildt SN, Kearns GL, Leeder JS, et al: Glucuronidation in humans: pharmacogenetic and developmental aspects. Clin Pharmacokinet 36:439–452, 1999

Frank D, Jaehde U, Fuhr U: Evaluation of probe drugs and pharmacokinetic metrics for CYP2D6 phenotyping. Eur J Clin Pharmacol 63:321–333, 2007

Grancharov K, Naydenova Z, Lozeva S, et al: Natural and synthetic inhibitors of UDP-glucuronosyltranferase. Pharmacol Ther 89:171–186, 2001

Hallberg P, Marten L, Wadelius M: Possible fluconazole-fentanyl interaction: a case report. Eur J Clin Pharmacol 62:491–492, 2006

Hawes EM: N+–Glucuronidation, a common pathway in human metabolism of drugs with a tertiary amine group. Drug Metab Dispos 26:830–837, 1998

Heiskanen T, Olkkola KT, Kalso E: Effects of blocking CYP2D6 on the pharmacokinetics and pharmacodynamics of oxycodone. Clin Pharmacol Ther 64:603–611, 1998

Holmes VF: Rifampin-induced methadone withdrawal in AIDS (letter). J Clin Psychopharmacol 10:443–444, 1990

Hutchinson MR, Meneaou A, Foster DJ, et al: CYP2D6 and CYP3A4 involvement in the primary oxidative metabolism of hydrocodone by human liver microsomes. Br J Clin Pharmacol 57:287–297, 2004

Iribarne C, Dreano Y, Bardou LG, et al: Interaction of methadone with substrates of human hepatic cytochrome P450 3A4. Toxicology 117:13–23, 1997

Jin M, Gock SB, Jannetto PJ, et al: Pharmacogenomics as molecular autopsy for forensic toxicology: genotyping cytochrome P450 3A4*1B and 3A5*3 for 25 fentanyl cases. J Anal Toxicol 29:590–598, 2005

Kaplan HL, Busto UE, Baylon GJ, et al: Inhibition of cytochrome P450 2D6 metabolism of hydrocodone to hydromorphone does not importantly affect abuse liability. J Pharmacol Exp Ther 281:103–108, 1997

Kato M, Satoh D, Okada Y, et al: Pharmacodynamics and pharmacokinetics or remifentanil: overview and comparison with other opioids. Masui 56:1281–1286, 2007

Kharasch ED, Russell M, Mautz D, et al: The role of cytochrome P450 3A4 in alfentanil clearance: implications for interindividual variability in disposition and perioperative drug interactions. Anesthesiology 87:36–50, 1997

Kharasch ED, Walker A, Isoherranen N, et al: Influence of CYP3A5 genotype on the pharmacokinetics and pharmacodynamics of the cytochrome P4503A probes alfentanil and midazolam. Clin Pharmacol Ther 82:410–426, 2007

Kreek MJ, Garfield JW, Gutjahr CL, et al: Rifampin-induced methadone withdrawal. N Engl J Med 294:1104–1106, 1976

Labroo RB, Thummel KE, Kunze KL, et al: Catalytic role of cytochrome P4503A4 in multiple pathways of alfentanil metabolism. Drug Metab Dispos 23:490–496, 1995

Labroo RB, Paine MF, Thummel KE, et al: Fentanyl metabolism by human hepatic and intestinal cytochrome P450 3A4: implications for interindividual variability in disposition, efficacy, and drug interactions. Drug Metab Dispos 25:1072–1080, 1997

Laugesen S, Enggaard TP, Pedersen RS, et al: Paroxetine, a cytochrome P450 2D6 inhibitor, diminishes the stereoselective O-demethylation and reduces the hypoalgesic effect of tramadol. Clin Pharmacol Ther 77:312–323, 2005

Lelas S, Wegert S, Otton SV, et al: Inhibitors of cytochrome P450 differentially modify discriminative-stimulus and antinociceptive effects of hydrocodone and hydromorphone in rhesus monkeys. Drug Alcohol Depend 54:239–249, 1999

Lotsch J, Geisslinger G: Morphine-6-glucuronide: an analgesic of the future? Clin Pharmacokinet 40:485–499, 2001

Lugo RA, Kern SE: The pharmacokinetics of oxycodone. J Pain Palliat Care Pharmacother 18:17–30, 2004

Niemi M, Tornio A, Pasanen MK, et al: Itraconazole, gemfibrozil and their combination markedly raise the plasma concentrations of loperamide. Eur J Clin Pharmacol 62:463–472, 2006

Numorphan (package insert). Chadds Ford, PA, Endo Pharmaceuticals Inc., 2002

Oda Y, Mizutani K, Hase I, et al: Fentanyl inhibits metabolism of midazolam: competitive inhibition of CYP3A4 in vitro. Br J Anaesth 82:900–903, 1999

Olkkola KT, Palkama VJ, Neuvonen PJ: Ritonavir's role in reducing fentanyl clearance and prolonging its half-life. Anesthesiology 91:681–685, 1999

Palkama VJ, Isohanni MH, Neuvonen PJ, et al: The effect of intravenous and oral fluconazole on the pharmacokinetics and pharmacodynamics of intravenous alfentanil. Anesth Analg 87:190–194, 1998a

Palkama VJ, Neuvonen PJ, Olkkola KT: The CYP 3A4 inhibitor itraconazole has no effect on the pharmacokinetics of i.v. fentanyl. Br J Anaesth 81:598–600, 1998b

Park HJ, Shinn HK, Ryu SH, et al: Genetic polymorphisms in the ABCB1 gene and the effects of fentanyl in Koreans. Clin Pharmacol Ther 81:539–546, 2007

Picard N, Cresteil T, Djebli N, et al: In vitro metabolism study of buprenorphine: evidence for new metabolic pathways. Drug Metab Dispos 33:689–698, 2005

Piscitelli SC, Kress DR, Bertz RJ, et al: The effect of ritonavir on the pharmacokinetics of meperidine and normeperidine. Pharmacotherapy 20:549–553, 2000

Poulsen L, Arendt-Nielsen L, Brosen K, et al: The hypoalgesic effect of tramadol in relation to CYP2D6. Clin Pharmacol Ther 60:636–644, 1996

Radominska-Pandya A, Czernik PJ, Little JM, et al: Structural and functional studies of UDP-glucuronosyltransferases. Drug Metab Rev 31:817–899, 1999

Ramirez J, Innocenti F, Schuetz EG, et al: CYP2B6, CYP3A4, and CYP2C19 are responsible for the in vitro N-demethylation of meperidine in human liver microsomes. Drug Metab Dispos 32:930–936, 2004

Saari TI, Laine K, Leino K, et al: Voriconazole, but not terbinafine, markedly reduces alfentanil clearance and prolongs its half-life. Clin Pharmacol Ther 80:502–508, 2006

Sadeque AJ, Wandel C, He H, et al: Increased drug delivery to the brain by P-glycoprotein inhibition. Clin Pharmacol Ther 68:231–237, 2000

Smith MT: Neuroexcitatory effects of morphine and hydromorphone: evidence implicating the 3-glucuronide metabolites. Clin Exp Pharmacol Physiol 27:524–528, 2000

Somogyi AA, Menelaou A, Fullston SV: CYP3A4 mediates dextropropoxyphenone N-demethylation to nordextropropoxyphene: human in vitro and in vivo studies and lack of CYP2D6 involvement. Xenobiotica 34:875–887, 2004

Stambaugh JE, Wainer IW, Schwartz I: The effect of phenobarbital on the metabolism of meperidine in normal volunteers. J Clin Pharmacol 18:482–490, 1978

Stone AN, Mackenzie PI, Galetin A, et al: Isoform selectivity and kinetics of morphine 3- and 6-glucuronidation by human UDP-glucuronosyltransferases: evidence for atypical glucuronidation kinetics by UGT2B7. Drug Metab Dispos 31:1086–1089, 2003

Subrahmanyam V, Renwick AB, Walters DG, et al: Identification of cytochrome P-450 isoforms responsible for cis-tramadol metabolism in human liver microsomes. Drug Metab Dispos 29:1146–1155, 2001

Susce MT, Murray-Carmichael E, de Leon J: Response to hydrocodone, codeine and oxycodone in a CYP2D6 poor metabolizer. Prog Neuropsychopharmacol Biol Psychiatry 30:1356–1358, 2006

Tateishi T, Krivoruk Y, Ueng YF, et al: Identification of human liver cytochrome P-450 3A4 as the enzyme responsible for fentanyl and sufentanil N-dealkylation. Anesth Analg 82:167–172, 1996

Totah RA, Sheffels P, Roberts T, et al: Role of CYP2B6 in steroselective human methadone metabolism. Anesthesiology 108:363–374, 2008

Trapnell CB, Klecker RW, Jamis-Dow C, et al: Glucuronidation of 3'-azido-3'-deoxythymidine (zidovudine) by human liver microsomes: relevance to clinical pharmacokinetic interactions with atovaquone, fluconazole, methadone, and valproic acid. Antimicrob Agents Chemother 42:1592–1596, 1998

Tyndale RF, Droll KP, Sellers EM: Genetically deficient CYP2D6 metabolism provides protection against oral opiate dependence. Pharmacogenetics 7:375–379, 1997

Ultram (package insert). Raritan, NJ, Ortho-McNeil Pharmaceutical Inc., 2000

Wandel C, Kim R, Wood M, et al: Interaction of morphine, fentanyl, sufentanil, alfentanil, and loperamide with the efflux drug transporter P-glycoprotein. Anesthesiology 96:913–920, 2002

Wright AW, Nocente ML, Smith MT: Hydromorphone-3-glucuronide: biosynthesis and preliminary pharmacological evaluation. Life Sci 63: 401–411, 1998

Yoo YC, Chung HS, Kim IS, et al: Determination of nalbuphine in drug abuser's urine. J Anal Toxicol 19:120–123, 1995

19

Psychiatry

Gary H. Wynn, M.D.

Neil Sandson, M.D.

Javier Muniz, M.D.

Reminder: This chapter is dedicated primarily to metabolic and transporter interactions. Interactions due to displaced protein-binding, alterations in absorption or excretion, and pharmacodynamics are not covered.

The last several decades have seen a steep rise both in the number of available psychotropic agents and the disorders treated by psychotropics. Until the 1980s, tricyclic antidepressants (TCAs), typical antipsychotics, lithium, and benzodiazepines were the mainstay of psychiatry. In the 1990s, methods of metabolic analysis were in their infancy, and U.S. Food and Drug Administration (FDA) requirements for understanding metabolism were less stringent. Even today there are significant gaps in the understanding of the metabolism and drug interactions of older medications. These gaps are slowly being addressed, and new data are frequently coming to light.

Currently, a body of literature exists that can help clinicians avoid some of the potential pitfalls of drug interactions. Recent prescribing habits show that psychiatric drugs are frequent offenders in this arena (Molden et al. 2005), and understanding these drugs and their potential for interactions is vital.

In this chapter, we review the psychotropic drugs in groups: antidepressants (selective serotonin reuptake inhibitors [SSRIs], other commonly used antidepressants, and tricyclics), anxiolytics and hypnotics (benzodiazepines and others), antipsychotics, drugs to treat attention-deficit/hyperactivity disorder (ADHD), and drugs for the treatment of addiction. Drugs used for the treatment of seizure disorders are often used as mood stabilizers, and they are discussed in Chapter 15, "Neurology."

Antidepressants

Selective Serotonin Reuptake Inhibitors

Each of the six SSRIs has unique P450 metabolisms and inhibitions. Their active metabolites also contribute to drug interactions (e.g., norfluoxetine is a mild to moderate inhibitor of 3A4, whereas its parent compound has less inhibitory ability). Each SSRI is reviewed for metabolism and inhibition. There is no known induction among the six SSRIs addressed here (see Table 19–1).

Citalopram and Escitalopram

Citalopram (Celexa) is a chiral drug with a more pharmacologically active enantiomer, escitalopram (*S*-citalopram; Lexapro). Both citalopram and escitalopram are metabolized via demethylation at 2C19, 2D6, and 3A4, and are mild inhibitors at 2D6 (Areberg et al. 2006; Preskorn et al. 2007; Rao 2007). In animal models, citalopram also appears to be a substrate and a weak inhibitor of P-glycoprotein (ABCB1) transport (Uhr and Grauer 2003; Weiss et al. 2003). Because the metabolic pathways are diverse and inhibition is mild, there are few drug interactions involving either citalopram or escitalopram.

Fluoxetine

Arguably the most famous antidepressant, fluoxetine (Prozac) is also a chiral drug. Fluoxetine and its long-lived metabolite norfluoxetine are substrates of

TABLE 19–1. Selective serotonin reuptake inhibitors

Drug	Metabolism/transport site(s)	Enzyme(s)/process inhibited[a]
Citalopram (Celexa)	2C19, 2D6, 3A4, ABCB1	2D6^c, ABCB1
Escitalopram (Lexapro)	2C19, 2D6, 3A4, ABCB1	2D6^c, ABCB1
Fluoxetine (Prozac)	2C9, 2C19, 2D6, 3A4, ABCB1	**2D6**, 2C9^b, 3A4^b, 2C19^b, 1A2^c, 2B6^b, ABCB1
Fluvoxamine (Luvox)	1A2, 2D6, ABCB1	**1A2**, **2C19**, 2B6^b, 2C9^b, 3A4^b, 2D6^b, ABCB1
Paroxetine (Paxil)	2D6, 3A4, ABCB1	**2D6**, **2B6**, 3A4^b, 1A2^c, 2C9^c, 2C19^c, ABCB1
Sertraline (Zoloft)	2B6, 2C9, 2C19, 2D6, 3A4, UGT2B7, UGT1A1, ABCB1	**2D6***, 2B6^b, 2C9^b, 2C19^b, 3A4^b, 1A2^c, ABCB1

Note.   Data presented relate to parent drug and metabolites combined.
*Sertraline is a potent 2D6 inhibitor at high concentrations (typically at dosages >200 mg/day).
[a]**Bold** type indicates potent inhibition.
[b]Moderate inhibition.
[c]Mild inhibition.

2C9, 2C19, 3A4, and 2D6 (Mandrioli et al. 2006; Ring et al. 2001). Together, they are potent inhibitors of 2D6 and mild to moderate inhibitors of 1A2, 2B6, 2C9, 2C19, and 3A4 (Bertelsen et al. 2003; Hesse et al. 2000; Ohno et al. 2007). Additionally, fluoxetine is both a substrate and an inhibitor of ABCB1 transport (Khairul et al. 2006). In light of this, fluoxetine can reasonably be considered a "pan-inhibitor."

Because of its multiple metabolic pathways, fluoxetine's plasma levels are not usually altered by P450 inhibitors. However, fluoxetine's metabolism can be vulnerable to "pan-inducers" such as older anticonvulsants. Additionally, fluoxetine's potent inhibition of 2D6 means patients taking fluoxetine and 2D6 substrates are particularly at risk for increased concentrations of 2D6 substrates (e.g., desipramine's [Norpramin's] concentration is increased fourfold [Preskorn et al. 1994]). Because fluoxetine inhibits other P450 enzymes in addition to 2D6, providers prescribing drugs metabolized via 2D6 or other affected P450 enzymes should monitor for increased side effects. When fluoxetine and risperidone (Risperdal) are coadministered, the concentration of risperidone increases 75% due to fluoxetine and norfluoxetine inhibition of 2D6 and 3A4 (Spina et al. 2002).

Fluvoxamine

Fluvoxamine (Luvox) is an achiral drug that has no significant metabolites. Fluvoxamine undergoes metabolism via 2D6 and 1A2 (Spigset et al. 2001) and is a substrate for ABCB1 transport (Fukui et al. 2007). Fluvoxamine is a potent inhibitor of 1A2 and 2C19 even at low doses (Christensen et al. 2002), a mild to moderate inhibitor of 2B6, 2C9, and 3A4, and a mild inhibitor of 2D6 (Hesse et al. 2000). ABCB1 transport also appears to be inhibited by fluvoxamine (Weiss et al. 2003). The possible drug interactions associated with fluvoxamine's P450 inhibitions are legion. Particularly important are interactions with drugs that have narrow therapeutic indices (Spina et al. 2007), such as clozapine (1A2), theophylline (1A2), warfarin (2C9), phenytoin (2C9 and 2C19), and the tertiary TCAs. When adding *any* drug to fluvoxamine therapy, clinicians are advised to be cautious—"starting low and going slow," measuring blood levels, and watching carefully for signs of toxicity. Besides affecting the plasma concentrations of other medications, fluvoxamine, a substrate of 1A2 and 2D6, can be altered in the presence of any inhibitor or inducer of 1A2 (e.g., the 1A2 inducer cigarette smoke) or 2D6 (e.g., the 2D6 inhibitor cimetidine).

Paroxetine

Paroxetine (Paxil) has no significant metabolites and is metabolized via 2D6 with minor contribution from 3A4 (Hemeryck and Belpaire 2002), as well as being a substrate for ABCB1 transport (Kato et al. 2008). Paroxetine is a potent inhibitor of 2B6 and 2D6 (Kuhn et al. 2007) and a mild inhibitor of 1A2, 2C9, 2C19, 3A4, and ABCB1 (von Moltke et al. 1996b; Weiss et al. 2003). Patients taking substrates of 2B6 and 2D6 may be at particular risk for significant drug interactions when these drugs are combined with paroxetine (e.g., desipramine concentrations increased 360% with coadministration of paroxetine [Alderman et al. 1997]). Paroxetine is vulnerable to drug interactions only with 2D6 inhibitors more potent than itself, such as quinidine, because paroxetine occupies the P450 site and "pushes off" other drugs with less affinity for 2D6 (Bloomer et al. 1992).

Sertraline

Sertraline (Zoloft) and its active metabolite desmethylsertraline are substrates of several P450 enzymes, including 2B6, 2C9, 2C19, 2D6, 3A4, UGT2B7, and UGT1A1 (Greenblatt et al. 1999; Hemeryck and Belpaire 2002; Obach et al. 2005; Wang et al. 2001), as well as being a substrate for ABCB1 transport (Wang et al. 2008). Sertraline inhibits 2D6 in a dose-dependent manner. At doses under 100 mg/day, sertraline may only mildly inhibit 2D6. At doses above 150 mg/day, 2D6 inhibition may become moderate to potent (Crewe et al.1992). Sertraline and desmethylsertraline are also modest inhibitors of 2C9, 2C19, and 3A4 (Greenblatt et al. 1999) and ABCB1 (Weiss et al. 2003), which accounts for scattered case reports of increased concentrations of phenytoin, warfarin, and cyclosporine when these drugs are combined with sertraline (Lill et al. 2000; Sayal et al. 2000). *In vitro* evidence supports moderate 2B6 inhibition by sertraline, but *in vivo* evidence is lacking (Walsky et al. 2006). Sertraline also appears to be a specific and potent inhibitor of glucuronidation (UGT1A4), which can precipitate toxic lamotrigine concentrations (Kaufman and Gerner 1998). Sertraline's complex, multi-enzyme metabolism is likely to prevent its being affected by the inhibition or induction of another drug.

Non–Selective Serotonin Reuptake Inhibitors

Bupropion

Bupropion (Wellbutrin) is primarily metabolized via 2B6 (Faucette et al. 2001) with minor contribution by 1A2, 2A6, 2C9, 2E1, 3A4, and glucuronidation (Wellbutrin 2006) (see Table 19–2). Unlike SSRIs, bupropion does not appear to be a substrate of ABCB1 transport (Wang et al. 2008). Neither smoking nor cimetidine alters bupropion's pharmacokinetics (Desai et al. 2001; Kustra et al. 1999). Ritonavir, efavirenz, nelfinavir, paroxetine, sertraline, and other 2B6 inhibitors may interact with bupropion (Hesse et al. 2002; Walsky et al. 2006), but clinical documentation is lacking. Because carbamazepine induces several P450 enzymes, bupropion concentrations are reduced with coadministration of carbamazepine (Popli et al. 1995).

Bupropion is a modest inhibitor of 2D6 and can increase concentrations of 2D6 substrates such as venlafaxine (Kennedy et al. 2002) and nortriptyline (Weintraub 2001). When added to treatment with paroxetine or fluoxetine (substrates of 2D6), sustained-release bupropion did not increase the concentration of either SSRI, likely because of the lower Ki (dissociation constant value) of SSRIs, which then occupy the P450 site and "push off" bupropion from 2D6 (Kennedy et al. 2002). When adding 2D6 substrates with narrow therapeutic indices to bupropion therapy, clinicians should consider starting with lower doses.

Duloxetine

Duloxetine (Cymbalta) received approval by the FDA for the treatment of depression and diabetic neuropathy in 2004 and has seen increasing use for the treatment of comorbid psychiatric and pain issues. Duloxetine is primarily metabolized via 1A2 and 2D6 (Caccia 2004; Lobo et al. 2008). Coadministration of 1A2 inhibitors including fluvoxamine can result in significantly increased levels of duloxetine. Similarly, 2D6 inhibitors (e.g., paroxetine, fluoxetine, and bupropion) can result in elevated levels of duloxetine (Cymbalta 2008). Although elevated levels of duloxetine may not result in significant morbidity, high plasma duloxetine levels will likely potentiate side effects. Duloxetine is also a moderate to potent inhibitor of 2D6 (Skinner et al. 2003).

TABLE 19–2. Other antidepressants

Drug	Metabolism/Transport site(s)	Enzyme(s) inhibited[a]
Bupropion (Wellbutrin)	2B6, 2D6, 1A2, 2A6, 2C9, 2E1, 3A4, glucuronidation	2D6[b]
Desvenlafaxine (Pristiq)	UGTs, 3A4	2D6[c]
Duloxetine (Cymbalta)	1A2, 2D6	2D6[b]
Mirtazapine (Remeron)	1A2, 2D6, 3A4, glucuronidation	None known
Nefazodone (Serzone)	3A4, 2D6	3A4
Trazodone (Desyrel)	3A4	None known
Venlafaxine (Effexor)	2D6, 2C19, 3A4	2D6[c]

Note. Data presented relate to parent drug and metabolites combined. UGT=uridine-5′-diphosphate glucuronosyltransferase.

[a]**Bold** type indicates potent inhibition.

[b]Moderate inhibition.

[c]Mild inhibition.

Mirtazapine

Mirtazapine (Remeron) is metabolized to several metabolites by 1A2, 2D6, 3A4, and glucuronidation (Timmer et al. 2000) and is not a potent P450 inhibitor. Because fluvoxamine is an inhibitor of all three of mirtazapine's P450 pathways, mirtazapine's concentration is increased up to 4-fold when the two drugs are combined (Anttila et al. 2001). Cimetidine, a modest P450 inhibitor, increases the concentration of mirtazapine by only 22% (Sitsen et al. 2000). Conversely, the P450 "pan-inducer" carbamazepine decreases mirtazapine's concentration significantly (Sitsen et al. 2001).

Nefazodone and Trazodone

Nefazodone (Serzone) is metabolized to three active metabolites via 3A4, including the anxiogenic metabolite m-chlorophenylpiperazine (m-CPP) (Kalgutkar et al. 2005). m-CPP is then further metabolized via 2D6 (von Moltke et al. 1999a). The concentration of m-CPP may be increased in poor metabolizers at 2D6 and by potent inhibitors of 2D6, such as fluoxetine and paroxetine. Metabolism of nefazodone and its metabolites are vulnerable to 3A4 inhibitors and inducers. Nefazodone is also a potent inhibitor of 3A4 and can increase concentrations of other 3A4 substrates, such as carbamazepine, simvastatin, cyclosporine, tacrolimus, alprazolam (Xanax), and triazolam (Halcion). Nefazodone is contraindicated with pimozide (Serzone 2005). Though Serzone, marketed by Bristol-Myers Squibb, was voluntarily withdrawn from the U.S. market in 2004, several generic forms of nefazodone are still available.

Trazodone (Desyrel) is also primarily metabolized via 3A4 to several metabolites including m-CPP (Wen et al. 2008). Trazodone's metabolism, like the metabolism of nefazodone, is vulnerable to 3A4 inducers and inhibitors. There is no known inhibition or induction caused by trazodone.

Venlafaxine

Venlafaxine (Effexor) is primarily metabolized to its active metabolite, O-desmethylvenlafaxine, by 2D6, while an inactive metabolite is handled by 3A4 and 2C19 (McAlpine et al. 2007). Venlafaxine also appears to be a substrate of ABCB1 transport (Gareri et al. 2008). Although drug interactions involving venlafaxine have not been well studied, inhibitors of 2D6—such as quinidine, paroxetine, diphenhydramine, and bupropion—have been shown to increase venlafaxine concentrations (Fogelman et al. 1999; Kennedy et al.

2002; Lessard et al. 2001). As a mild 2D6 inhibitor, venlafaxine has been shown to increase concentrations of imipramine (Tofranil) (Ball et al. 1997), desipramine, haloperidol (Haldol), and risperidone.

Tricyclic Antidepressants

TCAs can be divided into two groups. The tertiary TCAs—which include amitriptyline (Elavil), clomipramine (Anafranil), doxepin (Adapin, Sinequan), imipramine (Tofranil), and trimipramine (Surmontil)—contain a tertiary amine side chain. The secondary TCAs—desipramine (Norpramin), nortriptyline (Pamelor), and protriptyline (Vivactil)—contain a secondary amine side chain. The metabolism of TCAs is very complex. Tertiary TCAs are first demethylated into secondary TCAs: amitriptyline to nortriptyline, imipramine to desipramine, doxepin to desmethyldoxepin, trimipramine to desmethyltrimipramine, and clomipramine to desmethylclomipramine. These compounds are then hydroxylated and finally are conjugated with glucuronic acid. Secondary TCAs are hydroxylated and glucuronidated. Extensive enterohepatic "recycling" of the glucuronidated TCA compounds may occur. As a general rule, 2D6 is responsible for hydroxylations, and other P450 enzymes are responsible for demethylations. In some cases, 2D6 may also act as a minor pathway for demethylation, and 3A4 may act as a "sink" or "reservoir" for hydroxylations. TCA hydroxymetabolites exist as enantiomers and are psychoactive.

Before SSRIs entered the U.S. market, it had been well established that phenothiazines could increase TCA concentrations (e.g., thioridazine [Mellaril] increases nortriptyline levels [Jerling et al. 1994a]), but it was TCA-SSRI interactions that ushered in the modern age of physician interest in P450 enzymes. Because the hydroxylation step of TCA metabolism is rate limited, significant increases in TCA concentrations can occur when potent 2D6 inhibitors are coadministered (Leucht et al. 2000).

Clinicians should be aware that TCAs are P450 inhibitors. The number of potential interactions is quite large, with a myriad of reports in the literature (Gillman 2007). For example, imipramine and nortriptyline can increase concentrations of chlorpromazine (Thorazine) and other phenothiazines (Loga et al. 1981; Rasheed et al. 1994). Olanzapine (Zyprexa) concentrations are increased by 20% with the addition of imipramine (Callaghan et al. 1997). See Table 19–3 for specific metabolism and inhibition details regarding TCAs.

TABLE 19–3.   Tricyclic antidepressants

Drug	Major metabolism site(s)	Enzyme(s) inhibited
Amitriptyline (Elavil)	1A2, 2C19, 2D6, 3A4, UGT1A4	1A2, 2C19, 2D6
Clomipramine (Anafranil)	1A2, 2C19, 2D6, 3A4	2D6, 1A2, 2C19
Desipramine (Norpramin)	2D6	2D6, 2C19
Doxepin (Adapin, Sinequan)	1A2, 2D6, 2C19, 3A4, UGT1A4, UGT1A3	1A2, 2C19, 2D6
Imipramine (Tofranil)	1A2, 2C19, 2D6, 3A4, UGT1A4, UGT1A3	2C19, 2D6, 1A2, 3A4
Nortriptyline (Pamelor)	2D6	2D6, 2C19
Protriptyline (Vivactil)	?2D6	?
Trimipramine (Surmontil)	2C19, 2D6, 3A4	?

Note.   Data presented relate to parent drug and metabolites combined. ?=unknown.

Anxiolytics and Hypnotics

Benzodiazepines

Benzodiazepines (BZDs) are used in a wide variety of medical settings and in a broad range of therapeutic ways. BZDs vary in half-life, potency, and metabolism. For simplicity, BZDs can be divided into two main groups based on metabolism and interaction (see Table 19–4 for details on BZD metabolism). Triazolobenzodiazepines, including midazolam, triazolam and alprazolam, are dependent on 3A4 for metabolism (Ohno et al. 2007). This dependence on 3A4 for metabolism increases the likelihood of significant interactions with inhibitors (e.g., ketoconazole, erythromycin, and nefazodone) or inducers (e.g., carbamazepine, rifampin, and St. John's wort) of 3A4. Given the use of midazolam in surgical and intensive care environments, potent 3A4 inhibitors such as ketoconazole can result in prolonged sedation and difficulty extubating individuals administered midazolam. Grapefruit juice, a potent inhibitor of intestinal 3A4, significantly increases midazolam and triazolam levels while having no impact on alprazolam (Kupferschmidt et al. 1995; Lilja et al. 2000; Yasui et al. 2000). This difference is likely due to the dependence of alprazolam on hepatic 3A4 for metabolism, versus dependence of midazolam and triazolam on intestinal 3A4 for first-pass metabolism.

Other BZDs, including diazepam, lorazepam, oxazepam, and temazepam are metabolized through a variety of enzymes, including UGT 1A1, 1A3, 1A9, 2B7, and 2B15, with some contribution via CYP3A4 (see Table 19–4). Diazepam also experiences significant metabolism via 2C19. Clonazepam undergoes part of its metabolism via 3A4 and is then acetylated (Seree et al. 1993). The wide variety of possible metabolic pathways significantly decreases the likelihood of a drug interaction, but a "pan-inhibitor" or "pan-inducer" may cause clinically relevant alterations to multi-enzymatically metabolized medications.

Flunitrazepam, known by the street name "roofies," is a potent BZD not legally available in the United States. Flunitrazepam is primarily metabolized via 3A4 and 2C19 as well as being a potent inhibitor of UGT 1A1, 1A3, and 2B7 (Ziqiang et al.1998).

TABLE 19–4.   Benzodiazepines

Drug	Metabolism
Alprazolam (Xanax)	**3A4**, glucuronidation
Clonazepam (Klonopin)	**3A4**, acetylation
Diazepam (Valium)	2C19, 3A4, 2B6, 2C9, glucuronidation
Flunitrazepam (Rohypnol ["roofies"])	**2C19, 3A4**
Lorazepam (Ativan)	**UGT2B7**, ?other UGTs
Midazolam (Versed)	**3A4**, glucuronidation
Oxazepam (Serax)	*S*-Oxazepam: UGT2B15 *R*-Oxazepam: UGT1A9, UGT2B7
Temazepam (Restoril)	**UGT2B7**, ?other UGTs, 2C19, 3A4
Triazolam (Halcion)	**3A4**, glucuronidation

Note.   **Bold** type indicates major pathway.
UGT=uridine 5′-diphosphate glucuronosyltransferase.

Nonbenzodiazepines

None of the nonbenzodiazepine anxiolytics and hypnotics are P450 inhibitors, and all are partially or completely dependent on 3A4 for metabolism (see Table 19–5).

Buspirone

Buspirone (BuSpar) is metabolized by 3A4 to an active metabolite, 1-pyrimidinylpiperazine (BuSpar 2007). Inhibitors of 3A4 can increase the C_{max} of buspirone: nefazodone up to 20-fold, itraconazole up to 13-fold (Kivisto et al. 1997, 1999), erythromycin 5-fold (Kivisto et al. 1997), grapefruit juice and diltiazem 4-fold (Lamberg et al. 1998b; Lilja et al. 1998), verapamil 3-fold (Lamberg et al. 1998b), and fluvoxamine 2-fold (Lamberg et al. 1998a). Other potent 3A4 inhibitors may also have the same effect. 3A4 inducers such as rifampin can reduce the plasma concentration of buspirone by 85% (Kivisto et al. 1999), as can dexamethasone, the older pan-inducing anticonvulsants (BuSpar 2007), and other potent 3A4 inducers.

Eszopiclone

Eszopiclone (Lunesta) is the racemic isomer of zopiclone and has metabolic properties similar to that of zopiclone. See Zopiclone section below for details.

TABLE 19–5. Nonbenzodiazepine sedative/hypnotics

Drug	Metabolism
Buspirone (BuSpar)	**3A4**
Eszopiclone (Lunesta)	2C8, **3A4**
Ramelteon (Rozerem)	**1A2**, 3A4, 2C family
Zaleplon (Sonata)	**Aldehyde oxidase**, 3A4
Zolpidem (Ambien)	**3A4**, 1A2, 2C9
Zopiclone (Imovane)	2C8, **3A4**

Note. **Bold** type indicates major pathway.

Ramelteon

Ramelteon (Rozerem) is a hypnotic agent approved for use in insomnia. Ramelteon undergoes metabolism primarily via 1A2 with minor contribution by 3A4 and the CYP2C family (Rozerem 2006). Potent inhibitors of 1A2 (e.g., fluvoxamine), 3A4 (e.g., ketoconazole), and 2C9 (e.g., fluconazole) may result in significant elevation of ramelteon plasma levels' potentiating effect while inducers such as rifampin may decrease plasma levels and overall efficacy (Rozerem 2006). Ramelteon does not appear to cause inhibition or induction.

Zaleplon

Zaleplon (Sonata) is principally metabolized by aldehyde oxidase and, to a lesser extent, by 3A4 (Obach 2004; Renwick et al. 1998). Potent inducers of 3A4 such as rifampin can reduce zaleplon's maximum concentration by 80% and should be avoided (Sonata 2007). Other potent 3A4 inducers are likely to have similar actions. Potent 3A4 inhibitors such as erythromycin increase zaleplon concentration by 34% (Sonata 2007), and other potent 3A4 inhibitors may cause similar increases. Cimetidine, which inhibits both 3A4 and aldehyde oxidase, has been shown to increase zaleplon's maximum concentration by 85% (Renwick et al. 2002; Sonata 2007), and if the drugs are coadministered, zaleplon's dosing should be substantially reduced.

Zolpidem

Zolpidem (Ambien) is principally metabolized via 3A4; 1A2 and 2C9 also participate in its metabolism (von Moltke et al. 1999b). The most potent inhibitors of 3A4 reduce zolpidem clearance by as much as 70% (ritonavir, ketoconazole [Greenblatt et al. 1998]), and itraconazole). Sertraline, a less potent 3A4 inhibitor, increases the C_{max} of zolpidem 43% (Ambien 2008), and fluoxetine and cimetidine have minimal effects (Allard et al. 1998; Hulhoven et al. 1988). These interactions contrast with the much more significant interaction involving buspirone (see the earlier section Buspirone), which is dependent on 3A4 for its metabolism. Potent inducers of 3A4 reduce the efficacy of zolpidem (e.g., rifampin reduces the peak plasma concentration of zolpidem by 58% [Villikka et al. 1997c]).

Zopiclone

Zopiclone (Imovane) is a chiral drug used for the treatment of insomnia. It has a complicated metabolism and may be excreted through the kidneys, lungs, and liver, where it is a substrate of 2C8 and 3A4 (Becquemont et al. 1999; Fernandez et al. 1995). Potent 3A4 inhibitors such as nefazodone and erythromycin can significantly increase zopiclone's plasma concentration and thus the risk for side effects (Aranko et al. 1994). Itraconazole increases zopiclone's peak plasma concentration by 77% (Jalava et al. 1996). There is no information on possible drug interactions with other 2C8 substrates or inhibitors. Potent inducers of 3A4 reduce zopiclone's clinical effectiveness (e.g., rifampin reduces zopiclone levels to one-third of C_{max} [Villikka et al. 1997b]).

Antipsychotics

Atypical Antipsychotics

Aripiprazole

Aripiprazole (Abilify) is a quinolone derivative atypical antipsychotic with partial dopamine type 2 and serotonin type 1A receptor activity. Aripiprazole undergoes metabolism via 2D6 and 3A4 to an active metabolite, dehydroaripiprazole (Molden et al. 2006). According to its manufacturer, aripiprazole is vulnerable to 2D6 inhibitors (e.g., paroxetine, fluoxetine, quinidine), 3A4 inhibitors (e.g., ketoconazole), and 3A4 inducers (e.g., carbamazepine) (Abilify 2008). Aripiprazole does not appear to cause inhibition or induction (see Table 19–6).

TABLE 19–6. Atypical antipsychotics

Drug	Major metabolism site(s)	Enzyme(s) inhibited[a]
Aripiprazole (Abilify)	2D6, 3A4	None known
Clozapine (Clozaril)	1A2, 3A4, 2D6, 2C9, 2C19, FMO_3, UGT1A4, UGT1A3	2D6[c]
Olanzapine (Zyprexa)	1A2, 2D6, UGTs, ABCB1	None known
Paliperidone (Invega)	2D6, 3A4, primarily excreted unchanged	None known
Quetiapine (Seroquel)	3A4, sulfation, ABCB1	None known
Risperidone (Risperdal)	2D6, 3A4, ABCB1	2D6[b]
Ziprasidone (Geodon)	Aldehyde oxidase, 3A4, 1A2	None known

Note. Data presented relate to parent drug and metabolites combined. FMO_3 = flavin monooxygenase; UGT = uridine 5'-diphosphate glucuronosyltransferase.

[a]**Bold** type indicates potent inhibition.
[b]Moderate inhibition.
[c]Mild inhibition.

Clozapine

Clozapine (Clozaril) is principally metabolized to its active metabolite, nor-clozapine, via 1A2 and (to a lesser extent) 3A4, 2D6, 2C9, and 2C19 (de Leon et al. 2005). Clozapine is also metabolized to clozapine-*N*-oxide, an inactive metabolite, through 3A4 and possibly flavin-containing monooxygenase 3 (Linnet and Olesen 1997). Both active compounds are further glucuronidated by UGT1A4, UGT1A3, and possibly other UGTs (Breyer-Pfaff and Wachsmuth 2001; Green et al. 1998; Mori et al. 2005).

The most repeatedly reviewed drug interaction involves clozapine and flu-voxamine (Heeringa et al. 1999; Lu et al. 2001; Shader and Greenblatt 1998; Szegedi et al. 1999). This interaction is predictable because fluvoxamine is a potent inhibitor of 1A2 and 2C19 and a milder inhibitor of 2D6, 3A4, 2C9, and 2B6. Therefore, all of clozapine's P450 pathways are blocked. Although the extent of this interaction varies greatly because of interindividual variability in the amounts of P450 enzymes, the increase in clozapine levels may be as high as 2- to 3-fold (Wetzel et al. 1998). Other potent inhibitors of 2C19 can also decrease clozapine clearance (e.g., modafinil [Dequardo 2002]). As inhibitors of 2C19 and 1A2, oral contraceptives containing ethinyl estradiol may also affect clozapine metabolism, and at least one case of this interaction has been reported (Gabbay et al. 2002). Inhibitors of 1A2, such as caffeine and ciprofloxacin (see Carrillo and Benitez 2000; Markowitz et al. 1997; Raaska and Neuvonen 2000), can be expected to decrease clozapine clearance. Potent inhibitors of 2D6 (e.g., paroxetine and fluoxetine) have also been reported to variably decrease clozapine and norclozapine clearance.

Clinical studies of potent 3A4 inhibitors such as ketoconazole and grape-fruit juice (Lane et al. 2001), itraconazole (Raaska and Neuvonen 1998), and nefazodone (Taylor et al. 1999) have failed to show decreased clozapine clearance (a few cases of decreased clearance have been reported). At first glance, this is puzzling. A likely explanation involves the important distinction between low-capacity, high-affinity P450 enzymes (e.g., 2D6) and low-affinity, high-capacity P450 enzymes (e.g., 3A4, the reservoir of the P450 system). Because clozapine is a substrate of so many "possible" P450 enzymes—2D6, 2C9, and 2C19—the drug fills these high-affinity P450 enzymes first. 1A2, though it is only 10% of the total liver P450 capacity, "expands" when cigarette smokers induce it. Therefore, except in the case of a clozapine overdose, clozapine does not "fill" 3A4, and thus no drug interaction occurs.

There are a number of interactions involving clozapine that are not well documented. A single case has been reported of a plasma clozapine concentration increase associated with lamotrigine therapy (Kossen et al. 2001). Both drugs use UGT1A4, and competitive inhibition of the conjugate may occur. Case and clinical studies have found both increased and decreased clozapine clearance with the addition of valproate (Conca et al. 2000; Facciola et al. 1999); the mechanism remains obscure. Several cases have been reported of respiratory distress or delirium in association with coadministration of clozapine and lorazepam, clonazepam, or diazepam, but similar cases have been reported with administration of clozapine alone (Cobb et al. 1991).

Potent "pan-inducers" of P450 enzymes, such as phenobarbital, carbamazepine, rifampin, and phenytoin, reduce the concentration of clozapine (Facciola et al. 1998; Jerling et al. 1994b; Miller 1991). Cigarette smoking is known to increase clearance of clozapine (Mookhoek and Loonen 2002). When these potent P450 inducers are withdrawn, clozapine toxicity can result (Zullino et al. 2002). Finally, clozapine is a modest 2D6 inhibitor, and concentrations of 2D6 substrates such as nortriptyline can be increased in patients taking clozapine (Smith and Riskin 1994).

Olanzapine

Olanzapine (Zyprexa) is not a significant P450 inhibitor or inducer (Ring et al. 1996a), but undergoes complex metabolism via 1A2, 2D6, and glucuronidation (Spina et al. 2007) as well as being a substrate for ABCB1 transport (Bozina et al. 2008). Probenecid, a broadly potent UGT inhibitor, has been shown to decrease clearance of olanzapine (Markowitz et al. 2002). Carbamazepine has been shown to increase the clearance of the glucuronidation of olanzapine (Linnet and Olesen 2002). It is likely that many other UGT-based drug interactions involving olanzapine will be found.

The second most important pathway for metabolism of olanzapine is through 1A2 (Kassahun et al. 1997). Potent inhibitors of 1A2, such as ciprofloxacin and fluvoxamine, increase olanzapine concentrations (de Jong et al. 2001; Markowitz and DeVane 1999). Potent 1A2 inducers such as cigarette smoking can reduce olanzapine concentrations (Lucas et al. 1998; Skogh et al. 2002). Cessation of smoking in individuals receiving olanzapine will lead to gradual increases in olanzapine concentrations, with resultant side effects or toxicity (Zullino et al. 2002).

Although 2D6 is a minor pathway of olanzapine metabolism, potent inhibitors such as fluoxetine have been shown to increase olanzapine concentrations by about 30% (Gossen et al. 2002).

Paliperidone

Paliperidone (Invega) is the 9-hydroxy metabolite of risperidone. Hepatic metabolism via 2D6 and 3A4 play a limited role in paliperidone metabolism with the majority excreted unchanged in urine (Invega 2008). Paliperidone does not appear to cause inhibition or induction. Drug interactions are unlikely given paliperidone's pharmacokinetics.

Quetiapine

Quetiapine (Seroquel) is principally metabolized by 3A4 (Grimm et al. 2006) and is a substrate of ABCB1 transport (Wikinski 2005). Its metabolism is vulnerable to all potent 3A4 inhibitors: ketoconazole (which increases C_{max} more than 300% [Seroquel 2001]), erythromycin, clarithromcyin, diltiazem, nefazodone, and others. Two drugs with less inhibitory effect on 3A4 have minimal effects on quetiapine: cimetidine has modest effects on quetiapine clearance (Strakowski et al. 2002), and fluoxetine decreases quetiapine clearance by only 11% (Potkin et al. 2002a). Similarly, all potent 3A4 inducers, such as phenytoin (Wong et al. 2001) and carbamazepine, increase clearance of quetiapine (DeVane and Nemeroff 2001a).

Surprisingly, thioridazine has also been shown to increase clearance of quetiapine (Potkin et al. 2002b). The mechanism of this interaction is unknown. Thioridazine may induce a heretofore unknown quetiapine P450 pathway or may induce 3A4 (Potkin et al. 2002b) or ABCB1 transport. Further studies are needed.

Risperidone

Not a P450 inducer and only a modest 2D6 inhibitor (Shin et al. 1999), risperidone (Risperdal) is metabolized predominantly by 2D6 and 3A4 to its major metabolite, 9-hydroxyrisperidone (Eap et al. 2001; Wang et al. 2007). Risperdal is also a substrate of ABCB1 transport (Wikinski 2005). Risperidone clearance may be decreased by potent 2D6 inhibitors such as paroxetine and fluoxetine and potent 3A4 inhibitors such as erythromycin and nefazodone. As information about P450-based drug interactions has become more available, and clinicians have been able to predict such interactions, use of these combinations has become less likely. As a result, reports of interactions between risperidone and these potent inhibitors (e.g., nefazodone) are rare in the literature.

Inducers of 3A4, such as carbamazepine and phenytoin, can increase risperidone clearance, resulting in decreased efficacy. When treatment with these inducers is discontinued and risperidone therapy is continued, the concentration of risperidone increases over time, and side effects may occur (Takahashi et al. 2001).

Carbamazepine concentrations can be increased by risperidone, although the mechanism of this interaction is not yet understood (Besag et al. 2006). Concentrations of both risperidone and valproate, when coadministered, have been found to remain the same or to be increased or decreased (DeVane and Nemeroff 2001b). As with carbamazepine, the mechanism is obscure. However, the clinical take-home lesson is clear: When another drug is added to risperidone therapy, plasma concentrations should be measured if possible.

Ziprasidone

The major pathway for metabolism of ziprasidone (Geodon) is through a molybdoflavoprotein, aldehyde oxidase, with minor contribution by 3A4 and 1A2 (Spina et al. 2007). Other substrates of aldehyde oxidase include zaleplon, citalopram, famciclovir, acyclovir, quinine, methotrexate, cyclophosphamide, and nicotine (Al-Salmy 2001; Heydorn 2000; Krenitsky et al. 1984; Rochat et al. 1998; Wierzchowski et al. 1996). Ziprasidone has no inhibitory P450 activity. As a result, pharmacokinetic drug interactions through P450 pathways are restricted to coadministered potent 3A4 inhibitors and inducers (Caley and Cooper 2002). Ketoconazole, the most potent 3A4 inhibitor, increases the C_{max} of ziprasidone by only about 40% (Miceli et al. 2000b), and carbamazepine increases its clearance about 35% (Miceli et al. 2000a).

In vitro data suggest that coadministration of an aldehyde oxidase inhibitor (e.g., chlorpromazine, cimetidine, hydralazine, and methadone) does not appear to inhibit the reduction of ziprasidone to *S*-methyldihydroziprasidone (Obach et al. 2005). Although not reproduced *in vivo,* this finding suggests that ziprasidone may be at lower risk for drug interactions, which is supported by the lack of case reports in the literature.

Typical Antipsychotics

Because the typical antipsychotics are older, only a limited number of studies of phase I and phase II metabolism have been done since sophisticated technology became available. Table 19–7 summarizes current understanding of these drugs' pharmacokinetics.

Psychostimulants and Atomoxetine

Psychostimulants can be grouped into three categories: methylphenidates (Concerta, Focalin, Metadate ER, Metadate CR, Ritalin, Ritalin-SR, Ritalin-LA); amphetamines (Adderall, Adderall-XR, Dexedrine-SR, dextroamphetamine sulfate); and others, including pemoline (Cylert; removed from the U.S. market by the FDA in 2005), modafinil (Provigil), and armodafinil (Nuvigil). Atomoxetine (Strattera) is not a psychostimulant; it is a selective norepinephrine reuptake inhibitor, new to the U.S. market. It is the first nonstimulant drug to be approved by the FDA for treatment of ADHD (see Table 19–8)

Methylphenidate and Amphetamines

Methylphenidate (Ritalin) has a complex metabolism that has not been completely elucidated, but it is generally believed that most of its inactive metabolite, ritalinic acid, is produced by plasma esterases and other phase I and phase II systems. Although 2D6 has been postulated to be a partial source of methylphenidate's metabolism, a study failed to show that a potent 2D6 inhibitor, quinidine, had any effect on methylphenidate or ritalinic acid (De-Vane et al. 2000). In the past 30 years, there have been reports of drug interactions between methylphenidate and imipramine, haloperidol, phenytoin, and others, which suggests that methylphenidate may be a P450 inhibitor (Markowitz et al. 1999). Amphetamines also have a complex metabolism, and the major metabolite is benzoic acid. 2D6 has been shown to be involved in amphetamine's metabolism (Miranda et al. 2007). There is no clinical evidence that amphetamines are P450 inhibitors.

Modafinil and Armodafinil

Modafinil (Provigil) and its racemic isomer armodafinil (Nuvigil) are wakefulness-promoting agents whose mechanism of action remains unknown.

TABLE 19–7. Typical antipsychotics

Drug	Metabolism site(s)[1]	Enzyme(s) inhibited[a]	Enzyme(s) induced
Chlorpromazine (Thorazine)	2D6, 1A2, 3A4, UGT1A4, UGT1A3	2D6	None known
Fluphenazine (Prolixin)	2D6, 1A2	2D6, 1A2	None known
Haloperidol (Haldol)	2D6, 3A4, 1A2	2D6	None known
Loxapine (Loxitane)	?2D6, ?3A4, ?1A2, UGT1A4, UGT1A3	None known	None known
Mesoridazine (Serentil)	2D6, 1A2	?	?
Perphenazine (Trilafon)	2D6, 3A4, 1A2, 2C19	2D6, 1A2	None known
Pimozide (Orap)	3A4, 1A2	2D6, 3A4	None known
Thioridazine (Mellaril)	2D6, 1A2, 2C19, FMO₃	2D6	?3A4

Note. Data presented relate to parent drug and metabolites combined. FMO$_3$ = flavin monooxygenase.
[1]**Bold** type indicates major pathway.
[a]**Bold** type indicates potent inhibition.

TABLE 19–8. Psychostimulants and atomoxetine

Drug	Metabolism site(s)	Enzyme(s) inhibited	Enzyme(s) induced
Amphetamines	Complex, 2D6	None known	None known
Armodafinil (Nuvigil)	3A4, glucuronidation, ABCB1	2C19	3A4
Atomoxetine (Strattera)	2D6	None known	None known
Methylphenidate (Ritalin)	Complex, plasma esterases, ?2D6	Possible P450 inhibition	None known
Modafinil (Provigil)	3A4, glucuronidation	2C19	3A4

Modafinil and its isomer primarily undergo metabolism via 3A4 and glucuronidation as well as being substrates for ABCB1 (Provigil 2007). *In vivo* analysis revealed inhibition of 2C19 and induction of 3A4; thus, substrates of 3A4 (e.g., cyclosporine and triazolam) and 2C19 (e.g., phenytoin) may require dosing adjustments when coadministered with modafinil or armodafinil (Darwish et al. 2008).

Atomoxetine

Although it is not a P450 inhibitor, atomoxetine (Strattera) is principally metabolized by 2D6 (Michelson et al. 2007). Potent 2D6 inhibitors such as fluoxetine, paroxetine, and quinidine can increase the maximum plasma concentration of atomoxetine as much as 4-fold, and the manufacturer recommends that its dosing be adjusted upward only after 4 weeks instead of within the first week (Strattera 2008). Other potent 2D6 inhibitors should also be used with caution, and the dosing of atomoxetine should be reduced.

Substance Abuse/Dependence Treatments

Acamprosate

Acamprosate (Campral) does not undergo phase I or phase II metabolism and is excreted unchanged by the kidneys (see Table 19–9).

Buprenorphine

Buprenorphine is an opiate partial agonist and antagonist available in the United States for the treatment of pain since the 1980s under the trade name Buprenex. In 2002, buprenorphine was approved by the FDA for use in the treatment of opiate addiction under the trade names Suboxone (buprenorphine and naloxone in a 4:1 ratio) and Subutex (sublingual buprenorphine). Buprenorphine (Suboxone) and its active metabolite norbuprenorphine first undergo *N*-dealkylation and then are further glucuronidated prior to excretion (Cone et al. 1984). Dealkylation of buprenorphine via CYP3A4 accounts for 70%–90% of buprenorphine's metabolism (Iribarne et al. 1997; Picard et al. 2005). 3A4 inhibitors including azole antifungals, macrolide antibiotics, and HIV protease inhibitors (McCance-Katz et al. 2006) may increase buprenorphine blood levels and require dosing adjustments. *In vitro* studies suggest

TABLE 19–9. Substance abuse/dependence treatments

Drug	Metabolism site(s)	Enzyme(s) inhibited[a]
Acamprosate (Campral)	Excreted unchanged by the kidneys	None known
Buprenorphine (Suboxone)	3A4, UGT1A1, UGT2B7	**3A4**[b], 2D6[b]
Disulfiram (Antabuse)	Not yet fully elucidated	2E1[b], 1A2[b]
Naltrexone (ReVia, Vivitrol)	Dihydrodiol dehydrogenase, UGT2B7, UGT1A1	Unknown
Varenicline (Chantix)	Primarily excreted unchanged in urine, UGT2B7 (minor)	Unknown

[a]**Bold** type indicates potent inhibition.
[b]Moderate inhibition.
[c]Mild inhibition.

both buprenorphine and norbuprenorphine mildly inhibit CYP3A4 and CYP2D6, but *in vivo* studies have not shown this effect at standard clinical doses (Umeda et al. 2005; Umehara et al. 2002; Zhang et al. 2003).

Norbuprenorphine, buprenorphine's active metabolite, undergoes further metabolism via phase II sulfonation and glucuronidation, though only a small fraction occurs via sulfonation (Chang et al. 2006). Glucuronidation via UGT1A1 and UGT2B7 provide the primary mechanism for phase II metabolism (Cheng et al. 1998; King et al. 1996).

Disulfiram

Disulfiram (Antabuse) blocks oxidation of acetaldehyde by aldehyde dehydrogenase, causing unpleasant and sometimes serious symptoms when alcohol is consumed. Disulfiram metabolism is not fully understood; however, disulfiram is known to cause moderate to potent inhibition of 2E1 (Damkier et al. 1999) and possible 1A2 inhibition with chronic administration (Fryc et al. 2002). Further elucidation of the pharmacokinetics of disulfiram is needed.

Naltrexone

Naltrexone (ReVia; Vivitrol), a competitive opioid receptor blocker, undergoes metabolism via dihydrodiol dehydrogenase to its primary, and active, metabolite 6-β-naltrexol (Vivitrol 2007). Naltrexone and its metabolites also undergo glucuronidation via UGT2B7 and UGT1A1 (King et al. 1996). The CYP450 enzymes do not contribute to naltrexone's metabolism (Porter et al. 2000). Caution should be used when naltrexone is prescribed with other medications that are metabolized by UGT2B7, such as NSAIDs, temazepam, and valproate.

Varenicline

Varenicline (Chantix) undergoes minimal metabolism and is excreted 90% unchanged in the urine. The remaining portion of varenicline primarily undergoes *N*-carbamoylglucuronidation via UGT2B7 (Obach et al. 2006). Varenicline is not known to inhibit or induce the CYP450 system, and no drug interactions have been observed with this medication.

References

Abilify (package insert). Tokyo, Japan, Otsuka America Pharmaceuticals, Inc., 2008

Alderman J, Preskorn SH, Greenblatt DJ, et al: Desipramine pharmacokinetics when coadministered with paroxetine or sertraline in extensive metabolizers. J Clin Psychopharmacol 17:284–291, 1997

Allard S, Sainati S, Roth-Schechter B, et al: Minimal interaction between fluoxetine and multiple-dose zolpidem in healthy women. Drug Metab Dispos 26:617–622, 1998

Al-Salmy HS: Individual variation in hepatic aldehyde oxidase activity. IUBMB Life 51:249–253, 2001

Ambien (package insert). Bridgeport, NJ, Sanofi-Aventis Inc., 2008

Anttila AK, Rasanen L, Leinonen EV: Fluvoxamine augmentation increases mirtazapine concentrations three- to fourfold. Ann Pharmacother 35:1221–1223, 2001

Aranko K, Luurila H, Backman JT, et al: The effect of erythromycin on the pharmacokinetics and pharmacodynamics of zopiclone. Br J Clin Pharmacol 38:363–367, 1994

Areberg J, Christophersen JS, Poulsen MN, et al: The pharmacokinetics of escitalopram in patients with hepatic impairment. AAPS J 8:E14–E19, 2006

Bach MV, Coutts RT, Baker GB: Involvement of CYP2D6 in the in vitro metabolism of amphetamine, two N-alkylamphetamines and their 4-methoxylated derivatives. Xenobiotica 29:719–732, 1999

Ball SE, Ahern D, Scatina J, et al: Venlafaxine: in vitro inhibition of CYP2D6 dependent imipramine and desipramine metabolism: comparative studies with selected SSRIs, and effects on human hepatic CYP3A4, CYP2C9 and CYP1A2. Br J Clin Pharmacol 43:619–626, 1997

Becquemont L, Mouajjah S, Escaffre O, et al: Cytochrome P-450 3A4 and 2C8 are involved in zopiclone metabolism. Drug Metab Dispos 27:1068–1073, 1999

Bertelsen KM, Venkatakrishnan K, von Moltke LL, et al: Apparent mechanism-based inhibition of human CYP2D6 in vitro by paroxetine: comparison with fluoxetine and quinidine. Drug Metab Dispos 31:289–293, 2003

Besag FM, Berry D: Interactions between antiepileptic and antipsychotic drugs. Drug Saf 29:95–118, 2006

Bloomer JC, Woods FR, Haddock RE, et al: The role of cytochrome P4502D6 in the metabolism of paroxetine by human liver microsomes. Br J Clin Pharmacol 33:521–523, 1992

Bozina N, Kuzman MR, Medved V, et al: Associations between MDR1 gene polymorphisms and schizophrenia and therapeutic response to olanzapine in female schizophrenic patients. J Psychiatr Res 42:89–97, 2008

Breyer-Pfaff U, Wachsmuth H: Tertiary N-glucuronides of clozapine and its metabolite desmethylclozapine in patient urine. Drug Metab Dispos 29:1343–1348, 2001

Brosen K, Naranjo CA: Review of pharmacokinetic and pharmacodynamic interaction studies with citalopram. Eur Neuropsychopharmacol 11:275–283, 2001

BuSpar (package insert). Princeton, NJ, Bristol-Myers Squibb Company, 2007

Caccia S: Metabolism of the newest antidepressants: comparisons with related predecessors. IDrugs 7:143–150, 2004

Caley CF, Cooper CK: Ziprasidone: the fifth atypical antipsychotic. Ann Pharmacother 36:839–851, 2002

Callaghan JT, Cerimele BJ, Kassahun KJ, et al: Olanzapine: interaction study with imipramine. J Clin Pharmacol 37:971–978, 1997

Carrillo JA, Benitez J: Clinically significant pharmacokinetic interactions between dietary caffeine and medications. Clin Pharmacokinet 39:127–153, 2000

Chang Y, Moody DE, McCance-Katz EF: Novel metabolites of buprenorphine detected in human liver microsomes and human urine. Drug Metab Dispos 34:440–448, 2006

Cheng Z, Rios GR, King CD, et al: Glucuronidation of catechol estrogens by expressed human UDP-glucuronosyltransferases (UGTs) 1A1, 1A3, and 2B7. Toxicol Sci 45:52–57, 1998

Christensen M, Tybring G, Mihara K, et al: Low daily 10-mg and 20-mg doses of fluvoxamine inhibit the metabolism of both caffeine (cytochrome P4501A2) and omeprazole (cytochrome P4502C19). Clin Pharmacol Ther 71:141–152, 2002

Cobb CD, Anderson CB, Seidel DR: Possible interaction between clozapine and lorazepam (letter). Am J Psychiatry 148:1606–1607, 1991

Conca A, Beraus W, Konig P, et al: A case of pharmacokinetic interference in comedication of clozapine and valproic acid. Pharmacopsychiatry 33:234–235, 2000

Cone EJ, Gorodetzky CW, Yousefnejad D, et al: The metabolism and excretion of buprenorphine in humans. Drug Metab Dispos 12:577–581, 1984

Court MH, Duan SX, Guillemette C, et al: Stereoselective conjugation of oxazepam by human UDP-glucuronosyltransferases (UGTs): Soxazepam is glucuronidated by UGT2B15, while R-oxazepam is glucuronidated by UGT2B7 and UGT1A9. Drug Metab Dispos 30:1257–1265, 2002

Crewe HK, Lennard MS, Tucker GT, et al: The effect of selective serotonin re-uptake inhibitors on cytochrome P4502D6 (CYP2D6) activity in human liver microsomes. Br J Clin Pharmacol 34:262–265, 1992

Cymbalta (package insert). Indianapolis, IN, Eli Lilly & Co., 2008

Damkier P, Hansen LL, Brosen K: Effect of diclofenac, itraconazole, grapefruit juice and erythromycin on the pharmacokinetics of quinidine. Br J Clin Pharmacol 48:829–838, 1999

Darwish M, Kirby M, Robertson P Jr, et al: Interaction profile of armodafanil with medications metabolized by cytochrome P450 enzymes 1A2, 3A4, and 2C19 in healthy subjects. Clin Pharmacokinet 47:61–74, 2008

de Jong J, Hoogenboom B, van Troostwijk LD, et al: Interaction of olanzapine with fluvoxamine. Psychopharmacology (Berl) 155:219–220, 2001

de Leon J, Armstrong SC, Cozza KL: The dosing of atypical antipsychotics. Psychosomatics 46:262–273, 2005

Dequardo JR: Modafinil-associated clozapine toxicity (letter). Am J Psychiatry 159:1243–1244, 2002

Desai HD, Seabolt J, Jann MW: Smoking in patients receiving psychotropic medications: a pharmacokinetic perspective. CNS Drugs 15:469–494, 2001

DeVane CL, Nemeroff CB: Clinical pharmacokinetics of quetiapine: an atypical antipsychotic. Clin Pharmacokinet 40:509–522, 2001a

DeVane CL, Nemeroff CB: An evaluation of risperidone drug interactions. J Clin Psychopharmacol 21:408–416, 2001b

DeVane CL, Markowitz JS, Carson SW, et al: Single-dose pharmacokinetics of methylphenidate in CYP2D6 extensive and poor metabolizers. J Clin Psychopharmacol 20:347–349, 2000

Eap CB, Bondolfi G, Zullino D, et al: Pharmacokinetic drug interaction potential of risperidone with cytochrome p450 isozymes as assessed by the dextromethorphan, the caffeine, and the mephenytoin test. Ther Drug Monit 23:228–231, 2001

Effexor (package insert). Philadelphia, PA, Wyeth Laboratories, 2002

el-Yazigi A, Chaleby K, Gad A, et al: Steady-state kinetics of fluoxetine and amitriptyline in patients treated with a combination of these drugs as compared with those treated with amitriptyline alone. J Clin Pharmacol 35:17–21, 1995

Facciola G, Avenoso A, Spina E, et al: Inducing effect of phenobarbital on clozapine metabolism in patients with chronic schizophrenia. Ther Drug Monit 20:628–630, 1998

Facciola G, Avenoso A, Scordo MG, et al: Small effects of valproic acid on the plasma concentrations of clozapine and its major metabolites in patients with schizophrenic or affective disorders. Ther Drug Monit 21:341–345, 1999

Faucette SR, Hawke RL, Shord SS, et al: Evaluation of the contribution of cytochrome P450 3A4 to human liver microsomal bupropion hydroxylation. Drug Metab Dispos 29:1123–1129, 2001

Fernandez C, Martin C, Gimenez F, et al: Clinical pharmacokinetics of zopiclone. Clin Pharmacokinet 29:431–441, 1995

Fogelman SM, Schmider J, Venkatakrishnan K, et al: O- and N-demethylation of venlafaxine in vitro by human liver microsomes and by microsomes from cDNA-transfected cells: effect of metabolic inhibitors and SSRI antidepressants. Neuropsychopharmacology 20:480–490, 1999

Frye RF, Branch RA: Effect of chronic disulfiram administration on the activities of CYP1A2, CYP2C19, CYP2D6, CYP2E1, and N-acetyltransferase in healthy human subjects. Br J Clin Pharmacol 53:155–162, 2002

Fukui N, Suzuki Y, Sawamura K, et al: Dose-dependent effects of the 3435 C>T genotype of ABCB1 gene on the steady state plasma concerntration of fluvoxamine in psychiatric patients. Ther Drug Monit 29:185–189, 2007

Gabbay V, O'Dowd MA, Mamamtavrishvili M, et al: Clozapine and oral contraceptives: a possible drug interaction. J Clin Psychopharmacol 22:621–622, 2002

Gareri P, De Fazio P, Galelli L, et al: Venlafaxine-propafenone interaction resulting in hallucinations and psychomotor agitation. Ann Pharmacother 42:434–438, 2008

Geodon (package insert). New York, NY, Pfizer Inc., 2002

Gillman PK: Tricyclic antidepressant pharmacology and therapeutic drug interactions updated. Br J Pharmacol 151:737–748, 2007

Gossen D, de Suray JM, Vandenhende F, et al: Influence of fluoxetine on olanzapine pharmacokinetics. AAPS PharmSci 4:E11, 2002

Green MD, King CD, Mojarrabi B, et al: Glucuronidation of amines and other xenobiotics catalyzed by expressed human UDP-glucuronosyltransferase 1A3. Drug Metab Dispos 26:507–512, 1998

Greenblatt DJ, von Moltke LL, Harmatz JS, et al: Kinetic and dynamic interaction study of zolpidem with ketoconazole, itraconazole, and fluconazole. Clin Pharmacol Ther 64:661–671, 1998

Greenblatt DJ, von Moltke LL, Harmatz JS, et al: Human cytochromes and some newer antidepressants: kinetics, metabolism, and drug interactions. J Clin Psychopharmacol 19(suppl):23S–35S, 1999

Grimm SW, Richtand NM, Winter HR, et al: Effects of cytochrome P450 3A modulators ketoconazole and carbamazepine on quetiapine pharmacokinetics. Br J Clin Pharmacol 61:58–69, 2006

Heeringa M, Beurskens R, Schouten W, et al: Elevated plasma levels of clozapine after concomitant use of fluvoxamine. Pharm World Sci 21:243–244, 1999

Hemeryck A, Belpaire FM: Selective serotonin reuptake inhibitors and cytochrome P-450 mediated drug-drug interactions: an update. Curr Drug Metab 3:13–37, 2002

Hesse LM, Venkatakrishnan K, Court MH, et al: CYP2B6 mediates the in vitro hydroxylation of bupropion: potential drug interactions with other antidepressants. Drug Metab Dispos 28:1176–1183, 2000

Hesse LM, von Moltke LL, Shader RI, et al: Ritonavir, efavirenz, and nelfinavir inhibit CYP2B6 activity in vitro: potential drug interactions with bupropion. Drug Metab Dispos 29:100–102, 2002

Heydorn WE: Zaleplon: a review of a novel sedative hypnotic used in the treatment of insomnia. Expert Opin Investig Drugs 9:841–858, 2000

Holmgren P, Carlsson B, Zackrisson AL, et al: Enantioselective analysis of citalopram and its metabolites in postmortem blood and genotyping for CYD2D6 and CYP2C19. J Anal Toxicol 28:94–104, 2004

Hulhoven R, Desager JP, Harvengt C, et al: Lack of interaction between zolpidem and H2 antagonists, cimetidine and ranitidine. Int J Clin Pharmacol Res 8:471–476, 1988

Invega (package insert), Mountain View, CA, ALZA Corporation, 2008

Iribarne C, Picart D, Dréano Y, et al: Involvement of cytochrome P450 3A4 in N-dealkylation of buprenorphine in human liver microsomes. Life Sci 60:1953–1964, 1997

Jalava KM, Olkkola KT, Neuvonen PJ: Effect of itraconazole on the pharmacokinetics and pharmacodynamics of zopiclone. Eur J Clin Pharmacol 51:331–334, 1996

Jerling M, Bertilsson L, Sjoqvist F: The use of therapeutic drug monitoring data to document kinetic drug interactions: an example with amitriptyline and nortriptyline. Ther Drug Monit 16:1–12, 1994a

Jerling M, Lindstrom L, Bondesson U, et al: Fluvoxamine inhibition and carbamazepine induction of the metabolism of clozapine: evidence from a therapeutic drug monitoring service. Ther Drug Monit 16:368–374, 1994b

Johnson C, Stubley-Beedham C, Stell JG: Hydralazine: a potent inhibitor of aldehyde oxidase activity in vitro and in vivo. Biochem Pharmacol 34:4251–4256, 1985

Kalgutkar AS, Vaz AD, Lame ME, et al: Bioactivation of the nontricyclic antidepressant nefazodone to a reactive quinone-imine species in human liver microsomes and recombinant cytochrome P4503A4. Drug Metab Dispos 33:243–253, 2005

Kassahun K, Mattiuz E, Nyhart E Jr, et al: Disposition and biotransformation of the antipsychotic agent olanzapine in humans. Drug Metab Dispos 25:81–93, 1997

Kato M, Fukuda T, Serretti A, et al: ABCB1 (MDR1) gene polymorphisms are associated with the clinical response to paroxetine in patients with major depressive disorder. Prog Neuropsychopharmacol Biol Psychiatry 15:398–404, 2008

Kaufman KR, Gerner R: Lamotrigine toxicity secondary to sertraline. Seizure 7:163–165, 1998

Kennedy SH, McCann SM, Masellis M, et al: Combining bupropion SR with venlafaxine, paroxetine, or fluoxetine: a preliminary report on pharmacokinetic, therapeutic, and sexual dysfunction effects. J Clin Psychiatry 63:181–186, 2002

Khairul MF, Min TH, Nasriyyah CH, et al: Fluoxetine potentiates chloroquine and mefloquine effect on multi-drug resistant Plasmodium falciparum in vivo. Jpn J Infect Dis 59:329–331, 2006

Kilicarslan T, Haining RL, Rettie AE, et al: Flunitrazepam metabolism by cytochrome P450s 2C19 and 3A4. Drug Metab Dispos 29:460–465, 2001

King CD, Green MD, Coffman BL, et al: The glucuronidation of exogenous and endogenous compounds by stably expressed rat and human UDP-glucuronosyl-transferases. Arch Biochem Biophys 332:92–100, 1996

Kivisto KT, Lamberg TS, Kantola T, et al: Plasma buspirone concentrations are greatly increased by erythromycin and itraconazole. Clin Pharmacol Ther 62:348–354, 1997

Kivisto KT, Lamberg TS, Neuvonen PJ: Interactions of buspirone with itraconazole and rifampicin: effects on the pharmacokinetics of the active 1-(2-pyrimidinyl)-piperazine metabolite of buspirone. Pharmacol Toxicol 84:94–97, 1999

Kossen M, Selten JP, Kahn RS: Elevated clozapine plasma level with lamotrigine (letter). Am J Psychiatry 158:1930, 2001

Krenitsky TA, Hall WW, de Miranda P, et al: 6-Deoxyacyclovir: a xanthine oxidase-activated prodrug of acyclovir. Proc Natl Acad Sci U S A 81:3209–3213, 1984

Kuhn UD, Kirsch M, Merkel U, et al: Reboxetine and cytochrome P450: comparison with paroxetine treatment in humans. Int J Clin Pharmacol Ther 45:36–46, 2007

Kupferschmidt H, Ha H, Ziegler W, et al: Interaction between grapefruit juice and midazolam in humans. Clin Pharmacol Ther 58:20–28, 1995

Kustra R, Corrigan B, Dunn J, et al: Lack of effect of cimetidine on the pharmacokinetics of sustained-release bupropion. J Clin Pharmacol 39:1184–1188, 1999

Lamberg TS, Kivisto KT, Laitila J, et al: The effect of fluvoxamine on the pharmacokinetics and pharmacodynamics of buspirone. Eur J Clin Pharmacol 54:761–766, 1998a

Lamberg TS, Kivisto KT, Neuvonen PJ: Effects of verapamil and diltiazem on the pharmacokinetics and pharmacodynamics of buspirone. Clin Pharmacol Ther 63:640–645, 1998b

Lane HY, Chiu CC, Kazmi Y, et al: Lack of CYP3A4 inhibition by grapefruit juice and ketoconazole upon clozapine administration in vivo. Drug Metabol Drug Interact 18:263–278, 2001

Lessard E, Yessine MA, Hamelin BA, et al: Diphenhydramine alters the disposition of venlafaxine through inhibition of CYP2D6 activity in humans. J Clin Psychopharmacol 21:175–184, 2001

Leucht S, Hackl HJ, Steimer W, et al: Effect of adjunctive paroxetine on serum levels and side-effects of tricyclic antidepressants in depressive inpatients. Psychopharmacology (Berl) 147:378–383, 2000

Lilja JJ, Kivisto KT, Backman JT, et al: Grapefruit juice substantially increases plasma concentrations of buspirone. Clin Pharmacol Ther 64:655–660, 1998

Lilja JJ, Kivisto KT, Backman J, et al: Effect of grapefruit juice dose on grapefruit-triazolam interaction. Eur J Clin Pharmacol 56:411–415, 2000

Lill J, Bauer LA, Horn JR, et al: Cyclosporine-drug interactions and the influence of patient age. Am J Health Syst Pharm 57:1579–1584, 2000

Linnet K, Olesen OV: Metabolism of clozapine by cDNA-expressed human cytochrome P450 enzymes. Drug Metab Dispos 25:1379–1382, 1997

Linnet K, Olesen OV: Free and glucuronidated olanzapine serum concentrations in psychiatric patients: influence of carbamazepine comedication. Ther Drug Monit 24:512–517, 2002

Lobo ED, Bergstron RF, Reddy S, et al: In vitro and in vivo evaluations of cytochrome p4501a2 interactions with duloxetine. Clin Pharmacokinet 47:191–202, 2008

Loga S, Curry S, Lader M: Interaction of chlorpromazine and nortriptyline in patients with schizophrenia. Clin Pharmacokinet 6:454–462, 1981

Lu ML, Lane HY, Chang WH: Differences between in vitro and in vivo determinations of fluvoxamine-clozapine interaction (letter). J Clin Psychopharmacol 21:625–626, 2001

Lucas RA, Gilfillan DJ, Bergstrom RF: A pharmacokinetic interaction between carbamazepine and olanzapine: observations on possible mechanism. Eur J Clin Pharmacol 54:639–643, 1998

Mandrioli R, Forti GC, Raggi MA: Fluoxetine metabolism and pharmacologic interactions: the role of cytochrome P450. Curr Drug Metab 7:127–133, 2006

Markowitz JS, DeVane CL: Suspected ciprofloxacin inhibition of olanzapine resulting in increased plasma concentration (letter). J Clin Psychopharmacol 19:289–291, 1999

Markowitz JS, Gill HS, DeVane CL, et al: Fluoroquinolone inhibition of clozapine metabolism (letter). Am J Psychiatry 154:881, 1997

Markowitz JS, Morrison SD, DeVane CL: Drug interactions with psychostimulants. Int Clin Psychopharmacol 14:1–18, 1999

Markowitz JS, DeVane CL, Liston HL, et al: The effects of probenecid on the disposition of risperidone and olanzapine in healthy volunteers. Clin Pharmacol Ther 71:30–38, 2002

Massica A, Mayo G, Wilkinson G: In vivo comparison of constitutive cytochrome P4503A activity assessed by alprazolam, triazolam, and midazolam. Clin Pharmacol Ther 76:341–349, 2004

McAlpine DE, O'Kane DJ, Black JL, et al: Cytochrome P450 2D6 genotype variation and venlafaxine dosage. Mayo Clin Proc 82:1065–1068, 2007

McCance-Katz EF, Moody DE, Smith PF, et al: Interactions between buprenorphine and antiretrovirals, II: the protease inhibitors nelfinavir, lopinavir/ritonavir, and ritonavir. Clin Infect Dis 43 (suppl 4):S235–S246, 2006

Miceli JJ, Anziano RJ, Robarge L, et al: The effect of carbamazepine on the steady-state pharmacokinetics of ziprasidone in healthy volunteers. Br J Clin Pharmacol 49 (suppl 1):65S–70S, 2000a

Miceli JJ, Smith M, Robarge L, et al: The effects of ketoconazole on ziprasidone pharmacokinetics—a placebo-controlled crossover study in healthy volunteers. Br J Clin Pharmacol 49 (suppl 1):71S–76S, 2000b

Michelson D, Read HA, Ruff DD, et al: CYP2D6 and clinical response to atomoxetine in children and adolescents with ADHD. J Am Acad Child Adolesc Psychiatry 46:242–251, 2007

Miller DD: Effect of phenytoin on plasma clozapine concentrations in two patients. J Clin Psychiatry 52:23–25, 1991

Miranda GE, Sordo M, Salazar AM, et al: Determination of amphetamine, methamphetamine, and hydroxyamphetamine derivatives in urine by gas chromatography-mass spectrometry and its relation to CYP2D6 phenotype of drug users. J Anal Toxicol 31:31–36, 2007

Molden E, Garcia BH, Braathen P, et al: Co-prescription of cytochrome P450 2D6/3A4 inhibitor-substrate pairs in clinical practice. Eur J Clin Pharmacol 61:119–125, 2005

Molden E, Lunde H, Lunder N, et al: Pharmacokinetic variability of aripiprazole and the active metabolite dehydroaripiprazole in psychiatric patients. Ther Drug Monit 28:744–749, 2006

Mookhoek EJ, Loonen AJ: Does the change of omeprazole to pantoprazole affect clozapine plasma concentrations? Br J Clin Pharmacol 53:545, 2002

Mori A, Maruo Y, Iwai M, et al: UDP-glucuronosyltransferase 1A4 polymorphism in a Japanese population and kinetics of clozapine glucuronidation. Drug Metab Dispos 33:672–675, 2005

Mula M, Monaco F: Carbamazepine-risperidone interactions in patients with epilepsy. Clin Neuropharmacol 25:97–100, 2002

Obach RS: Potent inhibition of human liver aldehyde oxidase by raloxifene. Drug Metab Dispos 32:89–97, 2004

Obach RS, Walsky RL: Drugs that inhibit oxidation reactions catalyzed by aldehyde oxidase to not inhibit the reductive metabolism of ziprasidone to its major metabolite, S-methyldihydroziprasidone: an in vitro study. J Clin Psychopharmacol 25:605–608, 2005

Obach RS, Reed-Hagen AE, Krueger SS, et al: Metabolism and disposition of vareni-cline, a selective alpha4beta2 acetylcholine receptor partial agonist, in vivo and in vitro. Drug Metab Dispos 34:121–130, 2006

Obach RS, Cox LM, Tremaine LM: Sertraline is metabolized by multiple cytochrome P450 enzymes, monoamine oxidases, and glucuronyl transferases in human: an in vitro study. Drug Metab Dispos 33:262–270, 2005

Ohno Y, Hisaka A, Suzuki H: General framework for the quantitative prediction of CYP3A4-mediated oral drug interactions based on the AUC increase by coad-ministration of standard drugs. Clin Pharmacokinet 46:681–696, 2007

Picard N, Cresteil T, Djebli N, et al: In vitro metabolism study of buprenorphine: evidence for new metabolic pathways. Drug Metab Dispos 33:689–695, 2005

Popli AP, Tanquary J, Lamparella V, et al: Bupropion and anticonvulsant drugs. Ann Clin Psychiatry 7:99–101, 1995

Porter SJ, Somogyi AA, White JM: Kinetics and inhibition of the formation of 6 beta-naltrexol from naltrexone in human liver cytosol. Br J Clin Pharmacol 50:465–171, 2000

Potkin SG, Thyrum PT, Alva G, et al: Effect of fluoxetine and imipramine on the pharmacokinetics and tolerability of the antipsychotic quetiapine. J Clin Phar-macol 22:174–182, 2002a

Potkin SG, Thyrum PT, Alva G, et al: The safety and pharmacokinetics of quetiapine when coadministered with haloperidol, risperidone, or thioridazine. J Clin Psy-chopharmacol 22:121–130, 2002b

Preskorn SH, Alderman J, Chung M, et al: Pharmacokinetics of desipramine coad-ministered with sertraline or fluoxetine. J Clin Psychopharmacol 14:90–98, 1994

Preskorn SH, Greenblatt DJ, Flockhart D, et al: Comparison of duloxetine, escitalo-pram, and sertraline effects on cytochrome P450 2D6 in healthy volunteers. 27:28–34, 2007

Provigil (package insert), Frazer, PA, Cephalon Inc., 2007

Raaska K, Neuvonen PJ: Serum concentrations of clozapine and N-desmethylclozapine are unaffected by the potent CYP3A4 inhibitor itraconazole. Eur J Clin Pharmacol 54:167–170, 1998

Raaska K, Neuvonen PJ: Ciprofloxacin increases serum clozapine and N-desmethyl-clozapine: a study in patients with schizophrenia. Eur J Clin Pharmacol 56:585–589, 2000

Rao N: The clinical pharmacokinetics of escitalopram. Clin Pharmacokinet 46:281–290, 2007

Rasheed A, Javed MA, Nazir S, et al: Interaction of chlorpromazine with tricyclic anti-depressants in schizophrenic patients. J Pak Med Assoc 44:233–234, 1994

Renwick AB, Mistry H, Ball S, et al: Metabolism of zaleplon by human hepatic microsomal cytochrome P450 isoforms. Xenobiotica 28:337–348, 1998

Renwick AB, Ball SE, Tredger JM, et al: Inhibition of zaleplon metabolism by cimetidine in the human liver: in vitro studies with subcellular fractions and precision-cut liver slices. Xenobiotica 32:849–862, 2002

Ring BJ, Binkley SN, Vandenbranden M, et al: In vitro interaction of the antipsychotic agent olanzapine with human cytochromes P450 CYP2C9, CYP2C19, CYP2D6 and CYP3A. Br J Clin Pharmacol 41:181–186, 1996a

Ring BJ, Catlow J, Lindsay TJ, et al: Identification of the human cytochromes P450 responsible for the in vitro formation of the major oxidative metabolites of the antipsychotic agent olanzapine. J Pharmacol Exp Ther 276:658–666, 1996b

Ring BJ, Eckstein JA, Gillespie JS, et al: Identification of the human cytochromes P450 responsible for in vitro formation of R- and S-norfluoxetine. J Pharmacol Exp Ther 297:1044–1050, 2001

Rochat B, Kosel M, Boss G, et al: Stereoselective biotransformation of the selective serotonin reuptake inhibitor citalopram and its demethylated metabolites by monoamine oxidases in human liver. Biochem Pharmacol 56:15–23, 1998

Rozerem (package insert), Deerfield, IL, Takeda Pharmaceuticals America Inc., 2006

Sayal KS, Duncan-McConnell DA, McConnell HW, et al: Psychotropic interactions with warfarin. Acta Psychiatr Scand 102:250–255, 2000

Seree EJ, Pisano PJ, Placidi M, et al: Identification of the human and animal hepatic cytochromes P450 involved in clonazepam metabolism. Fundam Clin Pharmacol 7:69–75, 1993

Seroquel (package insert). Wilmington, DE, AstraZeneca Pharmaceuticals LP, 2001

Serzone (package insert). Princeton, NJ, Bristol-Myers Squibb Company, 2005

Shader RI, Greenblatt DJ: Clozapine and fluvoxamine, a curious complexity (editorial). J Clin Pharmacol 18:101–102, 1998

Shin JG, Soukhova N, Flockhart DA: Effect of antipsychotic drugs on human liver cytochrome P-450 (CYP) isoforms in vitro: preferential inhibition of CYP2D6. Drug Metab Dispos 27:1078–1084, 1999

Shin JG, Park JY, Kim MJ, et al: Inhibitory effects of tricyclic antidepressants (TCAs) on human cytochrome p450 enzymes in vitro: mechanism of drug interaction between TCAs and phenytoin. Drug Metab Dispos 30:1102–1107, 2002

Sitsen JM, Maris FA, Timmer CJ: Concomitant use of mirtazapine and cimetidine: a drug-drug interaction study in healthy male subjects. Eur J Clin Pharmacol 56:389–394, 2000

Sitsen J, Maris F, Timmer C: Drug-drug interaction studies with mirtazapine and carbamazepine in healthy male subjects. Eur J Drug Metab Pharmacokinet 26:109–121, 2001

Skinner MH, Kuan HY, Pan A, et al: Duloxetine is both an inhibitor and a substrate of cytochrome P4502D6 in healthy volunteers. Clin Pharmacol Ther 73:170–177, 2003

Skogh E, Reis M, Dahl ML, et al: Therapeutic drug monitoring data on olanzapine and its N-demethyl metabolite in the naturalistic clinical setting. Ther Drug Monit 24:518–526, 2002

Smith T, Riskin J: Effect of clozapine on plasma nortriptyline concentration. Pharmacopsychiatry 27:41–42, 1994

Sonata (package insert). Philadelphia, PA, Wyeth Laboratories, 2007

Spigset O, Axelsson S, Norstrom A, et al: The major fluvoxamine metabolite in urine is formed by CYP2D6. Eur J Clin Pharmacol 57:653–658, 2001

Spina E, Avenoso A, Facciola G, et al: Effect of fluoxetine on the plasma concentrations of clozapine and its major metabolites in patients with schizophrenia. Int Clin Psychopharmacol 13:141–145, 1998

Spina E, Avenoso A, Salemi M, et al: Plasma concentrations of clozapine and its major metabolites during combined treatment with paroxetine or sertraline. Pharmacopsychiatry 33:213–217, 2000

Spina E, Avenoso A, Facciola G, et al: Plasma concentrations of risperidone and 9-hydroxyrisperidone during combined treatment with paroxetine. Ther Drug Monit 23:223–237, 2001

Spina E, Avenoso A, Scordo MG, et al: Inhibition of risperidone metabolism by fluoxetine in patients with schizophrenia: a clinically relevant pharmacokinetic drug interaction. J Clin Psychopharmacol 22:419–423, 2002

Spina E, de Leon J: Metabolic drug interactions with newer antipsychotics: a comparative review. Basic Clin Pharmacol Toxicol 100:4–22, 2007

Strakowski SM, Keck PE Jr, Wong YW, et al: The effect of multiple doses of cimetidine on the steady-state pharmacokinetics of quetiapine in men with selected psychotic disorders. J Clin Psychopharmacol 22:201–205, 2002

Strattera (package insert). Indianapolis, IN, Eli Lilly and Company, 2008

Szegedi A, Anghelescu I, Wiesner J, et al: Addition of low-dose fluvoxamine to low-dose clozapine monotherapy in schizophrenia: drug monitoring and tolerability data from a prospective clinical trial. Pharmacopsychiatry 32:148–153, 1999

Takahashi H, Yoshida K, Higuchi H, et al: Development of parkinsonian symptoms after discontinuation of carbamazepine in patients concurrently treated with risperidone: two case reports. Clin Neuropharmacol 24:358–360, 2001

Taylor D, Bodani M, Hubbeling A, et al: The effect of nefazodone on clozapine plasma concentrations. Int Clin Psychopharmacol 14:185–187, 1999

Timmer CJ, Sitsen JM, Delbressine LP: Clinical pharmacokinetics of mirtazapine. Clin Pharmacokinet 38:461–474, 2000

Umeda S, Harakawa N, Yamamoto M, et al: Effect of nonspecific binding to microsomes and metabolic elimination of buprenorphine on the inhibition of cytochrome P4502D6. Biol Pharm Bull 28:212–216, 2005

Umehara K, Shimokawa Y, Miyamoto G: Inhibition of human drug metabolizing cytochrome P450 by buprenorphine. Biol Pharm Bull 25:682–685, 2002

Uhr M, Grauer MT: ABCB1AB P-glycoprotein is involved in the uptake of citalopram and trimipramine into the brain of mice. J Psychiatr Res 37:179–185, 2003

Villikka K, Kivisto KT, Backman JT, et al: Triazolam is ineffective in patients taking rifampin. Clin Pharmacol Ther 61:8–14, 1997a

Villikka K, Kivisto KT, Lamberg TS, et al: Concentrations and effects of zopiclone are greatly reduced by rifampicin. Br J Clin Pharmacol 43:471–474, 1997b

Villikka K, Kivisto KT, Luurila H, et al: Rifampin reduces plasma concentrations and effects of zolpidem. Clin Pharmacol Ther 62:629–634, 1997c

Vivitrol (package insert). Cambridge, MA, Alkermes, Inc., 2007

von Moltke LL, Manis M, Harmatz JS, et al: Inhibition of acetaminophen and lorazepam glucuronidation in vitro by probenecid. Biopharm Drug Dispos 14:119–130, 1993

von Moltke LL, Greenblatt DJ, Duan SX, et al: Phenacetin O-deethylation by human liver microsomes in vitro: inhibition by chemical probes, SSRI antidepressants, nefazodone and venlafaxine. Psychopharmacology (Berl) 128:398–407, 1996a

von Moltke LL, Greenblatt DJ, Harmatz JS, et al: Triazolam biotransformation by human liver microsomes in vitro: effects of metabolic inhibitors and clinical confirmation of a predicted interaction with ketoconazole. J Pharmacol Exp Ther 276:370–379, 1996b

von Moltke LL, Greenblatt DJ, Granda BW, et al: Nefazodone, meta-chlorophenylpiperazine, and their metabolites in vitro: cytochromes mediating transformation, and P450-3A4 inhibitory actions. Psychopharmacology (Berl) 145:113–122, 1999a

von Moltke LL, Greenblatt DJ, Granda BW, et al: Zolpidem metabolism in vitro: responsible cytochromes, chemical inhibitors, and in vivo correlations. Br J Clin Pharmacol 48:89–97, 1999b

von Moltke LL, Greenblatt DJ, Giancarlo GM, et al: Escitalopram (S-citalopram) and its metabolites in vitro: cytochromes mediating biotransformation, inhibitory effects, and comparison to R-citalopram. Drug Metab Dispos 29:1102–1109, 2001

Walsky RL, Astuccio AV, Obach RS: Evaluation of 227 drugs for in vitro inhibition of cytochrome P450 2B6. J Clin Pharmacol 46:1426–1438, 2006

Wang JH, Liu ZQ, Wang W, et al: Pharmacokinetics of sertraline in relation to genetic polymorphism of CYP2C19. Clin Pharmacol Ther 70:42–47, 2001

Wang JS, Zhu HJ, Gibson BB, et al: Sertraline and its metabolite desmethylsertraline, but not buproprion or its three major metabolites, have high affinity for P-glycoprotein. Biol Pharm Bull 31:231–234, 2008

Wang L, Yu L, Zhang AP, et al: Serum prolactin levels, plasma risperidone levels, polymorphism of cytochrome P450 2D6 and clinical response in patients with schizophrenia. J Psychopharmacol 21:837–842, 2007

Weintraub D: Nortriptyline toxicity secondary to interaction with bupropion sustained-release. Depress Anxiety 13:50–52, 2001

Weiss J, Dormann SM, Martin-Facklam M, et al: Inhibition of P-glycoprotein by newer antidepressants. J Pharmacol Exp Ther 305:197–204, 2003

Wellbutrin (package insert). Research Triangle Park, NC, GlaxoSmithKline, 2006

Wen B, Ma L, Rodrigues AD, et al: Detection of novel reactive metabolites of trazodone: evidence for CYP2D6-mediated bioactivation of m-chlorophenylpiperazine. Drug Metab Dispos 36:841–850, 2008

Wetzel H, Anghelescu I, Szegedi A, et al: Pharmacokinetic interactions of clozapine with selective serotonin reuptake inhibitors: differential effects of fluvoxamine and paroxetine in a prospective study. J Clin Psychopharmacol 18:2 9, 1998

Wierzchowski J, Wroczynski P, Interewicz E: Selective assay of the cytosolic forms of the aldehyde dehydrogenase in rat, with possible significance for the investigations of cyclophosphamide cytotoxicity. Acta Pol Pharm 53:203–208, 1996

Wikinski S: Pharmacokinetic mechanisms underlying resistance in psychopharmacological treatment: the role of P-glycoprotein. Vertex 16:438–441, 2005

Wong YW, Yeh C, Thyrum PT: The effects of concomitant phenytoin administration on the steady-state pharmacokinetics of quetiapine. J Clin Psychopharmacol 21:89–93, 2001

Yasui N, Kondo T, Furukori H, et al: Effects of repeated ingestion of grapefruit juice on single and multiple oral-dose pharmacokinetics and pharmacodynamics of alprazolam. Psychopharmacol 150:185–190, 2000

Ziqiang C, Rios G, King C, et al: Glucuronidation of catechol estrogens by expressed human UDP-glucuronosyltransferases (UGTs) 1A1, 1A3, and 2B7. Toxicol Sci 45:448–457, 1998

Zhang W, Ramamoorthy Y, Tyndale RF, et al: Interaction of buprenorphine and its metabolite norbuprenorphine with cytochromes p450 in vitro. Drug Metab Dispos 31:768–772, 2003

Zullino DF, Delessert D, Eap CB, et al: Tobacco and cannabis smoking cessation can lead to intoxication with clozapine or olanzapine. Int Clin Psychopharmacol 17:141–143, 2002

20

Transplant Surgery and Rheumatology

Immunosuppressants

Gary H. Wynn, M.D.

Reminder: *This chapter is dedicated primarily to metabolic and ABCB1 (P-gp) interactions. Interactions due to displaced protein-binding, alterations in absorption or excretion, and pharmacodynamics are not covered.*

Immunosuppressants are used to prevent organ transplant rejection and to treat patients with autoimmune disorders. They can be classified by their mechanism or mechanisms of actions. Many immunosuppressants are administered in other disease processes; for example, cyclophosphamide is used in cancer chemotherapy, and steroids are prescribed for numerous inflammatory medical conditions. Recent advances in immunosuppressant therapy have focused on interventions such as monoclonal antibodies. Frequently these new therapies do not interact with the enzymatic or transport systems covered in this text and are thus not included (see Table 20–1).

Glucocorticoids

The synthetic glucocorticoids methylprednisolone, prednisolone, and prednisone are all used for a variety of purposes, including immunosuppression in organ transplant recipients and in patients with autoimmune disorders. Prednisone is actually a pro-drug and is metabolized by the liver to prednisolone, the active glucocorticoid. Because these three drugs are very old, there is minimal information on metabolism. Methylprednisolone appears to be metabolized at least in part via 3A4 (Kotlyar et al. 2003), although additional pathways have yet to be determined. The naturally occurring glucocorticoids—cortisone and hydrocortisone—are metabolized in part by 3A4 (Lin et al. 1999) via 6β-hydroxylation, so it is believed that synthetic glucocorticoids are metabolized in a similar fashion.

Use of 3A4 inhibitors may alter the pharmacokinetics of glucocorticoids. Imani et al. (1999) demonstrated that when prednisone is used with the potent 3A4 inhibitor diltiazem, the latter actually enhances prednisone's effect—in part by inhibiting prednisone's active metabolite, prednisolone, from being metabolized by 3A4. This reaction could lead to toxic levels of prednisone or prednisolone. Phenobarbital and phenytoin, inducers of 3A4, have been noted in multiple case reports to decrease transplant survival and to necessitate higher doses of prednisolone to maintain efficacy (Gambertoglia et al. 1982; Wassner et al. 1977). The glucocorticoids are reportedly inducers of 3A4 themselves, but few cases of related interactions have been presented in the literature.

Cyclosporine

Cyclosporine (Sandimmune) is a cyclic polypeptide of 11 amino acids that is produced by a fungus (Sandimmune 2001). It suppresses immune responses by inhibiting the first phase of T-cell activation by binding with a natural immunosuppressant protein, cyclophilin. This complex then inhibits calcineurin, a calcium/calmodulin–activated phosphatase. It is considered a mainstay medication for preventing rejection of organ transplants.

TABLE 20–1. Immunosuppressants

Drug	Metabolism site(s)	Enzyme(s)/ Process(es) inhibited	Enzyme(s)/ Process(es) induced	Transporter substrate
Azathioprine (Imuran)[1]	Unknown enzymes metabolize to mercaptopurine	?	?	?
Cyclosporine (Sandimmune)	3A4	3A4[2], ABCB1	None	ABCB1
Mercaptopurine (Purinethol)	TPMT	?	?	?
Methotrexate	Non-P450 oxidative enzymes	?	?	?
Methylprednisolone	3A4	3A4[2]	3A4[4]	No
Mitoxantrone (Novantrone)	Unknown oxidative enzymes and UGTs	?	?	ABCB1, ABCC1, ABCG2
Muromonab-CD3 (Orthoclone OKT3)	Not phase I or phase II enzymes	?	None	?
Mycophenolate mofetil (CellCept)[1]	Hydrolysis, UGT1A7, UGT1A8, UGT1A9, UGT2B7	?	?	?
Prednisolone[1]	3A4	None	3A4[4]	?
Prednisone[1]	3A4	None	3A4[4]	?
Sirolimus (Rapamune)	3A4	?	None	ABCB1
Tacrolimus (Prograf)	3A4, UGT2B7	3A4,[3] UGT1A1,[4] UGT2B7, ABCB1	None	ABCB1

Note. ?=unknown; TPMT=thiopurine methyltransferase; UGT=uridine 5'-diphosphate glucuronosyltransferase.
[1]Pro-drug, requires enzymatic activity to produce active pharmacological compound. [2]Moderate inhibition. [3]Weak inhibition. [4]Possible inhibition.

Cyclosporine is given both intravenously and orally. A significant first-pass effect occurs with oral administration, primarily because the drug is metabolized and cleared extensively by 3A4 (Fahr 1993; Hu et al. 2006). Cyclosporine is also an inhibitor of 3A4, as evidenced by elevation in repaglinide plasma levels during coadministration (Kajosaari et al. 2005). This agent is also a substrate and inhibitor of ABCB1 (P-gp) transport (Amioka et al. 2007; Lo and Burckart 1999). Other enzymes may be involved in cyclosporine clearance, since multiple metabolites are found in the urine after oral administration, but these other enzymes are not well identified (Sandimmune 2001).

Because cyclosporine is dependent on 3A4 for clearance and is an ABCB1 substrate, there are numerous reports of drug interactions. The following agents have been shown to significantly inhibit cyclosporine metabolism (often *in vivo*) through inhibition of 3A4: clarithromycin (Biaxin), diltiazem (Cardizem), erythromycin, fluconazole (Diflucan), fluvoxamine (Luvox), grapefruit juice, indinavir (Crixivan), itraconazole (Sporanox), ketoconazole (Nizoral), nefazodone, quinupristin/dalfopristin (Syncrcid), ritonavir (Norvir), and troleandomycin (Tao). Indeed, the list of drugs that can increase cyclosporine levels reads nearly like a list of all known 3A4 inhibitors. In one case, a 20-mg dose of fluoxetine (Prozac) doubled cyclosporine levels (Horton and Bonser 1995). In another case, a 100-mg dose of fluvoxamine increased serum cyclosporine levels from 200 to 380 ng/mL (Vella and Sayegh 1998). Conversely, discontinuation of a 3A4 inhibitor can result in significantly lowered cyclosporine doses, placing patients at risk for organ rejection (Moore et al. 1996).

Because cyclosporine is a substrate of ABCB1, inhibitors of ABCB1 can increase cyclosporine levels as well. It can be difficult to pinpoint this mechanism, however, because there is so much overlap with 3A4 and ABCB1 substrates, inhibitors, and inducers. Additionally, cyclosporine inhibits ABCB1 (Krishna et al. 2007) and may be responsible for increases in levels of other drugs, such as cancer chemotherapy agents (Theis et al. 1998), but again, the overlap with 3A4 makes it difficult to draw definite conclusions in many cases.

Drugs that induce 3A4 (or ABCB1) can decrease cyclosporine levels and bring about transplant rejection. Well-known inducers of 3A4 have all been reported, in multiple cases, to significantly decrease levels of cyclosporine; such drugs include carbamazepine, phenobarbital, phenytoin, and rifampin. In one case, cyclosporine levels decreased 50% after 1 month of coadministration of the modest 3A4 inducer modafinil (Provigil) at 200 mg/day (Provigil 1999).

St. John's wort has been publicized as a safe and natural alternative for treating modest depression; therefore, its use in medically complicated recipients of organ transplants is not surprising. This herbal supplement is considered an inducer of 3A4 and ABCB1 (Zhou et al. 2004). Since the first report of heart transplant rejection from the use of the supplement with cyclosporine (Ruschitzka et al. 2000), multiple case reports have shown similar interactions in patients taking cyclosporine and St. John's wort, and thus individuals should be encouraged to avoid such a combination.

DNA Cross-Linking Agents

Cyclophosphamide

Cyclophosphamide is discussed in Chapter 16, "Oncology."

Mitoxantrone

Mitoxantrone (Novantrone) is a parenteral antineoplastic agent similar to anthracyclines such as doxorubicin. It is approved for use in treating a variety of neoplasms, and in 2000 the U.S. Food and Drug Administration (FDA) approved its use in multiple sclerosis. Mitoxantrone is cleared by oxidative metabolism, glucuronidation, and excretion (some of the drug is excreted unchanged) (Novantrone 2001). Mitoxantrone is a substrate of several transporters, including ABCB1 (P-gp), ABCC1 (MRP1), and ABCG2 (BCRP) (Pawarode et al. 2007), although the clinical importance of this is not yet fully understood. Overall, very little is known about any pharmacokinetic drug interactions involving mitoxantrone, although pharmacodynamic interactions are a known potential problem—particularly with other drugs that may affect bleeding times.

Folate Inhibitors

Methotrexate

Methotrexate is discussed in Chapter 16, "Oncology."

Inhibitors and Analogs of DNA Base Pairs

Azathioprine

Azathioprine (Imuran), first approved by the FDA in 1968, is used in a host of autoimmune disorders and is administered to prevent renal transplant rejection. Few drug interactions involving azathioprine are known (Haagsma 1998), but it is a pro-drug and is metabolized by unknown liver enzymes to its active purine analog, mercaptopurine (see next section). In this sense, azathioprine's drug interaction profile is similar to that of mercaptopurine.

Mercaptopurine

Mercaptopurine (6-MP; Purinethol) was approved by the FDA in 1953 and is used in a wide variety of disorders, including Crohn's disease, ulcerative colitis, and several forms of leukemia. Although a small portion of mercaptopurine is excreted unchanged, the majority is metabolized by thiopurine methyltransferase (TPMT), an oxidative enzyme in the liver that metabolizes other endogenous thiopurines (Gardiner et al. 2006). For years it has been known that TPMT is polymorphic and that 1 in 300 individuals essentially lacks the enzyme (Lennard et al. 1989). Normal doses of azathioprine or 6-MP in patients with low TPMT activity can cause severe toxicity, resulting in acute myelosuppression. It is current practice for genetic screening to be performed before azathioprine or 6-MP is prescribed, to determine whether the patient has diminished TPMT activity (Richard et al. 2007).

Mycophenolate Mofetil

Mycophenolate mofetil (CellCept) is a parenteral immunosuppressive agent that works by inhibiting an enzyme necessary for purine synthesis. It is actually a pro-drug and is quickly metabolized to mycophenolic acid. It is approved for use in a variety of organ transplantation procedures, for the purpose of preventing graft rejection. The pro-drug is rapidly hydrolyzed, and 5 minutes after intravenous injection, levels of mycophenolate mofetil are undetectable. Mycophenolic acid, the metabolite, is metabolized through glucuronidation via UGT1A7, UGT1A8, UGT1A9, and UGT2B7 (Inoue et al. 2007; Kagaya et al. 2007). Coadministration of rifampin with mycophenolic acid causes a reduction of mycophenolic acid levels, which appears to be caused by rifampin induction of glucuronidation (Naesens et al. 2006). Pharmacodynamic drug interactions involving mycophenolate mofetil are known, but there are few reports of pharmacokinetic drug interactions that relate to its rapid hydrolysis or conjugation (CellCept 2000). Further study is needed to elucidate these potential interactions.

Macrolides

Sirolimus

Sirolimus (Rapamune) is a macrolide immunosuppressant that inhibits the second phase of T-cell activation (Rapamune 2002). This mechanism is different from that of cyclosporine or tacrolimus (Prograf). Unlike cyclosporine and tacrolimus, sirolimus appears not to produce end-organ toxicity. Therefore, higher levels of sirolimus are rarely associated with cognitive impairment or nephrotoxicity.

Sirolimus is metabolized by 3A4 and is a substrate of ABCB1 transport (Anglicheau et al. 2006; Picard et al. 2007). Inhibitors of 3A4 (including cyclosporine, diltiazem, and ketoconazole) have been shown to increase sirolimus levels. Of greater concern is the possibility of a decrease in sirolimus levels when the drug is administered with a 3A4 or ABCB1 inducer. Concomitant administration of sirolimus and rifampin can result in significantly lowered plasma sirolimus levels requiring dose adjustments to achieve effective therapy (Boni et al. 2007). The manufacturer has indicated that other typical 3A4 inducers, such as carbamazepine, phenobarbital, and phenytoin, can decrease sirolimus levels (Rapamune 2002). Therefore, when sirolimus is used with any 3A4 inducer, sirolimus concentrations should be monitored closely to avoid the potential for graft rejection.

Tacrolimus

Tacrolimus (Prograf) is a macrolide immunosuppressant that inhibits the first phase of T-cell activation, a mechanism similar to that of cyclosporine (Prograf 1998). It appears to be a substrate of 3A4, UGT2B7, and ABCB1 (Op den Buijsch et al. 2007; Strassburg et al. 2001), as well as an inhibitor of ABCB1 (Naito et al. 1992; Tanaka et al. 2007). Tacrolimus has narrow safety and therapeutic windows. Excessive levels can lead to cognitive impairment and nephrotoxicity, and low levels can lead to organ transplant rejection.

Multiple cases have been reported of increased serum tacrolimus levels with coadministration of the 3A4 inhibitors clarithromycin, diltiazem, erythromycin, fluconazole, indinavir, itraconazole, ketoconazole, nefazodone, and ritonavir. Additionally, grapefruit juice (Fukatsu et al. 2006) and quinupristin (Synercid 2000) may significantly increase tacrolimus through inhibition of 3A4 as well as ABCB1 transport. Inhibitors or inducers of ABCB1 transport may similarly alter plasma concentrations and overall efficacy of tacrolimus.

Clotrimazole, a modest inhibitor of 3A4, was shown to significantly increase tacrolimus levels (Vasquez et al. 2005). Felodipine, a calcium channel blocker with modest 3A4 inhibition, increased tacrolimus levels in one case (Butani et al. 2002).

Homma et al. (2002) reported that a patient who was a poor metabolizer at 2C19 had increased levels of tacrolimus when prescribed lansoprazole (Prevacid), a proton pump inhibitor. Further study has confirmed that poor metabolizers at 2C19 who are coadministered tacrolimus and rabeprazole (Aciphex) or lansoprazole will likely require lower tacrolimus dosing due to inhibition of 3A4 (Miura et al. 2007).

Tacrolimus is a weak inhibitor of 3A4 (Niwa et al. 2007), with an inhibitory effect not likely to cause clinically significant interactions. Tacrolimus is also an inhibitor of glucuronidation, including UGT2B7 (Hara et al. 2007), and possibly UGT1A1 (Gornet et al. 2001).

Because of the narrow therapeutic index and complex metabolism and transport of tacrolimus, patients receiving this agent should be monitored closely and counseled extensively regarding the potential complications associated with tacrolimus drug interactions.

Monoclonal Antibodies

Three monoclonal antibodies are used as immunosuppressant agents in recipients of allografts: basiliximab (Simulect), daclizumab (Zenapax), and muromonab-CD3 (Orthoclone OKT3). Basiliximab and daclizumab are produced by DNA technology (Simulect 2001; Zenapax 1999). They are immunoglobulin G monoclonal antibodies with half-lives similar to those of endogenous immunoglobulin G (several weeks). Muromonab-CD3 is a monoclonal antibody of murine origin. It recognizes and binds to CD3 antigens of human T lymphocytes, inhibiting their function (Orthoclone OKT3 2001). Its clearance is dependent on the amount of available CD3 molecules. Because of the nature of monoclonal antibodies, few studies have evaluated their pharmacokinetics. Further study is needed, but currently no evidence exists for clinically meaningful interaction of monoclonal antibodies with enzymatic metabolism.

Summary

Potent inhibitors and inducers of 3A4 metabolism may greatly affect immunosuppressants, leading to nephrotoxicity or organ failure. The addition of a potent inhibitor may make it possible to decrease the dose of an administered immunosuppressant—and thus reduce the cost of immunosuppressant therapy—but the risks of exploitation of the drug interaction may outweigh the benefits. Most authors recommend serial monitoring of serum immunosuppressant and creatinine levels.

References

Amioka K, Kuzuya T, Kushihara H, et al: Carvedilol increases cyclosporin bioavailability by inhibiting P-glycoprotein-mediated transport. J Pharm Pharmacol 59:1383–1387, 2007

Anglicheau D, Pallet N, Rabant M, et al: Role of P-glycoprotein in cyclosporine cytotoxicity in the cyclosporine-sirolimus interaction. Kidney Int 70:1019–1025, 2006

Arima H, Yunomae K, Hirayama F, et al: Contribution of P-glycoprotein to the enhancing effects of dimethyl-beta-cyclodextrin on oral bioavailability of tacrolimus. J Pharmacol Exp Ther 297:547–555, 2001

Boni J, Leister C, Burns J, et al: Pharmacokinetic profile of temsirolimus with concomitant administration of cytochrome p450-inducing medications. J Clin Pharmacol 47:1430–1439, 2007

Butani L, Berg G, Makker SP: Effect of felodipine on tacrolimus pharmacokinetics in a renal transplant recipient. Transplantation 73:159–160, 2002

CellCept (package insert). Nutley, NJ, Roche Laboratories, 2000

Fahr A: Cyclosporin clinical pharmacokinetics. Clin Pharmacokinet 24:472–495, 1993

Fukatsu S, Fukudo M, Masuda S, et al: Delayed effect of grapefruit juice on pharmacokinetics and pharmacodynamics of tacrolimus in a living-donor transplant recipient. Drug Metab Pharmacokinet 21:122–125, 2006

Gambertoglia JG, Frey FJ, Holferd NH, et al: Prednisone and prednisolone bioavailability in renal transplant patients. Kidney Int 21:621–626, 1982

Gardiner SJ, Begg EJ: Pharmacogenetics, drug-metabolizing enzymes, and clinical practice. Pharmacol Rev 58:521–590, 2006

Gornet JM, Lokiec F, Duclos-Vallee JC, et al: Severe CPT-11 induced diarrhea in the presence of FK-506 following liver transplantation for hepatocellular injury. Anticancer Res 21:4203–4206, 2001

Haagsma CJ: Clinically important drug interactions with disease-modifying antirheumatic drugs. Drugs Aging 13:281–289, 1998

Hara Y, Nakajima M, Miyamoto K, et al: Morphine glucuronosyltransferase activity in human liver microsomes is inhibited by a variety of drugs that are co-administered with morphine. Drug Metab Pharmacokinet 22:103–112, 2007

Homma M, Itagaki F, Yuzawa K, et al: Effects of lansoprazole and rabeprazole on tacrolimus blood concentrations: case of a renal transplant recipient with CYP2C19 gene mutation. Transplantation 73:303–304, 2002

Horton RC, Bonser RS: Interaction between cyclosporin and fluoxetine. BMJ 311:422, 1995

Hu YF, Qiu W, Liu ZQ, et al: Effects of genetic polymorphisms of CYP3A4, CYP3A5, and MDR1 on cyclosporine pharmacokinetics after renal transplantation. Clin Exp Pharmacol Physiol 33:1093–1098, 2006

Imani S, Jusko WJ, Steiner R: Diltiazem retards the metabolism of oral prednisone with effects on T-cell markers. Pediatr Transplant 3:126–130, 1999

Inoue K, Miura M, Satoh S, et al: Influence of UGT1A7 and UGT1A9 intronic I399 genetic polymorphisms on mycophenolic acid pharmacokinetics in Japanese renal transplant recipients. Ther Drug Monit 29:299–304, 2007

Kagaya H, Inoue K, Miura M, et al: Influence of UGT1A8 and UGT2B7 genetic polymorphisms on mycophenolic acid pharmacokinetics in Japanese renal transplant recipients. Eur J Clin Pharmacol 63:279–288, 2007

Kajosaari LI, Niemi M, Neuvonen M, et al: Cyclosporine markedly raises the plasma concentrations of repaglinide. Clin Pharmacol Ther 78:388–399, 2005

Kotlyar M, Brewer ER, Golding M, et al: Nefazodone inhibits methylprednisolone disposition and enhances adrenal-suppressant effect. J Clin Psychopharmacol 23:652–656, 2003

Krishna R, Bergman A, Larson P, et al: Effect of a single cyclosporine dose on the single-dose pharmacokinetics of sitagliptin (MK-0431), a dipeptidyl peptidase-4 inhibitor, in healthy male subjects. J Clin Pharmacol 47:165–174, 2007

Lennard L, Van Loon JA, Weinshilboum RM: Pharmacogenetics of acute azathioprine toxicity: relationship to thiopurine methyltransferase genetic polymorphism. Clin Pharmacol Ther 46:149–154, 1989

Lin Y, Anderson GD, Kantor E, et al: Differences in the urinary excretion of 6-beta-hydroxycortisol/cortisol between Asian and Caucasian women. J Clin Pharmacol 39:578–582, 1999

Lo A, Burckart GJ: P-glycoprotein and drug therapy in organ transplantation. J Clin Pharmacol 39:995–1005, 1999

Miura M, Inoue K, Kagaya H, et al: Influence of rabeprazole and lansoprazole on the pharmacokinetics of tacrolimus in relation to CYP2C19, CYP3A5 and MDR1 polymorphisms in renal transplant recipients. Biopharm Drug Dispos 28:167–175, 2007

Moore LW, Alloway RR, Acchiardo SR, et al: Clinical observations of metabolic changes occurring in renal transplant recipients receiving ketoconazole. Transplantation 61:537–541, 1996

Naesens M, Kuypers DR, Streit F, et al: Rifampin induces alterations in mycophenolic acid glucuronidation and elimination: implications for drug exposure in renal allograft recipients. Clin Pharmacol Ther 80:509–521, 2006

Naito M, Oh-hara T, Yamazaki A, et al: Reversal of multidrug resistance by an immunosuppressive agent FK-506. Cancer Chemother Pharmacol 29:195–200, 1992

Niwa T, Yamamoto S, Saito M et al: Effect of cyclosporine and tacrolimus on cytochrome p450 activities in human liver microsomes. Yakugaku Zasshi 127:209–216, 2007

Novantrone (package insert). Seattle, WA, Immunex Corp., 2001

Op den Buijsch RA, Christiaans MH, Stolk LM, et al: Tacrolimus pharmacokinetics and pharmacogenetics: influence of adenosis triphosphate-binding cassette B1 (ABCB1) and cytochrome (CYP) 3A polymorphisms. Fundam Clin Pharmacol 21:427–435, 2007

Orthoclone OKT3 (package insert). Raritan, NJ, Ortho Biotech Products, 2001

Pawarode A, Shukla S, Minderman H, et al: Differential effects of the immunosuppressive agents cyclosporine A, tacrolimus and sirolimus on drug transport by multidrug resistant proteins. Cancer Chemother Pharmacol 60:179–188, 2007

Picard N, Djebli N, Sauvage FL, et al: Metabolism of sirolimus in the presence or absence of cyclosporine by genotyped human liver microsomes and recombinant cytochromes P4503A4 and 3A5. Drug Metab Dispos 35:350–355, 2007

Prograf (package insert). Deerfield, IL, Fujisawa Healthcare Inc., 1998

Provigil (package insert). West Chester, PA, Cephalon, Inc., 1999

Rapamune (package insert). Philadelphia, PA, Wyeth-Ayerst Pharmaceuticals Inc., 2002

Richard VS, Al-Ismail D, Salamat A: Should we test TPMT enzyme levels before starting azathioprine? Hematology 12:359–360, 2007

Ruschitzka F, Meier PJ, Turina M, et al: Acute heart transplant rejection due to Saint John's wort. Lancet 355:548–549, 2000

Sandimmune (package insert). East Hanover, NJ, Novartis Pharmaceuticals Corp., 2001

Simulect (package insert). East Hanover, NJ, Novartis Pharmaceuticals Corp., 2001

Strassburg CP, Barut A, Obermayer-Straub P, et al: Identification of cyclosporine A and tacrolimus glucuronidation in human liver and the gastrointestinal tract by a differentially expressed UDP-glucuronosyltransferase: UGT2B7. J Hapatol 34:865–872, 2001

Synercid (package insert). Bridgewater, NJ, Aventis Pharmaceuticals Products, Inc., 2000

Tanaka S, Hirano T, Saito T, et al: P-glycoprotein function in peripheral blood mononuclear cells of myasthenia gravis patients treated with tacrolimus. Biol Pharm Bull 30:291–296, 2007

Theis JG, Chan HS, Greenberg ML, et al: Increased systemic toxicity of sarcoma chemotherapy due to combination with the P-glycoprotein inhibitor cyclosporin. Int J Clin Pharmacol Ther 36:61–64, 1998

Vasquez EM, Shin GP, Sifontis N, et al: Concomitant clotrimazole therapy more than doubles the relative oral bioavailability of tacrolimus. Ther Drug Monit 27:587–591, 2005

Vella JP, Sayegh MH: Interactions between cyclosporine and newer antidepressant medications. Am J Kidney Dis 31:320–323, 1998

Wassner SJ, Malekzadeh MH, Dennis AJ, et al: Allograft survival in patients receiving anticonvulsant medications. Clin Nephrol 8:293–297, 1977

Zenapax (package insert). Nutley, NJ, Roche Laboratories, 1999

Zhou S, Chan E, Pan SQ, et al: Pharmacokinetic interactions of drugs with St. John's wort. J Psychopharmacol 18:262–276, 2004

PART IV

Practical Matters

Kelly L. Cozza, M.D., F.A.P.M., F.A.P.A.
Editor

21

Guidelines

Kelly L. Cozza, M.D., F.A.P.M., F.A.P.A.

This chapter offers guidelines for clinical reference in regard to three key issues: recognizing the basic patterns of drug interactions (Table 21–1), prescribing within a polypharmacy environment (Table 21–2), and assessing and managing drug interactions when they occur (Table 21–3).

Guidelines for Categorizing Drug Interactions

There are six easy-to-identify pharmacokinetic drug interaction patterns. Carefully reviewing the first section of this text has laid the foundation for understanding these principles. Table 21–1 reviews those concepts and presents them in an easy-to-remember format.

TABLE 21–1. Six patterns of pharmacokinetic drug interactions

Pattern 1: Inhibitor added to a substrate

This pattern generally results in increased substrate levels. If the substrate has a narrow margin of safety, toxicity may result unless care is exercised (such as close monitoring of blood levels and/or decreasing of substrate drug doses in anticipation of the interaction).

Example: Paroxetine (2D6 inhibitor) added to nortriptyline (2D6 substrate with a narrow therapeutic window), leading to tricyclic toxicity

Corollary: If a drug or substrate must be activated at a particular enzyme (i.e., is a pro-drug) and the necessary enzyme is inhibited, there will be a loss of efficacy (but no toxicity), because the active metabolite will not be produced (e.g., tramadol [pro-drug requiring activation at 2D6] administered with paroxetine [2D6 inhibitor]).

If the substrate compound has a narrow margin of safety and accumulates, toxicity may result (e.g., terfenadine [cardiotoxic 3A4 substrate pro-drug] administered with ketoconazole [3A4 inhibitor], leading to terfenadine's being unable to form the noncardiotoxic yet active fexofenadine).

Pattern 2: Substrate added to an inhibitor

This pattern may cause difficulties if the substrate has a narrow margin of safety and is titrated according to preset guidelines that do not take into account the presence of an inhibitor. However, if the substrate is titrated to specific blood levels or to therapeutic effect, or with appreciation that an inhibitor is present, toxicity is less likely.

Example: Nortriptyline (2D6 substrate) added to quinidine (2D6 inhibitor) rapidly (i.e., not slowly and at low doses, or without checking levels), leading to nortriptyline toxicity.

Pattern 3: Inducer added to a substrate

This pattern generally results in decreased substrate levels after 7–10 days; the decrease may lead to a loss of substrate efficacy unless levels are monitored and/or substrate doses are increased in anticipation of the interaction.

Example: Carbamazepine (3A4 inducer) added to oral contraceptives or cyclosporine (3A4 substrates), leading to contraceptive or organ transplant failure

Corollary: If an inducer is added to a compound with a more active or toxic metabolite, toxicity rather than loss of efficacy may result (e.g., acetaminophen is metabolized at 2E1 to a hepatotoxic metabolite that normally is detoxified by the liver; but if acute acetaminophen overdose is associated with chronic ethanol use or with use of another "pan-inducer," hepatotoxic effects are produced).

TABLE 21–1. Six patterns of pharmacokinetic drug interactions *(continued)*

Pattern 4: Substrate added to an inducer

This pattern may lead to ineffective dosing if preset dosing guidelines are followed that do not take into account the presence of a chronic inducer. If the substrate is titrated to a therapeutic blood level or to clinical effect, or with an appreciation that an inducer is present, dosing is more likely to be effective.

Example: Cyclosporine (3A4 substrate) added to St. John's wort (3A4 inducer), leading to subtherapeutic cyclosporine levels

Pattern 5: Reversal of inhibition

A substrate and an inhibitor are coadministered, equilibrium is achieved, and then treatment with the inhibitor is discontinued. This leads to a resumption of normal enzyme function and immediately results in lower levels of substrate, increased metabolite formation, and, possibly, loss of efficacy of the substrate.

Example: Cimetidine (pan-inhibitor; enzymes inhibited include 2D6) and nortriptyline (2D6 substrate) are simultaneously administered to good effect. When treatment with cimetidine is abruptly discontinued, subtherapeutic nortriptyline levels and loss of efficacy result.

Pattern 6: Reversal of induction

A substrate and an inducer are simultaneously administered, equilibrium is achieved, and then treatment with the inducer is abruptly discontinued. This gradually results (over 2–3 weeks) in decreased amounts of available enzyme, leading to increased substrate levels, decreased metabolite formation, slower metabolism, and possibly substrate toxicity.

Example: A chronic smoker (1A2 inducer in smoke) with therapeutic clozapine levels (1A2 substrate) stops smoking abruptly (during a hospitalization). Several weeks later, clozapine levels increase to the point of toxicity.

Source. Adapted from Sandson NB: *Drug-Drug Interaction Primer: A Compendium of Case Vignettes for the Practicing Clinician.* Washington, DC, American Psychiatric Publishing, 2007. Used with permission.

Guidelines for Prescribing in a Polypharmacy Environment

Given the modern polypharmacy environment, clinicians may feel hindered in making prescribing decisions. Because there are so many interactions and potential interactions to watch for, clinicians often feel "frozen" when facing polypharmacy. The use of multiple physicians and pharmacies by patients increases the risk of insufficient clinical coordination of complicated medication regimens. The five basic principles or guidelines shown in Table 21–2 can help prevent the occurrence of most metabolism-mediated drug interactions.

TABLE 21–2.   Guidelines for prescribing in a polypharmacy environment

If at all clinically possible:

1.   **Avoid** prescribing medications that either significantly inhibit or significantly induce enzymes.

2.   **Prescribe** medications that are eliminated by multiple pathways (phase I metabolism, phase II metabolism, and/or renal excretion).

   Principles 1 and 2 should be weighed concurrently. By following the first principle, a clinician can avoid trouble. The serum levels of medications a patient is already taking will not be altered appreciably by the newly introduced medication, and if the patient is later prescribed a new medication by another physician, the prescribing clinician can feel relatively confident that potential drug interactions are less likely.

   Similarly, if a clinician follows Principle 2 by prescribing a medication that is eliminated by multiple pathways, drugs later introduced—drugs with the potential to inhibit or induce one or two enzymes—will not cause as much of a problem with the original drug. In addition, phenotypic variability may be less of an issue; if the patient lacks full activity of one enzyme, other enzymes will help clear the drug.

3.   **Prescribe** medications that do not have serious consequences if their metabolism is prolonged (as may occur with concomitant use of a substrate inhibitor or if the patient being treated is a phenotypic poor metabolizer).

TABLE 21–2. Guidelines for prescribing in a polypharmacy environment *(continued)*

Some psychotropic drugs have narrow margins of safety. Pimozide, clozapine, tricyclic antidepressants, many typical antipsychotics, and some antiseizure medications may have serious untoward effects if their metabolisms are significantly prolonged by substrate inhibitors or because of phenotypic variability. Some of these serious effects may include cardiac toxicity, seizures, extreme sedation, and extrapyramidal symptoms.

4. **Monitor** serum drug levels often if you are concerned about potential metabolism-mediated interactions.

   Although it is not necessary to obtain serum drug levels at all times, drug level monitoring can be of great assistance when multiple drugs are being administered. Some examples of the practical use of serum drug levels are given in the study cases in Part II of this volume.

5. **Remind** patients to tell you when other physicians prescribe medications for them. Educating patients about the possibilities of drug interactions often reduces the risk of such interactions. *All patients receiving medications with narrow margins of safety or severe toxicities should be given this instruction.* A prescribing clinician can keep up with possible interactions by asking patients to contact him or her any time a new medication has been prescribed by another clinician. Patients appreciate being given the responsibility of monitoring the effects of their medications (with their providers' assistance), and side effects, toxicities, and potentially fatal outcomes have been prevented many, many times with this approach.

Two other issues important to consider in making prescribing decisions:

1. Metabolism by P450 enzymes does not always inactivate compounds. Some drugs are pro-drugs, meaning they are activated by enzymatic action to active or more active compounds.

   Tramadol and some alkylating agents are examples of such medications. Additionally, some drugs, such as bupropion, haloperidol, fluoxetine, and risperidone, are metabolized to equally effective pharmacologically active metabolites.

2. Generally, the older the drug (particularly if the drug was released in the United States before 1990), the less is known about its metabolism.

   Older information on drug-drug interactions is often based on case reports, whereas newer drugs have been better profiled by the manufacturers to improve their chances of approval by the U.S. Food and Drug Administration. Much of this change occurred after unexpected interactions associated with fluoxetine and with nonsedating antihistamines were reported in the 1980s and 1990s.

Algorithm for Assessing and Managing Drug Interactions

Polypharmacy cannot be avoided. It is important to remain watchful and to understand the steps for identification and management of drug interactions should they arise. A useful algorithm for attending to drug interactions is found in Table 21–3.

TABLE 21–3.   Algorithm for assessing and managing drug interactions

1. **Identify interaction**

    Consult pertinent resources: Resources include literature identified through PubMed or other citation databases, drug interaction Web sites, textbooks or newsletters, commercial drug interaction software programs, and abstracts from conferences. Important information may be obtained from researchers or from the manufacturer directly.

    Anticipate or predict likely interactions: On the basis of the pharmacology and pharmacokinetics of the suspected medications, would you anticipate a potential interaction? (For example, are the drugs metabolized by the same subset of P450 enzymes? Do they have enzyme-inhibiting or -inducing properties? Is drug absorption pH dependent on or susceptible to cation binding?)

2. **Verify existence of interaction**

    In the literature:

    How was the interaction described? Was it noted retrospectively, in a study of a single case, an *in vitro* study, preclinical testing, or a controlled pharmacokinetic human study? In healthy volunteers or in a target patient population?

    Can the data be applied to your patient population? Was the study population similar to your own? What were the doses and duration of the agent used? Did the subjects have coexisting disease states? Were the subjects taking concomitant medications?

    In a clinical situation:

    What is the time course of the interaction? How long will it take for the interaction to develop? What clinical consequences do you expect to see? Has the interaction already occurred?

    Do the **clinical signs and symptoms** support your assumptions?

    Is the **objective evidence,** such as drug concentrations, available?

    Have **other potentiating factors** been ruled out?

TABLE 21–3. Algorithm for assessing and managing drug interactions *(continued)*

3. Assess clinical significance of interaction

Do the agents involved have a narrow therapeutic index? Are the drugs associated with dose-related efficacy or toxicity?

Is there risk of therapeutic failure or of development of resistance?

4. Evaluate available therapeutic alternatives

Space doses: Can doses be spaced in a practical and/or convenient way for the patient?

Increase dose: Is increasing the dose affordable? Are appropriate dosage forms available?

Decrease dose: Are appropriate dosage forms available?

Discontinue administration of one drug: What are the therapeutic consequences of temporarily or permanently discontinuing treatment with one agent?

Change agent: What are the comparative efficacy, adverse effects, cost, availability, compliance issues, and drug interactions associated with a new agent?

Add another agent to counteract effect of interaction: What are the comparative efficacy, adverse effects, cost, availability, compliance issues, and drug interactions associated with a new agent?

Take no action: In certain situations (e.g., when the likelihood of an interaction occurring is low or when the clinical effect of a potential interaction would be minor or insignificant), the practitioner may want to maintain the patient's current regimen and monitor the patient's condition. Should evidence of a clinically significant interaction be detected, one of the above-mentioned management options may then be considered.

Source. Adapted from Tseng AL, Foisy MM: "Management of Drug Interactions in Patients With HIV." *Annals of Pharmacotherapy* 31:1040–1058, 1997. Used with permission.

22

Medicolegal Implications of Drug-Drug Interactions

David M. Benedek, M.D., D.F.A.P.A.

Ethical principles of beneficence and nonmalfeasance dictate that physicians keep abreast of research on drug-drug interactions so that they can provide the best care for their patients. Moreover, legally defined standards of care necessitate that physicians not only incorporate this knowledge into their clinical practice but also share this information with patients, so that patient and clinician can collaborate in decision making (Armstrong et al. 2002).

In this chapter, legal standards and landmark medicolegal cases defining the scope of these responsibilities are reviewed. The extent to which drug interactions have been implicated in malpractice cases and other forensic arenas is explored. Finally, guidelines for risk management are outlined, and case vignettes illustrate the potential for application of these recommendations.

Medication Therapy and Malpractice

A malpractice suit is an action in tort: a request for compensation for damage to one party caused by the tortious (noncriminal) act or omission of another. Tradition, common law, and case law have upheld the idea that a physician, by virtue of his or her fiduciary relationship with the patient, has special duties—beyond those reasonably expected between other members of society—to prevent harm. In a malpractice suit, a plaintiff must establish, by a preponderance of the evidence, dereliction (by act or omission) of this special duty—dereliction resulting in physical or emotional damage. Notably, the standard of proof (often defined as 51% of the evidence) is markedly lower than that necessary for determination of criminal culpability. Malpractice suits may result from various aspects of care, ranging from misdiagnosis to inadequate performance of surgical procedures to failure to establish or provide adequate follow-up. However, negligent use of medications (excessive or inadequate dosing, failure to monitor levels, failure to select appropriate medications) and failure to obtain appropriate informed consent are causes of action that may result from a physician's lack of awareness of drug interactions or failure to address this potential in treatment decisions.

Legal Principles and Landmark Cases

The courts determine the degree to which a physician's care represents appropriate treatment versus dereliction of duty by considering the standard of care as it applies to particular practices (acts or omissions). Exact definitions vary from jurisdiction to jurisdiction, but this benchmark is generally regarded as "the custom of the profession, as reflected in the standard of care common to other professionals of the practitioner's training and theoretical orientation" (Gutheil and Appelbaum 2000, p. 142). The courts most commonly rely on medical expert testimony to determine whether a specific practice represents a deviation from the standard of care. However, if a clinician chooses a medication for an unapproved use, prescribes medication in doses beyond those recommended, or fails to monitor levels as specifically required by the U.S. Food and Drug Administration or recommended by the manufacturer, these actions are viewed as prima facie evidence of negligence (under the legal principle of *res ipsa loquitur*—"the thing speaks for itself"). The burden of proof in such circumstances shifts from the plaintiff to the physician, who must now establish that his or her care was not negligent. In cases in which drug interactions are known to result in inhibition of a specific agent's metabolism, prescribing a medication in otherwise appropriate doses might be viewed by the court as excessive use. Failure to monitor drug levels appropriately (when therapeutic and toxic levels have been established) would similarly represent negligent care if an adverse drug reaction resulted from drug toxicity.

Cases involving death or permanent injury have resulted in multimillion-dollar awards for plaintiffs even when both agents involved in the interaction were prescribed within recommended dosing guidelines. In 1999, for example, the Oregon Court of Appeals awarded $23 million (in a case against a physician and a pharmaceutical company) when permanent brain damage resulted from seizures precipitated by theophylline toxicity 1 week after appropriately dosed ciprofloxacin was added to the patient's chronic asthma treatment (*Bocci v. Key Pharmaceuticals* 1999). Ciprofloxacin potently inhibits 1A2, an enzyme on which theophylline depends for clearance.

Legal decisions have also established more precisely the nature of the information that a physician must provide to the patient in obtaining informed consent for treatment (including medication management). If treatment is initiated without voluntary, competent, and knowing (informed) consent, this care would constitute an unconsented touching or battery as defined by common law, even if the intent was benevolent. This idea reflects the degree to which law attempts to protect bodily autonomy as a part of the fundamental right to liberty. In *Natanson v. Kline* (1960), the Kansas Supreme Court opined that "the physician should explain the nature of the ailment, the nature of the proposed treatment, the probability of success or alternatives, and perhaps, the risks of unfortunate results" (p. 1093) in response to a complaint in which the patient alleged negligent care after radiation therapy following radical mastectomy. The court was not sympathetic to Dr. Kline's argument that the prevailing standard of care did not require the provision of any information about the significant risks associated with the procedure.

Decisions in other jurisdictions have further amplified the need to inform patients of risk potential. After a plaintiff sought damages when a surgeon failed to inform him of the 1% risk of paraplegia associated with an orthopedic procedure, the U.S. Court of Appeals for the District of Columbia held that the scope of a physician's communications regarding potential risks must be "measured by the patient's need" (*Canterbury v. Spence* 1972, p. 773). This opinion, and subsequent decisions in other jurisdictions, established the idea that the patient must be informed of the material risks of treatment: those risks that are most common, and those that are most severe. In a 1996 case, after a subtotal thyroidectomy, a patient with Graves' disease received calcitriol (Rocaltrol; a synthetic vitamin D analog) and calcium carbonate (Os-Cal) for postoperative hypoparathyroidism and hypocalcemia. Shortly thereafter, he presented to the emergency room with decreased appetite, nausea, emesis, headache, ataxia, weakness, and lethargy. Then he experienced seizures, respiratory arrest, and loss of consciousness. In the subsequent malpractice suit, he alleged that he was improperly advised of his potential to develop hypocalcemia or early warning signs of hypocalcemia on this regimen and that his physician failed to monitor calcium levels properly. He was awarded more than $8 million in damages (*Brown v. Pepe et al.* 1996). Because patients must be warned about potentially severe reactions to their medications, informed consent for treatment with an agent known to interact adversely with another of the patient's medications must include provision of information about the risks of this interaction, if the adverse interaction is either common or severe. Multiple jurisdictions have upheld the principle that whereas others must exercise due care and diligence in the delivery of medications, the responsibility for warnings and informed consent rests primarily with the physician (*Kintigh v. Abbott Pharmacy* 1993; *Leesley v. West* 1988).

Drug Interactions, Adverse Outcomes, and Malpractice: Scope of the Problem

If physical or emotional damage results directly from a drug interaction that a physician knew or should have known about, or if an interaction-related injury was a material risk about which the patient should have been informed during the informed consent process, malpractice actions may ensue. The extent to which this type of malpractice occurs is not easy to establish. However, recent studies have shed light on the scope of these problems.

In August 2001, Kelly reviewed case reports of adverse drug reactions published in *Clin-Alert,* an established compendium of adverse drug reactions reported in medical, legal, and pharmacy journals. He found that 9% of the

1,520 reported events over a 20-year period resulted in a fatal outcome, a life-threatening event, or permanent disability. Although causality was not clearly established in all cases, and adverse drug reactions and allergies were the reported causes for the majority of adverse events, nearly one-third of the adverse drug reactions resulted from medication errors (dosing or administration) or drug interactions. Thirteen percent of the adverse drug reactions resulted in financial penalties (either via settlement or verdict) against health care providers, and the mean award amount was $3,127,890 (Kelly 2001). Data on awards stemming directly from drug-drug interactions were not reported. However, reasons for lawsuits were noted generally as overdoses, poor or no monitoring, or improper treatment (Kelly 2001). Drug interactions may figure into each of these reasons, because interactions affect dosing and monitoring. In cases in which there is known to be a high risk of a drug-drug interaction, even the choice of a medication that the U.S. Food and Drug Administration approved for use in a specific illness could constitute improper treatment (when more reasonable alternatives are available).

Actual medicolegal significance may be considerably greater than that indicated by the study. Kelly (2001) noted that *Clin-Alert* abstracted adverse drug reactions from only one legal journal, and it was not necessarily clear whether a particular lawsuit evolved from cases reported in the medical or the pharmacological journals. In addition, clinicians have noted experience in cases involving drug-interaction damage awards or settlements that have not been reported in a published format (R.C. Hall, personal communication, March 2001; R.S. Simon, personal communication, May 2001).

Preskorn (2002) reported on case series and case reports of deaths presumed to be due to intentional overdoses of psychotropic medications. He noted that in some of these cases, drug-drug interactions with other prescribed medications appeared to have resulted in truly accidental deaths. In other cases, the drug-drug interaction—rather than the toxic effects of the psychotropic agent intentionally taken in excess—actually caused the (intended) death by suicide. He concluded that in such cases, although the outcome is recognized as a suicide, the extent to which a drug-drug interaction contributed to the outcome may be underappreciated. The problem has also been increasingly recognized in international arenas (Yeom et al. 2005).

Although courts have traditionally been hesitant to recognize legal duties of pharmacists beyond the duty to dispense medications accurately, recent cases have established the principle that pharmacists may share some responsibility for informing a patient of potential interactions, particularly if pharmacists voluntarily assume such a burden by advertising a capability to detect them (Cacciatore 1996; Termini 1998). Nurses have been named as codefendants in cases involving failure to monitor drug levels with medications known to exhibit considerable drug-drug or drug-dietary interactions (Morris 2002). Nonetheless, the deference the courts have shown to the physician-patient relationship indicates that even in these circumstances, physicians will continue to bear much of the burden for knowing about potential drug interactions, avoiding them when prudent, or informing patients about them.

When a psychiatrist supervises another therapist and provides active guidance and professional direction to treatment, the doctrine of *"respondeat superior"*—let the master answer for the deeds of his servant—provides the legal basis for the assumption of liability by the supervisor in malpractice cases involving the clinician whom he or she supervises. This circumstance must be distinguished from the case where a psychiatrist provides a consultation to another clinician. Here, the latter may accept or reject the advice given and the psychiatrist is not legally considered to have assumed care for the patient. Most commonly, the relationship between the psychiatrist and the nonpsychiatrist clinician or therapist takes the form of ongoing collaboration. Here, professional liability is shared within the professional qualifications and limitations of the respective collaborators. In such cases American Psychiatric Association professional guidelines dictate that the psychiatrist (and collaborator) must periodically reassess the appropriateness of continued collaborative care and inform the patient if the collaboration is terminated (American Psychiatric Association 1980). Clear communication not only with the patient but between collaborators may prevent contradictory guidance or the unwitting prescription of interactive medications or dietary items (Winkelaar 1999).

Reducing Liability Risk

Clearly, the physician must keep abreast of data regarding dangerous drug interactions. Reviewing medical journals and bulletins for reports of drug-drug interactions is one way to keep pace with the knowledge that eventually shapes the standard of care. Taking part in continuing medical education programs that address these issues is certainly another. Technological advances such as computerized order entry and medication screening are gaining increased recognition as means to reduce medication errors, and it is likely they will soon represent the standard of care in hospitals. Pharmacy literature has advocated for such institutional changes for years (Cacciatore 1996). For individual providers, low-cost and free computer programs are available that can assist in rapidly establishing patient medication profiles for known interactions. The software includes capability for frequent Internet updates. Tens of thousands of physicians are availing themselves of these programs, but the extent to which appropriate use of such software may become the standard of care is an area provoking considerable speculation.

Sharing this increased knowledge of drug interactions with the patient when obtaining informed consent, and documenting such discussions in the treatment record in informed consent notes, will reduce the likelihood of liability in the event of an adverse drug interaction. Moreover, by virtue of the thorough review of medications these discussions entail, the discussions may actually modify treatment decisions and therefore decrease the potential for such interactions. Discussions with patients about the possibility that new data may emerge regarding the safety of potential medications, as well as reminders to consult clinicians and pharmacists before medication changes or use of over-the-counter medications, furthers patient confidence and trust in the care providers and decreases the risk of litigation when adverse events occur. Collaboration between the psychiatrist and other members of the patient's treatment team (e.g., psychotherapist, general practice physician) is also vital. While the advent of electronic medical records facilitates the sharing of knowledge between clinicians, the existence of such records may also increase legal expectations regarding what a competent psychiatrist *should* know about the care provided by other professionals. Encouraging patients to inform other physicians of any changes in their medication regime and to allow the sharing of necessary medical records among professionals involved in their care promotes treatment-team awareness of potentially dangerous drug interactions that could emerge as a result of prescriptions from different physicians.

The active development of an environment of trust and collaboration may reduce the patient's propensity for litigation even when adverse events result from clinician error. Finally, because patients with psychiatric illness are often at increased risk for suicide attempts, medication decisions must take into account not only the safety in overdose of the prescribed psychotropic agent, but also the potential for a drug-drug interaction resulting in death or disability when an otherwise "safe" psychotropic medication is taken in intentional overdose.

Case Examples

The following vignettes have been adapted from cases with which the author is familiar. The cases have been sufficiently altered to disguise facts and identities without altering the salient points with regard to medicolegal principles or risk management.

Case 1: Loss of the "Therapeutic Benefit" of a Drug-Drug Interaction

Mr. A, a 58-year-old man with chronic knee pain, was referred to a psychiatrist by his family physician after showing no improvement in depressed mood, irritability, and somatic complaints despite treatment with fluoxetine (Prozac) titrated to 80 mg/day over 8 weeks. When the psychiatrist obtained a history of worsening insomnia, increased migraine headaches, and gastrointestinal disturbance (heartburn), she recommended decreasing the fluoxetine dose to 40 mg/day to reduce activation and gastrointestinal side effects, adding amitriptyline (Elavil) 50 mg at night (to augment the selective serotonin reuptake inhibitor [SSRI], improve sleep, and potentially reduce the knee pain component of the patient's depressive syndrome), and adding cimetidine (Tagamet), 300 mg four times a day, to alleviate the patient's heartburn. The family physician followed these recommendations.

Three weeks later, Mr. A informed his family physician that he was indeed feeling significantly improved. His mood had lifted, his sleep had returned to normal, and his gastrointestinal symptoms had fully resolved. The family physician continued treatment with amitriptyline and fluoxetine but discontinued cimetidine therapy. Three weeks later, Mr. A was found at home dead. Autopsy revealed a toxic amitriptyline level, and the manner of death was considered suicide by overdose.

Mr. A's wife sued both physicians, alleging failure to monitor tricyclic antidepressant (TCA) levels and failure to warn of the potential interactions between the TCA and the SSRI *as well as the potential interaction between these medications and cimetidine.* In a deposition, an expert in clinical psychopharmacology opined that Mr. A's response to the combination of TCA and SSRI occurred at blood levels higher than expected from the dosing strategy, because of competitive inhibition of their hepatic metabolism by each other and cimetidine. He concluded that when treatment with cimetidine was discontinued, depressive symptoms returned because the antidepressant concentrations returned to subtherapeutic levels. The defendant physicians argued that the relatively low dose of amitriptyline did not necessitate therapeutic drug level monitoring and that fluoxetine levels are not routinely measured in clinical care. Both claimed that they warned the patient about the potential for reoccurrence of depressive symptoms, but both acknowledged that they did not specifically warn the patient that discontinuation of cimetidine therapy could result in a decrease in antidepressant blood levels or efficacy. The case was eventually settled out of court.

This vignette serves to illustrate several important points with regard to the medicolegal significance of drug-drug interactions. Although it is unclear whether the physicians involved recognized the potential "therapeutic benefit" of the cimetidine-TCA-SSRI interaction, available records and the physicians' testimony indicate that when making the decision to discontinue cimetidine therapy, they did not warn the patient about the potential effect of this interaction. Because the duty to obtain informed consent regarding treatment includes a requirement for the physician to provide information concerning the risk, benefits, and alternatives to treatment, it might have been successfully argued that failure to warn of the risks of discontinuing treatment breached this duty. If either physician had discussed the risk of discontinuation of cimetidine with the patient and documented the counseling in the medical records, this argument would be invalidated. The fact that the widow brought suit suggests that she believed that treatment decisions were not made in a collaborative and informed manner. There is no legal obligation to involve other family members in treatment decisions concerning competent patients. However, if patient consent can be obtained, informing family members regarding warning signs for reoccurrence of depression may facilitate return to care before an untoward outcome and may reduce the extent to which grieving family members, feeling isolated from the treatment process, seek to blame physicians in the face of tragedy. Finally, although the cause of death in this case was ruled suicide by intentional overdose, whether the overdose would have proven fatal if it were not for the inhibition of TCA metabolism resulting from the prescribed fluoxetine remains an unanswered question.

Case 2: Treatment Decisions Taking Into Account the Natural Course of Comorbid Conditions

Mrs. B, a 47-year-old woman with a history of alcohol dependence and hypertension, sought additional care for intermittent symptoms of anxiety from her internist at the recommendation of her Alcoholics Anonymous sponsor. Verapamil (Calan), 180 mg twice a day, had provided fair control of her hypertension for more than a year, and she was taking no other medications. Her doctor initiated treatment for panic disorder, prescribing sertraline (Zoloft), 50 mg each morning. Cognizant of the potential for insomnia early in treatment, but cautious because of her history of alcohol dependence, he prescribed a 7-day supply of triazolam (Halcion) 0.25 mg, to be used nightly as needed for sleep if the sertraline resulted in sleep disturbance. Days later, the patient relapsed in her alcohol use and called her sponsor while intoxicated. The sponsor drove to the patient's home and found her unconscious. The sponsor then summoned emergency services and the patient was hospitalized. The diagnosis was acute alcohol and benzodiazepine intoxication that resulted in respiratory compromise and transient coma.

On recovery, the patient sued her physician for failing to warn her about the potential interaction between the triazolobenzodiazepine triazolam and the calcium-channel blocker verapamil. She stated that had she been properly warned of this interaction, she would have continued taking verapamil but would have elected not to take triazolam. Her physician argued that her respiratory compromise and coma resulted from her voluntary alcohol ingestion rather than from the medications he prescribed, but the patient's attorney countered that the physician was aware of her alcohol dependence. Indeed, medical records indicated that the internist's decision to prescribe only a minimal quantity of triazolam supported the notion that he was aware of the patient's potential for addiction, and the court ruled in favor of the patient.

Although hepatic inhibition of 3A4 by verapamil contributed to the increased central nervous system toxicity of the triazolam and the subsequent respiratory collapse, so too did the patient's use of alcohol. The court, however, determined that the preponderance of evidence supported the patient's contention that she was improperly warned about the potential interactions between triazolam and verapamil and that she would not have taken the benzodiazepine if warned about this potentially serious interaction. Although the court recognized that the patient's own behavior contributed to the outcome, it apportioned a degree of responsibility to the physician in awarding damages. This vignette also illustrates the idea that when making medication decisions, physicians must remain aware of their patients' other illnesses (in Mrs. B's case, alcohol dependence), associated behavior (tendency toward relapse), and the potential effect of this behavior on drug-drug interactions. In this case, knowledge of the patient's potential for misuse of medications (given her history of alcohol dependence) and for use of medications in combination with alcohol necessitated consideration of alternative treatment choices for antidepressant-associated insomnia.

Summary

Keeping abreast of information regarding drug interactions poses significant challenges for the physician: new medications are rapidly introduced, and drug interactions are reported in a variety of forums, including legal journals, pharmacy journals, and pharmacological journals—sources to which clinicians do not frequently refer. Nonetheless, the law has established standards of physician conduct that necessitate assimilating this information into therapeutic decisions. Attention to these issues as they are reported in the medical literature, participation in relevant continuing medical education programs, use of technology that aids in identifying potential interactions, and respect for the patient's need to be aware of material risks when consenting to treatment are means by which the physician may reduce the risk of drug interaction–related liability.

References

American Psychiatric Association: American Psychiatric Association guidelines for psychiatrists in consultative, supervisory, or collaborative relationships with nonmedical therapists. Am J Psychiatry 137:1489–1491, 1980

Armstrong SC, Cozza KL, Benedek DM: Med-psych drug-drug interactions update. Psychosomatics 43:245–247, 2002

Cacciatore GG: Advertising, computers, and pharmacy liability: a Michigan court's decision has ramifications for pharmaceutical care. J Am Pharm Assoc 36:651–654, 1996

Gutheil TG, Appelbaum PS: Malpractice and other forms of liability, in Clinical Handbook of Psychiatry and the Law, 3rd Edition. Philadelphia, PA, Lippincott Williams & Wilkins, 2000, p 142

Kelly WN: Potential risks and prevention, part 4: reports of significant adverse drug events. Am J Health Syst Pharm 58:1406–1412, 2001

Morris K: What kinds of malpractice cases are happening around the state? Ohio Nurses Rev 77(8):16, 2002

Preskorn SH: Fatal drug-drug interactions as a differential consideration in apparent suicides. J Psychiatr Pract 8:233–238, 2002

Termini RB: The pharmacist duty to warn revisited: the changing role of pharmacy in health care and the resultant impact on the obligation of a pharmacist to warn, 4 Ohio N.U.L. Rev. 551–566, 1998

Winkelaar PG: Medicolegal file: tacit approval of alternative therapy. Can Fam Physician 45:905, 1999

Yeom JA, Park JS, OH O, et al: Computerized screening of DOR in Korea. Ann Pharmacother 39:1918–1923, 2005

Legal Citations

Bocci v Key Pharmaceuticals, 14 Pharmaceutical Litigation Reporter 13–14 (1999)

Brown v Pepe et al, Pharmaceutical Litigation Reporter 11781–11782 (1996)

Canterbury v Spence, 464 F2d 772 (DC Cir 1972)

Kintigh v Abbott Pharmacy, Mich Ct App, 503 NW2d 657, 661 (1993)

Leesley v West, Ill Ct App, 518 NE2d 758 (1988)

Natanson v Kline, 350 P2d 1093 (1960)

23

How to Retrieve and Review the Literature

Kelly L. Cozza, M.D., F.A.P.M., F.A.P.A.

We believe it is important to help readers update this Clinical Manual and maintain provider-specific lists themselves. Keeping abreast of the literature may seem daunting, but as providers in the setting of polypharmacy, physicians must do so.

Searching the Internet

The Internet is the most powerful, speedy, and all-encompassing way to keep up. Ways to interface with the Internet are changing daily. Many journals now provide "citation managers" that alert online subscribers when topics of interest to them are cited. Many of these services carry subscription fees.

Web browsers and Web sites are also evolving. Currently, our favorite search engine is PubMed (http://www.ncbi.nlm.nih.gov/pubmed). PubMed taps directly into MEDLINE. Dr. Oesterheld's Web site http://

mhc.daytondcs.com:8080/cgi-bin/ddiD4?ver=4&task=getDrugList has up-to-date P450, uridine 5′-diphosphate glucuronosyltransferase (UGT), and P-glycoprotein (ABCB1) lists, and the optional drug interaction program is thorough, is updated regularly, and provides individualized interaction checks. This program also makes predictions about potential drug interactions, even if there are no available published reports. This drug interaction tool provides selected references so that the reader can research why an interaction may occur.

Another interactive table is the Cytochrome P450 Drug Interaction Table (http://www.drug-interactions.com). This table is monitored by the University of Indiana Department of Medicine, is updated fairly often, and has hyperlinks to literature sources via PubMed. The table is restricted to P450-mediated interactions but is quite complete. *Physicians' Desk Reference* (2008) is online (http://www.pdr.net) and is a useful, though somewhat static, tool for reviewing drugs available in the United States. The drug interaction findings are limited to what is reported in the *Physicians' Desk Reference,* and drugs that are no longer patented are frequently not included. The U.S. Food and Drug Administration's Web site (http://www.fda.gov) provides the latest information on drugs, safety, and product approvals. Other helpful sites include http://www.medscape.com and http://www.drkoop.com. For HIV medication information, useful sites include http://www.hopkins-hivguide.org and http://www.hiv-druginteractions.org.

These sites have our seal of approval because rather than limiting the prescriber's options, they expand his or her knowledge and prescribing power. We believe that there are currently no personal digital assistant (PDA) programs with the power of these Web sites behind them. Most handheld programs do not provide the reasons or the references necessary to make an educated prescription choice. Many PDA programs simply state that an interaction exists, which may lead the user to *not* prescribe in situations in which it may be helpful to the patient to do so. Drug interaction software frequently leads to "prescriber paralysis" or avoidance of potentially helpful medications.

For a thorough search, we suggest going directly to one of the MEDLINE services mentioned earlier, which permits a review of the original research on one's own. We have found that the best MeSH terms to use to find literature on P450-mediated drug interactions are *pharmacokinetics, drug interaction, cytochrome,* and the names of the drugs being investigated. In a search for information on phase II–related interactions, using the drug name followed by the term *glucuronidation* or *metabolism* works well. Entering *P-glycoprotein, P-gp,* or *ABCB1* and the drug name yields references to literature on interactions involving P-glycoproteins. We search for articles published from 1985 on and in English, broadening the search by extending the year span and accepting publications in other languages only when several initial tries are fruitless or when we are researching an older drug. We restrict our searches to reports on human studies because findings of animal studies often are not translatable to humans or are too preliminary to be useful clinically. Additionally, we often "open up" the search and use large search engines like Google (http://www.google.com), which searches outside of the MEDLINE literature. Unfortunately, when using Google-like searches, you may find an obscure and helpful reference, but you also are led to manufacturer's, distributors', and blog sites. It is interesting to see what the lay population is reading about medications, however.

A "great" search, one that is broad enough in scope without being too large for adequate review, should provide the searcher with 20 to 35 titles. If a searcher retrieves 5 to 10 titles and they are exactly what was sought, that result is exceptional and a time-saver. We start with all the aforementioned terms, and if not enough titles are retrieved, we eliminate one term at a time, typically dropping *cytochrome* first, because few older reports include this word in the title or abstract. Expanding the range of years searched is necessary in the case of some older drugs, such as phenytoin, phenobarbital, and many oncology agents. Literature published before 1980 does not directly reflect an understanding of the P450 system, and the searcher may be left making a few leaps and assumptions. In summary, we like to perform a narrow search first. Beginning with a search that is too broad means wading through many titles, usually more than 35. Reading through titles and abstracts that may not be needed takes time.

Critically Reviewing the Literature

Case Reports

Individual cases are often the first clue to a potential drug interaction, and were particularly so in the days before human liver microsome studies. Although controls are not involved and confounding variables exist, individual cases have great merit, especially if they occur with enough frequency to lead to further study, as occurred with terfenadine. A strong case report includes information on pre- and postadministration serum levels of drugs, with corresponding clinical effects. Too often, serum levels are missing from case reports. In clinical practice, we find it helpful to routinely monitor serum levels of drugs with toxic metabolites or narrow margins of safety, especially before administering a new drug with potential interactions, and to obtain another measurement of levels (at the appropriate time) if a drug interaction is suspected. These data greatly strengthen a case report's credibility. Presentation of a finding of a serum level exceeding clinical standards after administration of a second drug, associated with a significant clinical event, is also common in a strong case report. Monahan et al. (1990), Armstrong and Schweitzer (1997), and Katial et al. (1998) have written excellent case reports involving drug interactions.

Reports of In Vitro Studies

In the United States, drugs must now be tested for P450-mediated interactions in the laboratory. This testing is done with human liver microsome studies, which permit thorough evaluation of the metabolisms of drugs at the level of the hepatocyte and determine which enzymes will be studied and probed in healthy volunteers and patient volunteers. Numerous drugs can be tested by these methods, quickly and without need for subjecting volunteers to drug effects. Zhao et al. (1999) conducted an excellent human liver microsome study. More recently, *Escherichia coli* has been used to manufacture recombinant human enzymes, thus fully automating *in vitro* assays. In this new manner, individual enzymes can be tested with drugs in the laboratory (McGinnity and Riley 2001). Human liver microsome studies and recombinant enzyme studies sometimes provide data that do not translate to clinical populations. This lack of clinical correlation is multifactorial and includes differences in culture concentrations and conditions compared with *in vivo* conditions (Venkatakrishnan et al. 2001).

Reports of In Vivo Studies

Studies involving human volunteers are the gold standard. Studies involving healthy volunteers are usually small ($N \leq 8$–15), randomized, placebo-controlled studies, sometimes crossover in design. These studies are helpful in testing drug interactions predicted by *in vitro* studies. The small size of these studies, the frequent use of the single-dose design, and the use of healthy volunteers may limit their generalizability to the general population. Studies involving actual patients, whose ages, diseases, and drug histories vary widely, are some of the strongest studies, particularly if confounding variables are well controlled. Drug concentration, area under the curve, and maximal drug concentration are typically measured in these studies. When these parameters as well as pharmacodynamic effect (e.g., psychomotor effects in studies of hypnotics) are studied, the clinical information gleaned can be very helpful. For example, Villikka et al. (1997) conducted an *in vivo* study with pharmacodynamic correlation, and Hesslinger et al. (1999) studied potential drug interactions in psychiatrically ill patients receiving multiple agents.

Review Articles

Review articles are necessary compilations of primary research reports. Reviews allow the sorting and evaluating of research necessary to understand a topic as a whole. These articles help introduce readers to an area of research and provide another viewpoint. A drawback of most reviews is the loss of specific data or of finer arguments and points within individual reports. Many reviews cover only a few drugs or a portion of the P450 system, so several reviews are necessary to compile a full picture. We recommend that reviews be used as guideposts in making clinical decisions. We postpone making final clinical decisions in critical cases until we have examined some of the more specific reports cited by the reviewers themselves or found in our own literature searches.

Jefferson and Greist (1996) wrote an excellent introduction to the P450 system. Other good reviews include one by Tredger and Stoll (2002) and a series of reviews for child psychiatrists by Oesterheld and Shader (1998) and Flockhart and Oesterheld (2000). More recently, Sandson at al. (2005) presented a comprehensive review of psychotropic drug interactions. Finch et al. (2002) provided a thorough review of rifampin and its metabolism, especially its induction of most metabolic enzymes. These authors included reviews and discussion of case reports, *in vitro* and *in vivo* data, and transporters across many drug classes. Cardiovascular and gastrointestinal medications have been reviewed in *Psychosomatics* (Williams et al. 2007; Wynn et al. 2007).

Epidemiological Reports

Population studies help guide the need for drug interaction research. Jankel and Fitterman (1993) and Hamilton et al. (1998) provided the big-picture reasoning for drug interaction study. Jankel and Fitterman (1993) found that 3% of hospitalizations each year are due to drug-drug interactions. Hamilton and colleagues (1998) confirmed that use of azole antifungals and rifamycins increases the risk of hospitalization because of these drugs' significant P450-mediated drug interactions. Hellinger et al. (2005) reviewed insurance claims of persons with HIV and reported significant inappropriate combinations of lipid-lowering drugs and antiretroviral therapy that resulted in significantly higher rates of myopathy, polyneuropathy, and myositis.

Manufacturer Package Inserts

Manufacturers are important sources of information about their products. *Physicians' Desk Reference* (2007) contains the texts of the package inserts of all drugs marketed in the United States and is also available online at http://www.pdr.net. Most newer drugs have discrete sections in the manufacturer's insert pertaining to their P450-mediated interactions. We have on many occasions called a manufacturer to obtain postmarketing or not yet published information about its drugs. Discussions with manufacturers' doctorate-level professionals are generally rewarding and informational.

Summary

A thorough review of the literature concerning a particular drug interaction usually involves reading the package insert and then a few review articles, followed by a more focused look at *in vitro* and *in vivo* data. All these research tools are necessary to develop a full understanding of the data available on any drug interaction. Because there are no perfect sources, a compilation similar to this Clinical Manual is what each provider should develop about his or her own frequently prescribed drugs.

References

Armstrong SC, Schweitzer SM: Delirium associated with paroxetine and benztropine combination (letter). Am J Psychiatry 154:581–582, 1997

Finch CK, Chrisman CR, Baciewicz AM, et al: Rifampin and rifabutin drug interactions: an update. Arch Intern Med 162:985–992, 2002

Flockhart DA, Oesterheld JR: Cytochrome P450-mediated drug interactions. Child Adolesc Psychiatr Clin N Am 9:43–76, 2000

Hamilton RA, Briceland LL, Andritz MH: Frequency of hospitalization after exposure to known drug-drug interactions in a Medicaid population. Pharmacotherapy 18:1112–1120, 1998

Hellinger FJ, Encinosa WE: Inappropriate drug combinations among privately insured patients with HIV disease. Med Care 43 (9 suppl):III53–62, 2005

Hesslinger B, Normann C, Langosch JM, et al: Effects of carbamazepine and valproate on haloperidol plasma levels and on psychopathologic outcome in schizophrenic patients. J Clin Psychopharmacol 19:310–315, 1999

Jankel CA, Fitterman LK: Epidemiology of drug-drug interactions as a cause of hospital admissions. Drug Saf 9:51–55, 1993

Jefferson JW, Greist JH: Brussels sprouts and psychopharmacology: understanding the cytochrome P450 enzyme system. Psychiatr Clin North Am 3:205–222, 1996

Katial RK, Stelzle RC, Bonner MW, et al: A drug interaction between zafirlukast and theophylline. Arch Intern Med 158:1713–1715, 1998

McGinnity DF, Riley RJ: Predicting drug pharmacokinetics in humans from in vitro metabolism studies. Biochem Soc Trans 29:135–139, 2001

Monahan BP, Ferguson CL, Killeavy ES, et al: Torsades de pointes occurring in association with terfenadine use. JAMA 264:2788–2790, 1990

Oesterheld JR, Shader RI: Cytochromes: a primer for child psychiatrists. J Am Acad Child Adolesc Psychiatry 37:447–450, 1998

Physicians' Desk Reference, 62nd Edition. Montvale, NJ, Medical Economics, 2008

Sandson NB, Armstrong SC, Cozza KL: An overview of psychotropic drug-drug interactions. Psychosomatics 46:464–494, 2005

Tredger JM, Stoll S: Cytochrome P450: their impact on drug treatment. Hospital Pharmacist 9:167–173, 2002

Venkatakrishnan K, von Moltke LL, Greenblatt DJ: Human drug metabolism and the cytochromes P450: application and relevance of in vitro models. J Clin Pharmacol 41:1149–1179, 2001

Villikka K, Kivisto KT, Luurila H, et al: Rifampin reduces plasma concentrations and effects of zolpidem. Clin Pharmacol Ther 62:629–634, 1997

Williams S, Wynn GH, Cozza KL, et al: Cardiovascular medications. Psychosomatics 48:537–547, 2007

Wynn GH, Sandson NB, Cozza KL: Gastrointestinal medications. Psychosomatics 48:79–85, 2007

Zhao XJ, Koyama E, Ishizaki T: An in vitro study on the metabolism and possible drug interactions of rokitamycin, a macrolide antibiotic, using human liver microsomes. Drug Metab Dispos 27:776–785, 1999

Appendix

Tables of Drug Interaction Pharmacokinetics

Metabolism, Inhibition, and Induction

Justin M. Curley, M.D.

TABLE 1. Internal medicine drug-drug interactions: allergy, cardiology, endocrine, gastrointestinal, infectious disease, pain (narcotics and nonnarcotics), oncology, and transplantation and rheumatology

Drug	Trade name	Metabolism site	Inhibition[a]	Induction[a]
A. Allergy				
Astemizole	Hismanal	3A4	None known	None known
Brompheniramine	—	Unknown	None known	None known
Cetirizine	Zyrtec	Unknown	None known	None known
Chlorpheniramine	Chlor-trimeton	2D6	2D6	None known
Desloratadine	Clarinex	3A4, 2D6	None known	None known
Diphenhydramine	Benadryl	2D6, 1A2, 2C9, 2C19	2D6	None known
Ebastine	Ebastel	3A4	None known	None known
Fexofenadine	Allegra	Excreted unchanged	None known	None known
Hydroxyzine	Atarax	2D6	2D6	None known
Levocetirizine	Xyzal	Excreted unchanged	None known	None known
Loratadine	Claritin, Alavert	3A4, 2D6	None known	None known
Montelukast	Singulair	3A4, 2C9	2C8	None known
Terfenadine	Seldane	3A4	None known	None known
Theophylline	Theo-Dur	1A2, 2E1	3A4	None known
Zafirlukast	Accolate	2C9, 3A4	2C9, 2C8, 3A4, 1A2	None known

TABLE 1. Internal medicine drug-drug interactions: allergy, cardiology, endocrine, gastrointestinal, infectious disease, pain (narcotics and nonnarcotics), oncology, and transplantation and rheumatology *(continued)*

Drug	Trade name	Metabolism site	Inhibition[a]	Induction[a]
B. Cardiology				
Amiodarone	Cordarone	2C8, 3A4	1A2, 2C9, 2D6, 3A4	None known
Amlodipine	Norvasc	3A4	None known	None known
Argatroban	Argatroban	3A4	None known	None known
Atorvastatin	Lipitor	3A4	2C9, 3A4	None known
Captopril	Capoten	Unknown	None known	None known
Carvedilol	Coreg	2D6, 1A2, 2C9, 2C19, 2E1, 3A4	None known	None known
Cilostazol	Pletal	3A4, 2C19	None known	None known
Clopidogrel	Plavix	3A4, 1A2, 2C9, 2C19	**2B6**, 2C19[b], 2C9[c]	None known
Digoxin	Lanoxin	Unknown	None known	None known
Diltiazem	Cardizem, Tiazac	3A4, 2D6, 2E1	3A4	None known
Dofetilide	Tikosyn	3A4	None known	None known
Enalapril	Vasotec	Unknown	None known	None known
Encainide	Enkaid	2D6	None known	None known
Felodipine	Plendil	3A4	None known	None known
Flecainide	Tambocor	2D6, 1A2	2D6	None known
Fluvastatin	Lescol	2C9, 2C8, 3A4	2C9	None known

TABLE 1. Internal medicine drug-drug interactions: allergy, cardiology, endocrine, gastrointestinal, infectious disease, pain (narcotics and nonnarcotics), oncology, and transplantation and rheumatology *(continued)*

Drug	Trade name	Metabolism site	Inhibition[a]	Induction[a]
B. Cardiology (continued)				
Ibutilide	Corvert	Unknown	None known	None known
Irbesartan	Avapro	2C9	None known	None known
Isradipine	Dynacirc	None known	None known	None known
Lidocaine	Xylocaine	1A2, 2B6, 3A4	1A2[c]	None known
Lisinopril	Prinivil	Unknown	None known	None known
Losartan	Cozaar	2C9	None known	None known
Lovastatin	Mevacor	3A4	None known	None known
Metoprolol	Lopressor	2D6	None known	None known
Mexiletine	Mexitil	2D6, 1A2	1A2	None known
Moricizine	Ethmozine	3A4	None known	1A2, 3A4
Nicardipine	Cardene	3A4	None known	None known
Nifedipine	Adalat, Procardia	3A4	None known	None known
Nimodipine	Nimotop	3A4	None known	None known
Nisoldipine	Sular	3A4, 2D6	None known	None known
Pravastatin	Pravachol	3A4	None known	None known
Propafenone	Rythmol	2D6, 1A2, 3A4	1A2, 2D6	None known
Propranolol	Inderal	2D6, 1A2, 2C19	2D6[b]	None known

TABLE 1. Internal medicine drug-drug interactions: allergy, cardiology, endocrine, gastrointestinal, infectious disease, pain (narcotics and nonnarcotics), oncology, and transplantation and rheumatology *(continued)*

Drug	Trade name	Metabolism site	Inhibition[a]	Induction[a]
B. Cardiology (continued)				
Quinidine	Quinidex	3A4	2D6	None known
Rosuvastatin	Crestor	2C9	None known	None known
Simvastatin	Zocor	3A4, 2D6	2C9	None known
Ticlopidine	Ticlid	?3A4	**2B6, 2C19, 2D6**[b] 1A2[c], 2C9[c]	None known
Tocainide	Tonocard	None known	1A2[c]	None known
Valsartan	Diovan	2C9	None known	None known
Verapamil	Calen, Covera, Verelan	3A4	3A4	None known
Warfarin	Coumadin	2C9, 1A2, 2C8, 2C18, 2C19, 3A4	None known	None known

TABLE 1.   Internal medicine drug-drug interactions: allergy, cardiology, endocrine, gastrointestinal, infectious disease, pain (narcotics and nonnarcotics), oncology, and transplantation and rheumatology *(continued)*

Drug	Trade name	Metabolism site	Inhibition[a]	Induction[a]
C. Endocrine				
Glimepiride	Amaryl	2C9	None known	None known
Glipizide	Glucotrol	2C9	None known	None known
Glyburide	Micronase	2C9, 2C19, 3A4	None known	None known
Pioglitazone	Actos	2C8	None known	None known
Repaglinide	Prandin	2C8, 3A4	None known	None known
Rosiglitazone	Avandia	2C8	None known	None known
Tolbutamide	Orinase	2C9	None known	None known
Troglitazone	Rezulin	2C8, 2C9	None known	3A4

TABLE 1. Internal medicine drug-drug interactions: allergy, cardiology, endocrine, gastrointestinal, infectious disease, pain (narcotics and nonnarcotics), oncology, and transplantation and rheumatology *(continued)*

Drug	Trade name	Metabolism site	Inhibition[a]	Induction[a]
D. Gastrointestinal				
Cimetidine	Tagamet	Renal	3A4, 2D6, 1A2, 2C9, 2C19	None known
Cisapride	Propulsid	3A4, 2A6	2D6, 3A4	None known
Esomeprazole	Nexium	2C19, 3A4	2C19	None known
Famotidine	Pepcid	Renal	None known	None known
Lansoprazole	Prevacid	2C19, 3A4	2C19, 2D6, 2C9, 3A4	1A2[c]
Metoclopramide	Reglan	2D6, 1A2	2D6	None known
Nizatidine	Axid	Renal	None known	None known
Omeprazole	Prilosec	2C19, 3A4	3A4, 2C19, 2C9	1A2
Pantoprazole	Protonix	2C19, 3A4	3A4	None known
Rabeprazole	Aciphex	2C19, 3A4	None known	None known
Ranitidine	Zantac	FMO3,5, 2C19, 1A2, 2D6	1A2, 2C9, 2C19[b], 2D6[c]	None known

TABLE 1. Internal medicine drug-drug interactions: allergy, cardiology, endocrine, gastrointestinal, infectious disease, pain (narcotics and nonnarcotics), oncology, and transplantation and rheumatology (continued)

Drug	Trade name	Metabolism site	Inhibition[a]	Induction[a]
E. Infectious disease				
Abacavir	Ziagen	Alcohol dehydrogenase, glucuronyl transferase	None known	None known
Amprenavir	Agenerase	3A4	3A4[b]	Possible 3A4
Atazanavir	Reyataz	3A4	3A4, 1A2, 2C9, UGT1A1	None known
Azithromycin	Zithromax	3A4	3A4[c]	None known
Ciprofloxacin	Cipro	3A4	1A2, 3A4	None known
Clarithromycin	Biaxin	3A4	3A4	None known
Delavirdine	Rescriptor	3A4, 2D6, 2C9, 2C19	3A4, 2C9, 2C19, 2D6	None known
Didanosine	ddI, Videx	Purine nucleoside phosphorylase	None known	None known
Dirithromycin	Dynabac	3A4	None known	None known
Efavirenz	Sustiva	3A4, 2B6	3A4, 2C9, 2C19, 2D6, 1A2	3A4[b], 2B6[c]
Emtricitabine	Emtriva	Excreted unchanged	None known	None known
Enoxacin	Penetrex	?	1A2	None known
Erythromycin	E-mycin	3A4	3A4	None known
Fluconazole	Diflucan	3A4	2C9, 3A4[b]	None known

TABLE 1. Internal medicine drug-drug interactions: allergy, cardiology, endocrine, gastrointestinal, infectious disease, pain (narcotics and nonnarcotics), oncology, and transplantation and rheumatology (continued)

Drug	Trade name	Metabolism site	Inhibition[a]	Induction[a]
E. Infectious disease (continued)				
Indinavir	Crixivan	3A4	3A4	None known
Isoniazid	—	?	2C9, 2E1	2E1
Itraconazole	Sporanox	3A4	3A4	None known
Ketoconazole	Nizoral	3A4	3A4	None known
Lamivudine	3TC, Epivir	Renal	None known	None known
Levofloxacin	Levaquin	Renal	None known	None known
Lopinavir	Kaletra (with ritonavir)	3A4	3A4[b], 2D6	Glucuronidation (phase II)
Miconazole	Micatin	3A4	3A4	None known
Moxifloxacin	Avelox	None known	None known	None known
Nelfinavir	Viracept	3A4, 2C19	3A4[c], 1A2, 2B6[b]	Possible 2C9
Nevirapine	Viramune	3A4, 2B6	None known	3A4[b], 2B6[b]
Norfloxacin	Noroxin	Renal	1A2, 3A4	None known
Ofloxacin	Floxin	Renal	1A2[c]	None known
Rifabutin	Mycobutin	3A4	None known	3A4, 1A2, 2C9, 2C19
Rifampin	Rifadin	3A4	None known	3A4, 1A2, 2C9, 2C19
Rifapentine	Priftin	3A4/other	None known	3A4, 2C9

TABLE 1.   Internal medicine drug-drug interactions: allergy, cardiology, endocrine, gastrointestinal, infectious disease, pain (narcotics and nonnarcotics), oncology, and transplantation and rheumatology *(continued)*

Drug	Trade name	Metabolism site	Inhibition[a]	Induction[a]
E. Infectious disease (continued)				
Ritonavir	Norvir	3A4, 2D6	3A4, 2D6, 2C9, 2C19, 2B6	3A4, 1A2[b], 2C9[b], 2C19
Rokitamycin	Ricamycin	3A4	3A4	None known
Saquinavir	Invirase	3A4	3A4[c]	None known
Sparfloxacin	Zagam	3A4, phase II	1A2, 3A4[c]	None known
Stavudine	d4T, Zerit	None known	None known	None known
Telithromycin	Ketek	3A4, others	3A4	None known
Tenofovir disoproxil fumarate	Viread	Renal	1A2	None known
Terbinafine	Lamisil	?	2D6	None known
Troleandomycin	Tao	3A4	3A4	None known
Trovafloxacin	Trovan	Phase II	None known	None known
Zalcitabine	ddC, Hivid	Renal	None known	None known
Zidovudine	AZT, Retrovir	UGT2B7	None known	None known

TABLE 1. Internal medicine drug-drug interactions: allergy, cardiology, endocrine, gastrointestinal, infectious disease, pain (narcotics and nonnarcotics), oncology, and transplantation and rheumatology *(continued)*

Drug	Trade name	Metabolism site	Inhibition[a]	Induction[a]
F. Pain: narcotics				
Alfentanil	Alfenta	3A4	None known	None known
Buprenorphine	Buprenex	3A4, UGT2B7, UGT1A3	None known	None known
Codeine		3A4 (10%), 2D6 (5%), UGT2B7 (80%)	UGT2B7	None known
Diphenoxylate	Lomotil	?	None known	None known
Fentanyl	Duragesic	3A4	Possible 3A4	None known
Hydrocodone	Vicodin (with acetaminophen)	2D6, 3A4	None known	None known
Hydromorphone	Dilaudid	2D6, 3A4, UGT1A3, UGT2B7	None known	None known
Loperamide	Imodium	2C8, 3A4	Possible 3A4	None known
Meperidine	Demerol	2B6, 3A4, 2C19, hydrolysis, UGTs	None known	None known
Methadone		3A4, 2D6, 2B6, UGT1A3, UGT2B7	3A4, UGT2B7	None known

TABLE 1. Internal medicine drug-drug interactions: allergy, cardiology, endocrine, gastrointestinal, infectious disease, pain (narcotics and nonnarcotics), oncology, and transplantation and rheumatology *(continued)*

Drug	Trade name	Metabolism site	Inhibition[a]	Induction[a]
F. Pain: narcotics (continued)				
Morphine		UGT2B7, UGT1A3, small amounts by 2D6 and 3A4	UGT2B7	None known
Nalbuphine	Nubain	?Unspecified P450 enzymes	None known	None known
Naloxone	Narcan	UGT2B7	None known	None known
Naltrexone	ReVia	?Unspecified P450 enzymes, UGT2B7	None known	None known
Oxycodone	OxyContin	2D6, 3A4	None known	None known
Oxymorphone	Numorphan	UGT2B7	None known	None known
Propoxyphene	Darvon	3A4	None known	None known
Remifentanil	Ultiva	Esterases	None known	None known
Sufentanil	Sufenta	3A4, 2D6	None known	None known
Tramadol	Ultram	2D6, 3A4, 2B6, UGTs	None known	None known

TABLE 1. Internal medicine drug-drug interactions: allergy, cardiology, endocrine, gastrointestinal, infectious disease, pain (narcotics and nonnarcotics), oncology, and transplantation and rheumatology *(continued)*

Drug	Trade name	Metabolism site	Inhibition[a]	Induction[a]
G. Pain: nonnarcotics				
Acetaminophen	Tylenol	2E1, 3A1, 1A2, UGT1A1, UGT1A6, UGT1A9, UGT2B15, SULT1A1	None known	None known
Acetylsalicylic acid	Aspirin	Deacetylation	None known	None known
Celecoxib	Celebrex	2C9, UGTs	2D6[c]	None known
Diclofenac	Voltaren	2C9, 3A4, UGT1A9, UGT2B7	UGT2B7[b]	None known
Diflunisal	Dolobid	UGT1A3, UGT1A9	UGT1A9, UGT2B7, SULTs	None known
Etodolac	Lodine	2C9, UGT1A9	2C9[c]	None known
Fenoprofen	Nalfon	UGT2B7	UGT2B7	None known
Flurbiprofen	Ansaid	2C9, UGT1A3, UGT1A9, UGT2B4, UGT2B7	None known	None known
Ibuprofen	Advil	2C8 (*R*-ibuprofen), 2C9 (*S*-ibuprofen), UGT1A1, UGT1A9, UGT2B4, UGT2B7	UGT2B7	None known

TABLE 1. Internal medicine drug-drug interactions: allergy, cardiology, endocrine, gastrointestinal, infectious disease, pain (narcotics and nonnarcotics), oncology, and transplantation and rheumatology *(continued)*

Drug	Trade name	Metabolism site	Inhibition[a]	Induction[a]
G. Pain: nonnarcotics (continued)				
Indomethacin	Indocin	2C9, 2C19, UGT1A1, UGT1A3, UGT1A9, UGT2B7	UGT2B7	None known
Ketoprofen	Orudis	?2C9, UGT1A3, UGT1A9, UGT2B4, UGT2B7	UGT2B7	None known
Ketorolac	Toradol	Unspecified P450 enzyme	None known	None known
Meclofenamate	Meclomen	2C9, UGT1A9, UGT2B7	SULTs, UGT1A9, UGT2B7	None known
Mefenamic acid	Ponstel	2C9, UGT1A9, UGT2B7	SULTs, UGT1A9, UGT2B7	None known
Meloxicam	Mobic	2C9, 3A4	2C9, 3A4[c]	None known
Nabumetone	Relafen	Unspecified phase I enzyme, UGTs	None known	None known
Naproxen	Naprosyn	2C9, 2C8, 1A2, UGT1A3, UGT1A9, UGT2B4, UGT2B7	UGT2B7	None known

TABLE 1. Internal medicine drug-drug interactions: allergy, cardiology, endocrine, gastrointestinal, infectious disease, pain (narcotics and nonnarcotics), oncology, and transplantation and rheumatology *(continued)*

Drug	Trade name	Metabolism site	Inhibition[a]	Induction[a]
G. Pain: nonnarcotics (continued)				
Oxaprozin	Daypro	Unspecified P450 and UGT enzymes	None known	None known
Piroxicam	Feldene	2C9, ?others	None known	None known
Rofecoxib	Vioxx	None	None known	None known
Salicylic acid	Salacid	UGT 2B7, glycine conjugation	SULTs, UGT2B7, β-oxidation	None known
Salsalate	Disalcid	Unspecified UGTs, esterases	SULTs, UGT2B7	None known
Sulindac	Clinoril	UGT1A1, UGT1A3, UGT1A9, UGT2B7, FMO₃	SULT1E1	None known
Tolmetin	Tolectin	3A4, UGTs	None known	None known
Valdecoxib	Bextra	3A4, 2C9, UGTs	2C19, 2C9, 3A4	None known

TABLE 1.   Internal medicine drug-drug interactions: allergy, cardiology, endocrine, gastrointestinal, infectious disease, pain (narcotics and nonnarcotics), oncology, and transplantation and rheumatology (*continued*)

Drug	Trade name	Metabolism site	Inhibition[a]	Induction[a]
H. Oncology				
Aminoglutethimide	Cytadren	Unknown	None known	None known
Anastrazole	Arimidex	3A4	1A2, 2C9, 3A4, 2C8	None known
Bicalutamide	Casodex	3A4 (*R*), glucuronidation (*S*)	2D6, 2C19, 2C9, 3A4	None known
Bleomycin	—	Renal	None known	None known
Busulfan	Myleran	3A4	None known	None known
Carmustine (BCNU)	—	2C9	None known	None known
Chlorambucil	Leukeran	Unknown	None known	None known
Cisplatin	Platinol	Unknown	None known	None known
Cyclophosphamide	Cytoxan	3A4, 2B6	None known	3A4, 2B6
Dacarbazine	—	1A2, 1A1, 2E1	None known	None known
Dasatinib	Sprycel	3A4	3A4	None known
Daunorubicin	Cerubidine	Unknown	None known	None known
Docetaxel	Taxotere	3A4	3A4	None known
Doxorubicin	Adriamycin	3A4	None known	None known
Epirubicin	Ellence	UGT2B7	None known	None known
Erlotinib	Tarceva	3A4, 1A2	3A4	None known

TABLE 1. Internal medicine drug-drug interactions: allergy, cardiology, endocrine, gastrointestinal, infectious disease, pain (narcotics and nonnarcotics), oncology, and transplantation and rheumatology *(continued)*

Drug	Trade name	Metabolism site	Inhibition[a]	Induction[a]
H. Oncology (continued)				
Etoposide	Toposar	3A4, 1A2, 2E1	None known	3A4
Exemestane	Aromasin	3A4	None known	None known
Flutamide	Eulexin	1A2, 3A4	1A2	None known
Gefitinib	Iressa	3A4, 2D6,2C19	2D6	None known
Idarubicin	Idamycin	2D6, 2C9	2D6[b]	None known
Ifosfamide	Ifex	3A4, 2B6	None known	3A4[b], 2C9
Imatinib	Gleevec	3A4, 2D6, 2C19	3A4	None known
Irinotecan	Camptosar	3A4, UGT1A1	3A4	None known
Lapatinib	—	Unknown	None known	None known
Letrozole	Femara	2A6, 3A4	2A6, 2C19	None known
Lomustine	Belustine	2D6	Possible 2D6	None known
Melphalan	Alkeran	Unknown	None known	None known
Nilotinib	—	Unknown	None known	None known
Paclitaxel	Taxol	3A4, 2C8	None known	2C8[b], 3A4[b]
Sorafenib	Nexavar	UGT1A9, 3A4	None known	None known
Streptozocin	Zanosar	Unknown	None known	None known
Sunitinib	Sutent	3A4	None known	None known

TABLE 1. Internal medicine drug-drug interactions: allergy, cardiology, endocrine, gastrointestinal, infectious disease, pain (narcotics and nonnarcotics), oncology, and transplantation and rheumatology *(continued)*

Drug	Trade name	Metabolism site	Inhibition[a]	Induction[a]
H. Oncology (continued)				
Tamoxifen	Soltamox	2D6, 2C9, 3A4, 1A2, 2A6, 2B6, 2E1	3A4, 2C9	3A4
Temozolomide	Temodar	Unknown	None known	None known
Teniposide	Vumon	3A4	3A4	None known
Thiotepa	Testamine	3A4, 2B6	2B6	None known
Topotecan	Hycamtin	3A4	None known	None known
Toremifene	Fareston	3A4, 1A2	None known	None known
Trofosfamide	Ixoten	3A4, 2B6	None known	None known
Vandetanib	Zactima	Unknown	None known	None known
Vinblastine	—	3A4	None known	None known
Vincristine	—	3A4	None known	None known
Vindesine	—	3A4	None known	None known
Vinorelbine	Navelbine	3A4	None known	None known

TABLE 1.   Internal medicine drug-drug interactions: allergy, cardiology, endocrine, gastrointestinal, infectious disease, pain (narcotics and nonnarcotics), oncology, and transplantation and rheumatology *(continued)*

Drug	Trade name	Metabolism site	Inhibition[a]	Induction[a]
I. Transplantation and rheumatology				
Azathioprine	Imuran	Unknown enzymes metabolize to mercaptopurine	?	?
Cyclophosphamide	Cytoxan	3A4, 2B6	?	2B6, 3A4
Cyclosporine	Sandimmune	3A4	3A4[b]	None known
Mercaptopurine	Purinethol	TPMT	?	?
Methotrexate	—	Non-P450 oxidative enzymes	?	?
Methylprednisolone	—	3A4	3A4[b]	3A4[b]
Mitoxantrone	Novantrone	Unknown oxidative enzymes, UGTs	?	?
Muromonab-CD3	Orthoclone OKT3	Not phase I or II enzymes	?	None known
Mycophenolate mofetil	CellCept	Hydrolysis, UGT1A7, UGT1A8, UGT1A9, UGT2B7	?	?

TABLE 1. Internal medicine drug-drug interactions: allergy, cardiology, endocrine, gastrointestinal, infectious disease, pain (narcotics and nonnarcotics), oncology, and transplantation and rheumatology *(continued)*

Drug	Trade name	Metabolism site	Inhibition[a]	Induction[a]
I. Transplantation and Rheumatology (continued)				
Prednisolone	—	3A4	None known	3A4
Prednisone	—	3A4	None known	3A4
Sirolimus	Rapamune	3A4	None known	None known
Tacrolimus	Prograf	3A4, UGT2B7	3A4, UGT1A1, UGT2B7	None known

Note. ABCB1 = P-glycoprotein; SULT = sulfotransferase; TPMT = thiopurine methyltransferase; UGT = uridine 5'-diphosphate glucuronosyltransferase; — = none; ? = unknown.

[a]**Bold** type indicates potent inhibition or induction.
[b]Moderate inhibition or induction.
[c]Mild inhibition or induction.

TABLE 2. Psychiatry drug-drug interactions

Drug	Trade name	Metabolism site	Inhibition[a]	Induction
Acamprosate	Campral	Excreted unchanged	None known	None known
Alprazolam	Xanax	3A4, UGT	None known	None known
Amitriptyline	Elavil	1A2, 2C19, 2D6, 3A4, UGT1A4	1A2, 2C19, 2D6	None known
Aripiprazole	Abilify	2D6, 3A4	None known	None known
Atomoxetine	Strattera	2D6	None known	None known
Bupropion	Wellbutrin	2B6, 2D6, 1A2, 2A6, 2C9, 2E1, 3A4, glucuronidation	**2D6**	None known
Buspirone	BuSpar	3A4	None known	None known
Chlorpromazine	Thorazine	2D6, 1A2, 3A4, UGT1A4, UGT1A3	**2D6**	None known
Citalopram	Celexa	2C19, 2D6, 3A4	2D6[c]	None known
Clomipramine	Anafranil	1A2, 2C19, 2D6, 3A4	2D6, 1A2, 2C19	None known
Clonazepam	Klonopin	3A4, acetylation	None known	None known
Clozapine	Clozaril	1A2, 3A4, 2D6, 2C9, 2C19, FMO₃, UGT1A4, UGT1A3	2D6[c]	None known
Desipramine	Norpramin	2D6	2D6, 2C19	None known
Desvenlafaxine	Pristiq	UGTs, 3A4	2D6[c]	None known
Diazepam	Valium	2C19, 3A4, 2B6, 2C9, UGTs	None known	None known
Disulfiram	Antabuse	Unknown	2E1[b], 1A2[b]	None known

TABLE 2. Psychiatry drug-drug interactions *(continued)*

Drug	Trade name	Metabolism site	Inhibition[a]	Induction
Doxepin	Adapin, Sinequan	1A2, 2D6, 2C19, 3A4, UGT1A4, UGT1A3	1A2, 2C19, 2D6	None known
Duloxetine	Cymbalta	1A2, 2D6	2D6[b]	None known
Escitalopram	Lexapro	2C19, 2D6, 3A4	2D6[c]	None known
Eszopiclone	Lunesta	2C8, 3A4	None known	None known
Flunitrazepam	Rohypnol	2C19, 3A4	None known	None known
Fluoxetine	Prozac	2C9, 2C19, 2D6, 3A4	**2D6**, 2C9[b], 3A4[b], 2C19[b], 1A2[c], 2B6[b]	None known
Fluphenazine	Prolixin	2D6, 1A2	2D6, 1A2	None known
Fluvoxamine	Luvox	1A2, 2D6	**1A2, 2B6**[b], 2C9[b], **2C19, 2D6**[b], 3A4[b]	None known
Haloperidol	Haldol	2D6, 3A4, 1A2	2D6	None known
Imipramine	Tofranil	1A2, 2C19, 2D6, 3A4, UGT1A4, UGT1A3	2C19, 2D6, 1A2, 3A4	None known
Lorazepam	Ativan	UGT2B7, possibly other UGTs	None known	None known
Loxapine	Loxitane	Possibly 2D6, 3A4, 1A2, and UGT1A4, UGT1A3	None known	None known
Mesoridazine	Serentil	2D6, 1A2	None known	None known
Methylphenidate	Ritalin	Possibly 2D6, plasma esterases	Possibly P450	None known
Midazolam	Versed	3A4, UGTs	None known	None known
Mirtazapine	Remeron	1A2, 2D6, 3A4, UGTs	None known	None known

TABLE 2. Psychiatry drug-drug interactions *(continued)*

Drug	Trade name	Metabolism site	Inhibition[a]	Induction
Modafinil	Provigil	3A4, glucuronidation	2C19	3A4
Nefazodone	Serzone	3A4, 2D6	3A4	None known
Nortriptyline	Pamelor	2D6	2D6, 2C19	None known
Olanzapine	Zyprexa	1A2, 2D6, UGTs	None known	None known
Oxazepam	Serax	*S*-Oxazepam: UGT2B15 *R*-Oxazepam: UGT1A9, UGT2B7	None known	None known
Paliperidone	Invega	2D6, 3A4, primarily excreted unchanged	None known	None known
Paroxetine	Paxil	2D6, 3A4	1A2[c], **2B6**, 2C9[c], 2C19[c], **2D6**, 3A4[b]	None known
Perphenazine	Trilafon	2D6, 3A4, 1A2, 2C19	**2D6**, 1A2	None known
Pimozide	Orap	3A4, 1A2	**2D6**, 3A4	None known
Protriptyline	Vivactil	?2D6	None known	None known
Quetiapine	Seroquel	3A4. sulfation	None known	None known
Ramelteon	Rozerem	1A2, 3A4, 2C family	None known	None known
Risperidone	Risperdal	2D6, 3A4	2D6[b]	None known
Sertraline	Zoloft	2B6, 2C9, 2C19, 2D6, 3A4, UGT2B7, UGT1A1	1A2[c], 2B6[b], 2C9[b], 2C19[b], 2D6[a,b,*], 3A4[b]	None known
Temazepam	Restoril	2C19, 3A4, UGT2B7	None known	None known

TABLE 2. Psychiatry drug-drug interactions *(continued)*

Drug	Trade name	Metabolism site	Inhibition[a]	Induction
Thioridazine	Mellaril	2D6, 1A2, 2C19, FMO_3	2D6	Possibly 3A4
Trazodone	Desyrel	3A4	None known	None known
Triazolam	Halcion	3A4, UGTs	None known	None known
Trimipramine	Surmontil	2C19, 2D6, 3A4	None known	None known
Varenicline	Chantix	UGT2B7, primarily excreted unchanged	None known	None known
Venlafaxine	Effexor	2D6, 2C19, 3A4	2D6[c]	None known
Zaleplon	Sonata	3A4, aldehyde oxidase	None known	None known
Ziprasidone	Geodon	3A4, 1A2, aldehyde oxidase	None known	None known
Zolpidem	Ambien	3A4, 1A2, 2C9	None known	None known
Zopiclone	Inovane	2C8, 3A4	None known	None known

Note. FMO_3 = flavin monooxygenase; UGT = uridine 5'-diphosphate glucuronosyltransferase.

*Sertraline is a potent 2D6 inhibitor at high concentrations (typically at dosages >200 mg/day).

[a]**Bold** type indicates potent inhibition.

[b]Moderate inhibition.

[c]Mild inhibition.

TABLE 3.   Neurology drug-drug interactions

Drug	Trade name	Metabolism site	Inhibition	Induction[a]
Almotriptan	Axert	MAO A, 3A4, 2D6	None known	None known
Bromocriptine	Parlodel	3A4, Possibly others	3A4[b]	None known
Carbamazepine	Tegretol	3A4, 2B6, 2C8, 2E1, 2C9, 1A2, UGT2B7	Possible 2C19	3A4, 2B6, 2C8, 2C9, 1A2, UGT1A4
Carbidopa-levodopa	Sinemet	Carbidopa: excreted unchanged Levodopa: aromatic amino acid decarboxylase	None known	None known
Donepezil	Aricept	2D6, 3A4, unspecified UGTs	None known	None known
Eletriptan	Relpax	3A4	None known	None known
Entacapone	Comtan	UGT1A subfamily	P450[c]	None known
Ethosuximide	Zarontin	3A4, phase II	None known	Possible pan-inducer
Felbamate	Felbatol	3A4, 2E1	2C19	3A4
Frovatriptan	Frova	1A2	None known	None known
Gabapentin	Neurontin	Excreted unchanged	None known	None known
Galantamine	Reminyl	2D6, 3A4, unspecified UGT(s)	None known	None known
Lamotrigine	Lamictal	UGT1A4	None known	UGT1A4[c]
Levetiracetam	Keppra	Hydrolysis	None known	None known
Memantine	Nemenda	Primarily excreted unchanged	2B6	None known
Methsuximide	Celontin	3A4, phase II	None known	Possible pan-inducer
Naratriptan	Amerge	P450, MAO A	None known	None known

TABLE 3. Neurology drug-drug interactions *(continued)*

Drug	Trade name	Metabolism site	Inhibition	Induction[a]
Oxcarbazepine	Trileptal	3A4	2C19	3A4[b], UGT1A4[b]
Pergolide	Permax	3A4, phase II	3A4	None known
Phenobarbital		2C9, 2C19, 2E1	3A4, possible phase II enzymes	UGTs, 3A4, 2C9, 2C19, 1A2, possibly others
Phenytoin	Dilantin	2C9, 2C19, UGT1A subfamily	None known	3A4, 2C9, 2C19, 2B6, UGT1A1, UGT1A4
Pramipexole	Mirapex	Excreted unchanged	None known	None known
Primidone	Mysoline	2C9, 2C19, 2E1	3A4	UGTs, 3A4, 2C9, 2C19, 1A2, possibly others
Rasagiline	Azilect	1A2	None known	None known
Rivastigmine	Exelon	Local cholinesterases	None known	None known
Rizatriptan	Maxalt	MAO A	None known	None known
Ropinirole	Requip	1A2, 3A4	1A2[c]	None known
Selegiline	Eldepryl	2B6, 3A4, 2A6	2C19	None known
Sumatriptan	Imitrex	MAO A	None known	None known
Tacrine	Cognex	1A2, 2D6	1A2	None known
Tiagabine	Gabitril	3A4, Unspecified UGTs	None known	None known
Tolcapone	Tasmar	UGT1A9, COMT, 3A4, 2A6	2C9[c]	None known
Topiramate	Topamax	Primarily excreted unchanged	2C19[b]	None known; decreases ethinyl estradiol levels

TABLE 3. Neurology drug-drug interactions *(continued)*

Drug	Trade name	Metabolism site	Inhibition	Induction[a]
Valproic acid	Depakote	Complex: 2C9, 2C19, 2A6, UGT1A6, UGT1A9, UGT2B7, β-oxidation	2D6, 2C9, UGT1A4, UGT1A9, UGT2B7, UGT2B15, epoxide hydroxylase	Possible 3A4
Vigabatrin	Sabril	Primarily excreted unchanged	None known	None known
Zolmitriptan	Zomig	1A2, MAO A	None known	None known
Zonisamide	Zonegran	3A4, acetylation, sulfonation	None known	None known

Note. COMT = catechol *O*-methyltransferase; UGT = uridine 5′-diphosphate glucuronosyltransferase; MAO A = monoamine oxidase A.

[a]**Bold** type indicates potent induction.
[b]Moderate inhibition or induction.
[c]Mild inhibition or induction.

Index

*Page numbers printed in **boldface** type refer to tables or figures.*